Recombinant Proteins from Plants

METHODS IN BIOTECHNOLOGY™

John M. Walker, SERIES EDITOR

Preface

Recombinant Proteins from Plants is one of the most exciting and fastest developing areas in biology. The latest molecular techniques are being applied to the exploitation of plants as novel expression systems for the production and overproduction of heterologous and native proteins. Transgenic plant technology is currently used in three broad areas: the expression of recombinant proteins to improve crop quality by increasing disease/pest resistance or increasing tolerance to stress, optimizing plant productivity and yield by the genetic manipulation of metabolic pathways, and the large-scale cost-effective production of recombinant proteins for use as specialist industrial or therapeutic biomolecules.

The intention of *Recombinant Proteins from Plants* is to provide comprehensive and detailed protocols covering all the latest molecular approaches. Because the production of transgenic plants has become routine in many laboratories, coverage is also given to some of the more "classical" approaches to the separation, analysis, and characterization of recombinant proteins. The book also includes areas of research that we believe will become increasingly important in the near future: efficient transformation of monocots with *Agrobacterium tumifaciens,* optimizing the stability of recombinant proteins, and a section highlighting the immunotherapeutic potential of plant-expressed proteins.

The contents bring together a broad range of knowledge and expertise from around the world; each chapter is written by research workers with hands-on experience of the protocols they describe. The book opens with chapters covering the introduction of transgenes into monocot and dicot species, and includes the expression and characterization of reporter genes and the application of recombinant proteins in crop protection. This is followed by sections that describe the expression of cytokines, antibodies/antibody fragments, and use of recombinant plant viral particles as vaccines. The later part of the book deals with the expression of proteins by plant suspension culture/protoplasts, explores the latest strategies to increase the yield of recombinant proteins by controlling the cellular location of protein folding, and the current protocols available for the molecular manipulation of photosynthetic pathways. The final chapters of the book outline techniques specifically designed to quantify and characterize plant-derived proteins and include: glycosylation state, ELISA, and two-dimensional electrophoresis.

We hope that we have provided a detailed handbook of essential approaches and methodologies that will be of benefit both to people new to plant protein biotechnology and to established laboratories in the field.

Charles Cunningham
Andrew J. R. Porter

Contents

Contributors

ILLIMAR ALTOSAAR • *Department of Biochemistry, University of Ottawa, Ottawa, Ontario, Canada*

OLGA ARTSAENKO • *IPK Gatersleben, Gatersleben, Germany*

KRISTIAN ASPEGREN • *VTT Biotechnology and Food Research, Finland*

ADRIAN D. BAVAGE • *Department of Brassica and Oilseeds Research, John Innes Centre, Norwich, UK*

ANNE-MARIE BRUYNS • *Laboratorium voor Genetica, Department of Genetics, Flanders Interuniversity Institute for Biotechnology, University of Gent, Belgium*

XIONGYING CHENG • *Department of Biochemistry, University of Ottawa, Ottawa, Ontario, Canada*

STEVEN P. COLLIVER • *Biochemistry and Physiology Department, IACR-Rothamsted, Harpenden, UK*

UDO CONRAD • *IPK Gatersleben, Gatersleben, Germany*

GEERT DE JAEGER • *Department of Genetics, Flanders Interuniversity Institute for Biotechnology, University of Gent, Belgium*

MYRIAM DE NEVE • *Department of Genetics, Flanders Interuniversity Institute for Biotechnology, University of Gent, Belgium*

CHRIS DE WILDE • *Department of Genetics, Flanders Interuniversity Institute for Biotechnology, University of Gent, Belgium*

LISE-ANNE DENMAT • *Laboratoire des Transports Intracellulaires, Université de Rouen, Faculté des Sciences, Mont-Saint-Aigen, France*

ANN DEPICKER • *Department of Genetics, Flanders Interuniversity Institute for Biotechnology, University of Gent, Belgium*

JÜRGEN DROSSARD • *Institut fur Biologie I (Botanik/Molekulargenetik), Rheinisch-Westfalische Technische Hochschule Aachen, Germany*

ANIL DUDANI • *Ottawa Blood Centre, CRCS, Ottawa, Ontario, Canada*

LOÏC FAYE • *Laboratoire des Transports Intracellulaires, Université de Rouen, Faculté des Sciences, Mont-Saint-Aigen, France*

ULRIKE FIEDLER • *IPK Gatersleben, Gatersleben, Germany*

RAINER FISCHER • *Institut fur Biologie I (Botanik/Molekulargenetik), Rheinisch-Westfalische Technische Hochschule Aachen, Germany*

ANNE-CATHERINE FITCHETTE-LAINÉ • *Laboratoire des Transports Intracellulaires, Faculté des Sciences, Université de Rouen, Mont-Saint-Aigen, France*

PETER R. GANZ • *Ottawa Blood Centre, CRCS, Ottawa, Ontario, Canada*

VÉRONIQUE GOMORD • *Laboratoire des Transports Intracellulaires, Université de Rouen, Faculté des Sciences, Mont-Saint-Aigen, France*

M. JACOBS • *Department of Plant Genetics, St. Genesius Rode, Belgium*

IRINA KORSCHHINECK • *Centre of Applied Biochemistry, University of Agriculture, Vienna, Austria*

PATRICE LEROUGE • *Centre Regional Universitaire de Spectroscopie, Faculté des Sciences, Université de Rouen, Mont-Saint-Aigen, France*

YU-CAI LIAO • *Institut fur Biologie I (Botanik/Molekulargenetik), Rheinisch-Westfalische Technische Hochschule Aachen, Germany*

KEITH LINDSEY • *Department of Biological Sciences, University of Durham, UK*

PIPPA MADGWICK • *Biochemistry and Physiology Department, IACR-Rothamsted, Harpenden, UK*

LEENA MANNONEN • *VTT Biotechnology and Food Research, Finland*

DOMINIQUE MICHAUD • *Centre de Recherche en Horticulture, Université Laval, Quebec, Canada*

JOHNATHAN A. NAPIER • *IACR Long Ashton Research Station, Department of Agricultural Sciences, University of Bristol, Bristol, UK*

MARTIN A. J. PARRY • *Biochemistry and Physiology Department, IACR-Rothamsted, Harpenden, UK*

MATTHEW J. PAUL • *Biochemistry and Physiology Department, IACR-Rothamsted, Harpenden, UK*

JULIAN PHILLIPS • *IPK Gatersleben, Gatersleben, Germany*

DIRK PRÜFER • *Max-Plank-Institut fur Zuchtungsforschung, Koln, Germany*

ANNELI RITALA • *VTT Biotechnology and Food Research, Finland*

STEPHEN J. REINL • *Biosource Technologies Inc, Vacaville, CA*

N. P. ESWARRA REDDY • *Department of Plant Genetics, University of Brussels, Belgium*

GAELLE RICHARD • *IACR Long Ashton Research Station, Department of Agricultural Sciences, University of Bristol, Bristol, UK*

MARK P. ROBBINS • *Cell Manipulation Group, Institute of Grassland and Environmental Research, Ceredigion, UK*

PIERRE ROUZÉ • *Laboratoire Associé de la Recherche Agronomique, University of Gent, Belgium*

RAVINDER K. SARDANA • *Department of Biochemistry, University of Ottawa, Ontario, Canada*

STEFAN SCHILLBERG • *Institut fur Biologie I (Botanik/Molekulargenetik), Rheinisch-Westfalische Technische Hochschule Aachen, Germany*

DETLEF SCHUMANN • *Institut fur Biologie I (Botanik/Molekulargenetik), Rheinisch-Westfalische Technische, Aachen, Germany*

PETER R. SHEWRY • *IACR Long Ashton Research Station, Department of Agricultural Sciences, University of Bristol, Bristol, UK*

HANNA SIMOLA • *VTT Biotechnology and Food Research, Finland*

LEIF SKØT • *Institute of Grassland and Environmental Research, Ceredigion, UK*

HERTA STEINKELLNER • *Centre of Applied Genetics, University of Agriculture, Vienna, Austria*

EILEEN S. TACKABERRY • *Ottawa Blood Centre, CRCS, Ottawa, Ontario, Canada*

TEEMU H. TEERI • *VTT Biotechnology and Food Research, Finland*

THOMAS H. TURPEN • *Biosource Technologies Inc, Vacaville, CA*

HELENA VAN HOUDT • *Department of Genetics, Flanders Interuniversity Institute for Biotechnology, University of Gent, Belgium*

MARC VAN MONTAGU • *Department of Genetics, Flanders Interuniversity Institute for Biotechnology, University of Gent, Belgium*

THIERRY C. VRAIN • *Centre de Recherche en Horticulture, Université Laval, Quebec, Canada*

K. JUDITH WEBB • *Institute of Grassland and Environmental Research, Ceredigion, UK*

JOY WILKINSON • *Plant Science, Unilever Research Colworth Laboratory, Bedfordshire, UK*

SABINE ZIMMERMANN • *Institut fur Biologie I (Botanik/Molekulargenetik), Rheinisch-Westfalische Technische, Aachen, Germany*

1

Rice Transformation by *Agrobacterium* Infection

Xiongying Cheng, Ravinder K. Sardana, and Illimar Altosaar

1. Introduction

Efficient production of transgenic plants is a prerequisite for the production of recombinant proteins in plants. In many dicotyledonous plants, *Agrobacterium tumefaciens*-mediated transformation has been well established. High efficacy and low cost make this the method of choice in many research laboratories for studying gene expression in plants *(1)*.

On the other hand, the monocotyledons and particularly the graminaceous species, including many of the most important food crops, such as wheat, rice, corn, and barley, are recalcitrant to *A. tumefaciens* infection. Alternative techniques have been developed to produce transgenic graminaceous plants (*see* **ref.** *2* for a recent review). With these methods, foreign DNA (e.g., a gene construct) is either taken up by plant cells (usually protoplasts) by electroporation or polyethylene glycol (PEG)-mediated transformation, or it is bombarded into plant cells via high-velocity microparticles (Biolistic or gene gun method). Transgenic plants are subsequently recovered through an in vitro selection and regeneration process *(3–5)*. Major problems, however, are associated with these methods, such as high cost and low transformation efficiency.

A. tumefaciens is a soil bacterium, capable of transferring a defined piece of DNA (T-DNA) from its tumor-inducing (Ti) plasmid into the genome of a large number of gymnosperms and angiosperms. However, only very few monocotyledonous plants in the families Liliaceae and Amaryllidaceae have been reported to be weakly susceptible to *A. tumefaciens* infection. Despite the earlier intensive efforts on *Agrobacterium*-mediated transformation in graminaceous plants *(6–9)*, success with *Agrobacterium* in these plants had been limited and remained controversial.

From: *Methods in Biotechnology, Vol. 3:*
Recombinant Proteins from Plants: Production and Isolation of Clinically Useful Compounds
Edited by: C. Cunningham and A. J. R. Porter © Humana Press Inc., Totowa, NJ

Recently, a major advance in rice transformation using *Agrobacterium* has been reported by Hiei and his co-workers *(10)*, who described an efficient transformation procedure for japonica rice cultivars. They also documented the stable integration, expression, and inheritance of the transgene in R_0–R_2 generations. We describe here an improved protocol modified from Hiei's for the production of transgenic rice plants, using embryogenic callus as explant. Using this procedure, it is possible to produce transgenic rice plants in 3–4 mo.

2. Materials

2.1. Plant Material

A suitable source for the preparation of embryogenic rice callus is healthy mature rice seeds or immature rice embryos. Seeds from both greenhouse and field-grown plants can be used for this purpose. Mature dry rice seeds can be stored at –20°C or 4°C for a long time (up to 5 yr). Before in vitro culture, they should be dehusked manually and checked individually to ensure that they are fully developed (discard small and shrunken seeds and seeds with brown or disfigured embryos) and free of pathogen infection (no trace of infection lesion). It is also important that each dehusked seed contains an intact embryo at its proximal end (*see* **Note 1**). For immature embryos, the caryopses, containing endosperm at the milky dough stage (10–14 d postanthesis), should be collected from healthy rice plants (*see* **Note 2**).

2.2. Bacterial Strains

Many *A. tumefaciens* strains are available for use in transformation. However, most of them have not yet been tested for rice transformation. Two strains that have been proven to work successfully in rice are LBA4404 *(11)* and EHA101 *(12)* (*see* **Note 3**).

2.3. Binary Vectors

A limited number of binary vectors have been tested for rice transformation. A Bin19-derived binary vector, pKGH4 *(13)*, was used as the major backbone vector in our experiments. This 13-kb vector contains three plant-expressible selection markers in the T-DNA region: *hph* gene for hygromycin-resistance and *gui*dA for β-glucuronidase (GUS), both driven by the CaMV 35S promoter, and the neomycin phosphotransferase II (*npt*II) gene for kanamycin-resistance, driven by the nopaline synthase promoter (Nos-promoter). A unique *Hin*dIII site in this region facilitates the insertion of the gene or gene cassette of interest (*see* **Note 4**).

2.4. Special Laboratory Tools and Materials

1. Laminar airflow hood (Canadian Cabinets, Ottawa).
2. Shaker (Lab-line, Baxter, Toronto).
3. Phytacon™ vessels (Sigma, St. Louis, MO).
4. Dissection microscope (Olympus, Tokyo).
5. Gallon pot (blow-moulded nursery container, Plant Products, Toronto; *see* **Note 5**).

2.5. Reagents

1. Acetosyringone (Aldrich, Milwaukee, WI).
2. Bleach (Javex™, Colgate-Palmolive Canada, Toronto).
3. Tween-20 (Sigma).
4. Hygromycin B (Boehringer Mannheim, Mannheim, Germany).
5. Ticarcillin disodium (SmithKline Beecham Pharma, Worthing, UK).

2.6. Media and Other Solutions

All media are autoclaved before use. They are made up from the components or commercially available stocks. Antibiotics are filter-sterilized and added to the media after autoclaving and cooling to 60°C.

1. YEP Medium: 10 g/L Bacto-peptone, 10 g/L bacto-yeast extract, 5 g/L NaCl, 15 g/L agar (Gibco, Gaithersburg, MD, for solid medium).
2. AB medium *(14)*: For preparation, prepare as two stock solutions: 20X AB buffer (60 g/L K_2HPO_4 and 20 g/L NaH_2PO_4) and 20X AB salts (20 g/L NH_4Cl, 6 g/L $MgSO_4 \cdot 7H_2O$, 3 g/L KCl, 0.2 g/L $CaCl_2$, and 50 mg/L $FeSO_4 \cdot 7H_2O$), and autoclave separately. For 1 L of medium, mix 5 g glucose and 900 mL distilled water, and autoclave. Add 50 mL each of the AB buffer and AB salts stock.
3. MS *(15)* callus induction medium: Murashige and Skoog basal medium (Sigma), 30 g/L sucrose, 2 mg/L 2,4-dichlorophenoxyacetic acid and 3 g/L PhytaGel™ (Sigma). Adjust pH before autoclaving to 5.8 with $1N$ NaOH.
4. 2N6-AS medium: N_6 major salts, N_6 minor salts, N_6 vitamins (Sigma), 1 g/L casamino acids, 30 g/L sucrose, 10 g/L glucose, 100 μM acetosyringone, 2 mg/L 2,4-dichlorophenoxyacetic acid, and 3 g/L PhytaGel, pH 5.2.
5. LHT medium: MS major and minor salts (Sigma), LS *(16)* vitamins (1 mg/L thiamine · HCl, 100 mg/L myo-inositol), 30 g/L sucrose, 2 mg/L 2.4-dichlorophenoxyacetic acid, 50 mg/L hygromycin, 200 mg/L ticarcillin disodium, and 3 g/L PhytaGel, pH 5.8.
6. LRHT regeneration medium: MS major and minor salts (Sigma), LS vitamins *(13)* (1 mg/L thiamine · HCl, 100 mg/L myo-inositol), 20 g/L sucrose, 2 mg/L kinetin, 0.5 mg/L indole-3-acetic acid (IAA), 50 mg/L hygromycin, 100 mg/L ticarcillin disodium, and 3 g/L PhytaGel, pH 5.8.

3. Methods

3.1. Callus Induction

In this procedure, embryogenic callus induced from mature and immature rice seeds (embryos) is used as the target tissue for the gene transfer. The following protocol applies to 100 seeds or embryos and can be scaled up or down, if necessary. During most of the following operations, standard sterile procedures should be followed.

1. To sterilize seeds (cv. Nipponbare), place 100 dehusked mature rice seeds, or immature rice caryopses collected fresh from plants, in 20 mL 50% Javex solution, with a drop of Tween-20, in a 100-mL sterile glass beaker covered with aluminium foil. Shake it on a shaker operated at 40 rpm for 30 min or longer (but no longer than 1 h). Rinse them thoroughly (at least three changes) with sterile distilled water in a sterile hood (*see* **Note 6**).
2. Press the sterilized immature seeds on the hull just outside the embryo with thumb and first finger to force the embryos out. Collect the excised embryos in 1 mL of sterile distilled water in a 3-cm Petri dish. After excision, wash the embryos a few times by pipeting in and out to change the water a few times, to remove the contaminated starch and other tissues (*see* **Note 7**).
3. Place 15 sterilized seeds or embryos on the surface of 20 mL solidified MS callus induction medium in 9-cm Petri dishes (9 × 1.5 cm). For mature seeds, anchor the endosperm end into the medium. Avoid direct embryo contact with the medium. For immature embryos, turn the scutellum surface upward with the aid of a dissection microscope.
4. Wrap the plates with parafilm and place them into an unlit growth room or growth chamber at 26°C. Callus becomes visible in a week and is ready for transformation in 3–4 wk after culture (*see* **Note 8**).

3.2. Agrobacterium *Cocultivation*

The transfer of T-DNA from *Agrobacterium* to plant cells takes place during the cocultivation period, when the bacteria are allowed to grow freely in contact with rice callus. Most of the DNA transfer is believed to occur in the surface cells of the callus. Prolonged exposure of the callus to the bacteria should be avoided, because it will cause tissue damage. These approaches were first described in some detail by Hiei et al. *(10)*, and we have found that the following adaptations have further increased the frequency of transformation. It is reassuring to note that this combination of starting materials, culture conditions, bacterial strains, and vectors has recently led to the high-efficiency transformation of maize, as well *(17)*.

1. Introduce Ti-plasmid-based vectors into *A. tumefaciens* (LBA4404 or EHA101) by the freeze–thaw method *(18)*. Plate the transformed cells on solidified YEP

medium with 50 mg/L kanamycin and 50 mg/L hygromycin, and grow at 28°C for 48–72 h, to allow colonies to appear (*see* **Note 9**).

2. Using a flamed and cooled bacterial loop, transfer a single colony from the plate to 5 mL AB medium, containing the same selective antibiotics, in a capped glass tube. Shake overnight at 28°C, using an orbital shaker with environmental control, at a shake rate of about 200 rpm.
3. Pour the overnight culture into 50 mL AB medium containing the same selective antibiotics and 100 μ*M* acetosyringone in a 500-mL flask. Shake overnight under the same conditions as the small culture above.
4. Collect the bacteria by centrifugation in a 50-mL sterile plastic tube, in a Beckman TJ-6 centrifuge or similar model, at 3000 rpm for 10 min. Pour off the supernatant. Wash the bacteria once by resuspending the pellet in 20 mL of MS callus induction medium containing 100 m*M* acetosyringone, and centrifuging again at 3000 rpm for 10 min. Discard the supernatant.
5. Dilute the bacteria with the same medium, to a density of about 10^8 cells/mL, and pour 3–5 mL into 6-cm sterile plastic Petri dishes for the callus immersion.
6. Select embryogenic callus from 3–4-wk-old callus cultures. Place 200–400 mg of vigorously growing callus pieces (3–5 mm in diameter) into the bacterial suspension and immerse them for 30 min, with occasional shaking, in a sterile laminar airflow hood.
7. Remove excess suspension from the calli by placing them on a pad of dry, sterile tissue paper. Transfer the inoculated calli to 2N6-AS medium in 9-cm Petri dishes (50–100 mg/dishes) and incubate at 28°C in darkness for 2 d. During the cocultivation period, check plates regularly for bacterial overgrowth.

3.3. Selection and Regeneration of Transformed Plants

After the cocultivation, the bacterial growth must be inhibited to allow for the preferential growth of the transformed plant cells, and their subsequent selection. This is achieved by including in the culture media two types of antibiotics: one for inhibiting (killing) the bacteria (ticarcillin disodium), the other for inhibiting (killing) the untransformed plant cells (hygromycin), thus allowing transformed cells to grow into colonies. The selected colonies then regenerate into plants in a regenerating medium.

1. Collect the cocultivated calli in a 50-mL sterile plastic tube. Wash the tissues twice with gentle shaking, using 30 mL of MS callus induction medium. Dry them on a pad of sterile tissue paper to remove excess surface water (*see* **Note 10**).
2. Transfer the callus pieces onto the surface of LHT medium in 9-cm Petri dishes (50–100 mg/dish) for bacteriostasis and the selection of transformed cells. Incubate the sealed dishes in darkness at 26°C. Check plates frequently for bacterial overgrowth. If bacterial contamination takes place, transfer uncontaminated callus into fresh dishes, or discard the whole contaminated dish if the contamination has spread.

3. Renew the culture medium every 2–3 wk by transferring the calli to fresh medium. Begin the visual selection of putative hygromycin-resistant microcalli 1 mo after culture on the LHT medium.
4. Excise the resistant microcalli from the co-cultivated callus and transfer them to fresh LHT medium for resistance confirmation and tissue proliferation. For this purpose, no more than 20 tentative resistant microcalli should be placed in a 9-cm Petri dish (*see* **Notes 11** and **12**).
5. Transfer the calli that grow well in the fresh medium to regeneration medium (LRHT) in 9-cm Petri dishes. Place the dishes at 25°C under a photoperiod of 16 h day/8 h dark (about 4000 lx) at 25°C in a controlled-temperature culture room (*see* **Note 13**).
6. Transfer the regenerating callus to 100 mL fresh LRHT medium in Phytacon vessels. Continue these culture conditions until the plantlets reach the top of the containers.
7. Gently remove well-developed plantlets, when they are about 10 cm in height, from the medium and immediately put them in tap water in a beaker or glass tray, to avoid desiccation (*see* **Note 14**).
8. Transplant the plantlets to soil in pots. Cover the plants in the first week after transplantation with a 1- or 2-L glass beaker to reduce water evaporation from the plants. Culture them as seed-grown plants in a greenhouse or growth chamber, with frequent watering.

4. Notes

1. Although old rice seeds (more than 1 yr old) may be used, freshly harvested seeds (within a year) are preferable, because they usually produce callus that has a better regeneration ability.
2. The genotype of rice can have a great influence on the transformation and plant regeneration. Japonica rice is usually superior in tissue culture to indica rice. So far, *Agrobacterium*-mediated transgenic plant production is mainly limited to japonica and javonica varieties *(19–23)*.
3. It seems that LBA4404 causes less tissue damage than EHA101 while their transformation efficiency remains similar. A kanamycin-sensitive derivative of EHA101, called EHA105, could also be used when kanamycin is required to select bacteria containing plasmid.
4. Although a large number of binary vectors are available, most of them have been developed for dicotyledonous plants and their plant selection marker is based on resistance to kanamycin. For rice or other monocot plants, which usually have a high background level of kanamycin resistance, other resistance markers, such as hygromycin and basta resistance, are necessary for effective selection.
5. This protocol simply needs general equipment (sterile loop, forceps, and spatula) for plant tissue culture; no other specialized utensils should be required under normal circumstances.
6. Trace residual amount of the Javex solution may cause severe damage to isolated immature embryos, resulting in greatly reduced callusing ability. Complete

removal of the sterilizing solution is very important. Soaking the rinsed caryopses in plenty of sterile distilled water for 30–60 min before embryo isolation will be helpful.

7. Wear an autoclaved examination glove (Premium Latex Examination Glove, DiaMed Lab Supplies, Mississauga, Ontario) while isolating the embryos, to eliminate contamination.

8. After 3 wk, there should be sufficient friable and embryogenic callus for transformation. Features of an embryogenic callus include the presence of distinct somatic embryos having a smooth epidermal surface and loose morphology, which make it easy to tease callus apart with forceps. Subculture of the embryogenic callus may be performed to provide a continuous supply of the callus for transformation. However, old callus (more than 4 mo old) may have reduced regeneration ability.

9. During the cocultivation period, bacteria grow rapidly. By 48 h, bacteria should become visible on the medium, but not on the callus. Excess growth of bacteria is undesirable, because it will be difficult later to inhibit their growth even under high concentration of antibiotics.

10. Before transfer to LHT medium, the co-cultivated callus may first be cultured in LHT medium without hygromycin for a week, to allow better recovery of the transformed rice cells.

11. Although resistant microcalli may become visible 1 mo after transfer to selection medium, most of them will appear later. Keep culturing the callus that has produced resistant colonies for two to three passages to allow more resistant microcalli to develop.

12. Resistant callus has a dry and compact morphology and grows very fast in LHT medium; tolerant or sensitive callus looks watery and soft, and it either turns brown or grows more slowly. A more reliable way to check if it is transgenic is to stain a part of it in X-gluc solution (Jersey Lab Supply, Livingston, NJ). Transgenic tissue will turn blue in minutes to hours at 37°C.

13. Plant differentiation occurs 2–3 wk after culture in the regeneration medium. For some cell lines, several subcultures may be required to induce shoot differentiation. Multipe-shoot formation is very common in rice. Shoots from the same callus should be recorded, accordingly, to allow for their proper identification.

14. Care should be taken not to damage the roots of plants. The best soil in which to grow rice plants is paddy soil from the field. Compost made up of vermiculite, top soil, and peat moss (Premier peat moss, Plant Product, Brampton, Ontario) (60:20:20) can also be used for growing rice plants.

Acknowledgments

This work was supported by The Rockefeller Foundation, and in part by the Natural Sciences and Engineering Research Council of Canada. We would like to thank T. Candresse for the pKGH4 vector and E. E. Hood for *A. tumefaciens* strains.

References

1. Robert, L. S., Donaldson, P. A., Ladaique, C., Altosaar, I., Arnison, P. G., and Fabijanski, S. F. (1990) Antisense RNA inhibition of β-glucuronidase gene expression in transgenic tobacco can be transiently overcome using a heat-inducible β-glucuronidase gene construct. *Bio/Technology* **8,** 459–464.
2. Vain, P., De Buyser, J., Bui Trang, V., Haicour, R., and Henry, Y. (1995) Foreign gene delivery into monocotyledonous species. *Biotechnol. Adv.* **13,** 653–671.
3. Shimamoto, K., Tereda, R., Izawa, T., and Fujimoto, H. (1989) Fertile transgenic rice plants regenerated from transformed protoplasts. *Nature* **338,** 274–276.
4. Datta, S. K., Peterhans, A., Datta, K., and Potrykus, I. (1990) Genetically engineered fertile indica-rice recovered from protoplasts. *Bio/Technology* **8,** 737–740.
5. Peng, J., Kononowicz, H., and Hodge, T. K. (1992) Transgenic indica rice plants. *Theor. Appl. Genet.* **83,** 855–863.
6. Chen, M.-T, Lee, T.-M., and Chang, H.-H. (1992) Transformation of indica rice (*Oryza sativa* L.) mediated by *Agrobacterium tumefaciens*. *Plant Cell Physiol.* **33,** 577–583.
7. Gould, J., Devery, M., Hasegawa, O., Ulian, E. C., Peterson, G., and Smith, R. H. (1991) Transformation of *Zea mays* L. using *Agrobacterium tumefaciens* and the shoot apex. *Plant Physiol.* **95,** 426–434.
8. Mooney, P. A., Goodwin, P. B., Dennis, E. S., and Llewellyn, D. J. (1991) *Agrobacterium tumefaciens*-gene transfer into wheat tissues. *Plant Cell Tissue Organ Cult.* **25,** 209–218.
9. Raineri, D. M., Bottino, P., Gordon, M. P., and Nester, E. W. (1990) *Agrobacterium*-mediated transformation of rice (*Oryza sativa* L.). *Bio/Technology* **8,** 33–38.
10. Hiei, Y., Ohta, S., Komari, T., and Kumashiro, T. (1994) Efficient transformation of rice (*Oryza sativa* L.) mediated by *Agrobacterium* and sequence analysis of the boundaries of the T-DNA. *Plant J.* **6,** 271–282.
11. Hoekema, A., Hirsch, P. R., Hooykaas, P. J. J., and Schilperoort, R. A. (1983) A binary plant vector strategy based on separation of vir- and T-region of the *Agrobacterium tumefaciens* Ti-plasmid. *Nature* **303,** 179–180.
12. Hood, E. E., Helmer, G. L., Fraley, R. T., and Chilton, M.-D. (1986) The hypervirulence of *Agrobacterium tumefaciens* A281 is encoded in a region of pTiBo542 outside of T-DNA. *J. Bacteriol.* **168,** 1291–1301.
13. Le Gall, O., Torregrosa, L., Danglot, Y., Candresse, T., and Bouquet, A. (1994) *Agrobacterium*-mediated genetic transformation of grapevine somatic embryos and regeneration of transgenic plants expressing the coat protein of grapevine chrome mosaic nepovirus (GCMV). *Plant Sci.* **102,** 161–170.
14. Chilton, M.-D., Currier, T. C., Farrand, S. K., Bendich, A. J., Gordon, M. P., and Nester, E. W. (1974) *Agrobacterium tumefaciens* DNA and PS8 bacteriophage DNA not detected in crown gall tumors. *Proc. Natl. Acad. Sci. USA* **71,** 3672–3676.
15. Murashige, T. and Skoog, F. (1962) A revised medium for rapid growth and bio-assays with tobacco tissue culture. *Physiol. Plantarum* **15,** 473–497.

16. Linsmaier, E. M. and Skoog, F. (1965) Organic growth factor requirements of tobacco tissue culture. *Physiol. Plantarum* **18,** 100–127.
17. Ishida, Y., Saito, H., Ohta, S., Hiei, Y., Komari, T., and Kumashiro, T. (1996) High efficiency transformation of maize (*Zea mays* L.) mediated by *Agrobacterium tumefaciens. Nature Biotechnol.* **14,** 745–750.
18. Holsters, M., de Waele, D., Depicker, A., Messens, E., Van Montagu, M., and Schell, J. (1978) Transfection and transformation of *A. tumefaciens. Mol. Gen. Genet.* **163,** 181–187.
19. Dong, J. J., Teng, W. M., Bucholz, W. G., and Hall, T. C. (1996) *Agrobacterium*-mediated transformation of javanica rice. *Mol. Breed.* **2,** 267–276.
20. Aldemita, R. R. and Hodges, T. K. (1996) *Agrobacterium tumefaciens*-mediated transformation of japonica and indica rice varieties. *Planta* **199,** 612–617.
21. Rashid, H., Yokoi, S., Toriyama, K., and Hinata, K. (1996) Transgenic plant production mediated by *Agrobacterium* in indica rice. *Plant Cell Rep.* **15,** 727–730.
22. Park, S. H., Pinson, S. R. M., and Smith, R. H. (1996) T-DNA integration into genomic DNA of rice following *Agrobacterium* inoculation of isolated shoot apices. *Plant Mol. Biol.* **32,** 1135–1148.
23. Yokoi, S., Tsuchiya, T., Toriyama, K., and Hinata, K. (1997) Tapetum-specific expression of the *osg6b* promoter-beta-glucuronidase gene in transgenic rice. *Plant Cell Rep.* **16,** 363–367.

2

Barley as a Producer of Heterologous Protein

Leena Mannonen, Kristian Aspegren, Anneli Ritala,
Hanna Simola, and Teemu H. Teeri

1. Introduction

Recent developments in genetic engineering and molecular biology have enabled the production of transgenic plants with improved characteristics, or cell and tissue cultures with altered or improved metabolic activity. However, no universally applicable gene transfer method has hitherto been established for plant transformation. The monocotyledonous plants, including grasses and cereals, have proved to be recalcitrant to genetic engineering. They do not belong to the natural host range of the efficient gene vector *Agrobacterium*, which is widely used for transformation of dicotyledonous plants. Particle bombardment, developed by Sanford et al. *(1)*, is a physical gene transfer method without target material limitations. Using this method, transgenic plants have been produced successfully from all major cereals *(2–8)*. Protoplast-mediated gene transfer methods, e.g., electroporation or polyethelene glycol (PEG) treatment, have long been hampered by the recalcitrance of important cereal crop species to the regeneration of plants from these protoplasts. However, success has recently been reported in this field, even with elite cultivars of barley *(9)*.

In this chapter, particle bombardment is used as a means to obtain stable transgenic barley cultures, which can be used for production of a heterologous enzyme, i.e., the endo-β-glucanase (EGI) of *Trichoderma reesei*, in amounts sufficient to conduct studies on its properties. Particle bombardment results in a mixed population of transgenic and untransformed cells, which leads to the need for efficient selection and screening in order to isolate a pure and stable clone. In this respect, a protoplast-mediated gene transfer method would offer an attractive alternative. All regenerated callus clusters would originate from a

From: *Methods in Biotechnology, Vol. 3:*
Recombinant Proteins from Plants: Production and Isolation of Clinically Useful Compounds
Edited by: C. Cunningham and A. J. R. Porter © Humana Press Inc., Totowa, NJ

single protoplast, which is already a pure clone. However, the applicability of protoplasts is hindered when the enzyme under study possesses activity that may disturb the cell wall synthesis needed for regeneration of productive cell lines, as may be the case in our particular example *(10)*.

This chapter presents a method for production of transgenic cell cultures and purification of a heterologous protein for evaluation of its properties. The β-glucanase activity of the model enzyme does not significantly affect the growth of the barley cell culture chosen as target for gene transfer through bombardment *(11)*. The procedure described may be used generally for production of foreign proteins in plants. Suspension cultures of barley cells are active protein secretors and proteins can be purified from the growth medium rather easily. Barley cell cultures may therefore offer an alternative to microbial expression systems, especially when posttranslational modifications, i.e., proper folding or glycosylation of produced proteins, are needed to retain their full activity. Plant cells are not yet comparable to microbial organisms as feasible production systems. There is no reason to doubt, however, that optimization of culture conditions and further development of plant expression vectors would not make plant cell cultures a competent production system, at least for specific proteins. The method presented in this paper also reveals the advantage of plant cell cultures in providing valuable information on the expression of the heterologous genes long before transgenic plants are available. Furthermore, the transformation method presented here is applicable to the production of transgenic barley plants when only the target tissue is chosen to bear embryogenic capacity *(6)*.

2. Materials

2.1. Plant Material

A cell culture of barley (*Hordeum vulgare* L.) var. Pokko was initiated from embryos on 211 medium (*see* **Subheading 2.4.**). A stable and fast-growing suspension culture is needed as the target for gene transfer and for production of heterologous proteins. The callus formed from the embryos in medium 211 was collected and suspended in the same medium. The culture stabilized during the first 5–10 passages; whereafter it has been maintained as a suspension culture in the same medium with weekly subculturing. The suspension, referred to as P1, is grown at 25°C in a 16/8 h light/dark regime under a white fluorescent lamp, 30 W (Airam) and Fluora 36 W (Osram), 1:1. The culture has been stable for 8 yr (*see* **Note 1**).

2.2. DNA Material

The barley culture is co-transformed with two plasmids. The DNA stock solution concentrations are 1.0 μg/μL.

1. Plasmid pHTT303 *(13)* contains the selectable marker *npt*II for antibiotic resistance under the control of the constitutive CaMV promoter.
2. Plasmid pKAH21 *(10)* carries the cDNA of *egl*1 for the production of heat-stable endo-β-glucanase (EGI, EC 3.2.1.4) of *T. reesei* under the control of the constitutive CaMV 35S promoter (*see* **Note 2**).

2.3. Special Laboratory Equipment

1. Rotary shaker (Infors, Bottmingen, Switzerland).
2. Biolistic® PDS 1000 He gene transfer device (Bio-Rad, Hercules, CA).
3. Programmable freezing unit for cryopreservation (Kryo 10, Planer Biomed, Sunbury-on-Thames, UK), cryotubes (Nunc, Roskilde, Denmark).
4. Glass-fiber filters (13400-47-S, Sartorius, Goettingen, Germany).
5. Gel filtration unit packed with Biogel 10 DG (Bio-Rad).
6. Column for hydrophobic chromatography, phenyl Sepharose FF (Pharmacia, Sollentuna, Sweden).
7. Gel filtration column packed with Sephadex G-25 (Pharmacia).
8. Anion exchange unit packed with DEAE-Sephadex (Pharmacia).
9. Cation exchange unit packed with CM Sepharose FF (Pharmacia).
10. Peristaltic pump P-1 (Pharmacia).
11. Fraction collector Frac-100 (Pharmacia).
12. UV detector single path monitor UV 1 (Pharmacia).
13. Spectrophotometer (Hitachi, Tokyo, Japan).
14. Power supply (Consort E722) 20–2000 V, 0–300 W, 0–200 mA (Turnhout, Belgium).
15. Midget electrophoresis unit (LKB 2050, Pharmacia).
16. Transblot shell (Bio-Rad).
17. Nitrocellulose filter paper (Schleicher&Schuell BA 85, Dassel, Germany).
18. MultiphorII electrophoresis unit (LKB 2117, Bromma, Sweden).
19. GelBond PAG film (LKB, Pharmacia).
20. pH Surface electrode (Ingold, Willmington, MA).

2.4. Growth Media, Buffers, and Reagents

2.4.1. Tissue Culture

1. Medium 211 is a modified B5 medium described by Séquin-Schwarz et al. *(12)*. All reagents are of analytical grade unless otherwise stated. In all solutions, UHP (ultrahigh pure water, Elga) is used. Before sterilization by autoclaving, the pH is adjusted to 5.8.
   a. Macrosalts: 270 mg/L KH_2PO_4, 2020 mg/L KNO_3, 400 mg/L NH_4NO_3, 370 mg/L $MgSO_4 \cdot 7H_2O$, and 290 mg/L $CaCl_2 \cdot 2H_2O$.
   b. Microsalts: 10 mg/L $MnSO_4 \cdot H_2O$, 3 mg/L H_3BO_3, 2 mg/L $ZnSO_4 \cdot 7H_2O$, 0.25 mg/L $Na_2MoO_4 \cdot 2H_2O$, 0.025 mg/L $CuSO_4 \cdot 5H_2O$, 0.025 mg/L $CoCl_2 \cdot 6H_2O$, 0.75 mg/L KI, 27.8 mg/L $FeSO_4 \cdot 7H_2O$, 37.3 mg/L $Na_2EDTA \cdot 2H_2O$.
   c. Vitamins: 100 mg/L myo-inositol, 1 mg/L nicotinic acid, 10 mg/L thiamine Cl-HCl, 1 mg/L pyridoxine-HCl.

   d. 30 mg/L Sucrose (food grade).
   e. 4 mg/L 2,4-Dichlorophenoxyacetic acid (2,4-D, Sigma D-2128).
   f. 3 g/L Gellan gum for solid medium (Phytagel, Sigma P-8169, St. Louis, MO).

2.4.2. Particle Bombardment

1. Tungsten particles M20 (1.32 μm, Bio-Rad), sterilized in 70% ethanol, as a 60 mg/mL suspension in sterile UHP water (*see* **Note 3**).
2. 2.5M CaCl$_2$ in UHP water, sterilized by autoclaving.
3. 0.1M Spermidine (Sigma S-2626) in UHP water, filter-sterilized.
4. Geneticin® G418 (Sigma G-9516), stock solution of 25 mg/mL, filter-sterilized.

2.4.3. Cryopreservation

1. 2M Sorbitol in growth medium 211 (sterile).
2. 4M Sorbitol in water (sterile).
3. Dimethylsulfoxide (DMSO, Merck 9678, Darmstadt, Germany).

2.4.4. Chromatographic Methods

1. Acetate buffer I: 20 mM NaAc, pH 4.5, 0.5M (NH$_4$)$_2$SO$_4$, conductivity 78 mS/cm.
2. Acetate buffer II: 20 mM NaAc, pH 4.5, 0.25M (NH$_4$)$_2$SO$_4$, conductivity 42 mS/cm.
3. Acetate buffer III: 20 mM NaAc, pH 4.5, 0.1M (NH$_4$)$_2$SO$_4$, conductivity 24 mS/cm.
4. Acetate buffer IV: 5 mM NaAc, pH 4.5.
5. Acetate buffer V: 200 mM NaAc, pH 5.5.
6. Acetate buffer VI: 8 mM NaAc, pH 5.5.
7. Acetate buffer VII: 10 mM NaAc, pH 5.5, conductivity 0.73 mS/cm.
8. Acetate buffer VIII: 100 mM NaAc, pH 3.8.
9. Acetate buffer IX: 5 mM NaAc, pH 3.8.
10. Acetate buffer X: 7 mM NaAc, pH 3.8, conductivity 0.13 mS/cm.
11. Acetate buffer XI: 20 mM NaAc, pH 5.0.

2.4.5. Biochemical Assays

1. Buffer I: 50 mM Na-citrate buffer, pH 4.8.
2. McIlvaine buffers (*14*): 0.2M Na$_2$HPO$_4$ adjusted to pH 2.5–8.0 with 0.1M citric acid.
3. 1% (w/v) Barley β-glucan (Megazyme, Bray, Ireland) in appropriate buffer.
4. DNS reagent (*15*): 10 g/L 3,5-dinitrosalisylic acid, 16 g/L NaOH, and 30 g/L K/Na-tartrate in water. Clarify by filtration. The reagent must be further filtered before each use.
5. Glucose standard stock solution of 10 mM glucose in water is diluted in sodium citrate buffer to yield a standard series of 2.5–10 mM glucose.
6. Dye reagent for protein assays (Bio-Rad 500-0006).
7. Protein standard stock solution of 20 mg/mL bovine serum albumin in water is diluted in water to yield a standard series of 0–20 μg/mL.

2.5. Electrophoretic Methods

2.5.1. SDS-Polyacrylamide Gel Electrophoresis (SDS-PAGE) and Western Blotting

1. Running gel: 10% (w/v) polyacrylamide (a mixture of 97% acrylamide and 3% bisacrylamide) in 375 mM Tris-HCl, pH 8.8, 0.1% (w/v) SDS, 0.03% (w/v) ammonium persulphate (APS), 0.67 µL/mL N,N,N',N'tetramethylene diamine (TEMED).
2. Upper gel: 5% (w/v) polyacrylamide (acryl-*bis*-acrylamide mixture) in 125 mM Tris-HCl, pH 6.8, 0.1% (w/v) SDS, 0.03% (w/v) APS, 1 µL/mL TEMED.
3. Running buffer: 25 mM Tris, 192 mM glycine, 0.01% (w/v) SDS, pH 8.3.
4. SDS-Mix: 367 mM Tris-HCl, pH 6.8, 30% (v/v) mercaptoethanol, 30% (v/v) glycerol, 0.003% bromophenol blue, 4% (w/v) SDS.
5. Low mol wt calibration kit (Pharmacia) in the range of 14.4–94 kDa.
6. Blotting buffer: 25 mM Tris, 192 mM glycine, 10% (v/v) methanol, pH 8.3.
7. Ponceau S (Sigma P-3504) in 3% (w/v) TCA.
8. Blocking buffer: 1% (w/v) ovalbumin or powdered skim milk in TBST buffer.
9. TBST buffer: 10 mM Tris-HCl, pH 8.0, 150 mM NaCl, 0.05% (v/v) Tween-20.
10. Primary antibody: Polyclonal rabbit antiserum against EGI of *T. reesei*, diluted 1:2000 in blocking buffer.
11. Secondary antibody: Goat antirabbit IgG-alkaline phosphatase conjugate (Bio-Rad) diluted 1:1000 in TBST buffer.
12. Staining reagent (Promega): 6.6 µL/mL NBT (nitroblue tetrazolium), 3.3 µL/mL BCIP (5-bromo-4-chloroindol-3-ylphosphate) in AP buffer.
13. AP buffer: 100 mM Tris-HCl, pH 9.5, 100 mM NaCl, 5 mM MgCl$_2$.

2.5.2. Isoelectric Focusing

1. Gel: 7.5% (w/v) polyacrylamide (acryl-bisacrylamide mixture) in water, 0.06 mL/mL ampholine mixture pH 3.5–5 (LKB 80-1125-89), 0.05% (w/v) APS, 0.5 µL/mL TEMED.
2. 0.5M Acetic acid.
3. 0.5M NaOH.
4. Staining for activity: 0.1% (w/v) barley β-glucan (Biocon) and 1% (w/v) agarose (indubiose A37 HAA IBF, Biosepra, France) in 0.1M Na-acetate buffer, pH 5.0.
5. 0.1% (w/v) Congo Red (Merck).

3. Methods

3.1. Gene Transfer and Selection of Transformants

Particle bombardment is performed according to the method of Ritala et al. *(13)*. Use the suspension culture of P1 1 d after subculture, when the cells are at the early acceleration phase of growth.

1. Sieve the suspension through a 1000 µm net and spread samples of about 200 mg fresh weight on a nylon membrane to cover an area of 2 cm in diameter. Place the membrane with cells on normal growth medium (211) solidified with 0.3% (w/v) gellan gum in a 5-cm diameter Petri dish.

2. Prepare the particles to be used in bombardment by mixing 100 µL of sterile tungsten suspension in an Eppendorf tube with 25 µL of DNA solution (12.5 µg pHTT303 and 12.5 µg pKAH21).

3. Precipitate the DNA onto the tungsten by adding 100 µL $CaCl_2$ and 40 µL spermidine, carefully mixing between additions. Allow the particles to settle to the bottom of the tube and, after settling, remove the supernatant completely.

4. Wash the particles with 500 µL of absolute ethanol and resuspend in 120 µL absolute ethanol.

5. After thorough mixing, apply 10 µL of the suspension carefully on the macrocarriers. The macrocarriers may be stored in a desiccator and used 1–3 h after preparation.

6. The Biolistic PDS 1000 He device is used for bombardment. Use the following conditions *(13)*: pressure in sample chamber 3.1 KPa, helium shock wave 9.0 MPa. The distances between macrocarrier and stopping screen, and between macrocarrier and rupture disk, are 6 mm and 9 mm, respectively. The distance between the stopping screen and sample is 6 cm (*see* **Note 4**).

7. After bombardment, allow the cells to recover for 2 d on the nylon net on solid medium, in their normal growth conditions.

8. Spread the cells on fresh medium supplemented with 100 µL/mL geneticin for selection of antibiotic resistant calli. Plate the cells every 2 wk on fresh growth medium with 100 µg/mL geneticin, for a minimum of eight passages.

9. In order to obtain pure transgenic clones, divide the calli into small clusters of 80–120 cells, and cultivate them under selection. After four passages with division, suspend the clusters in the normal liquid medium (211) with 75 µg/mL geneticin (*see* **Note 5**).

10. After a week of cultivation, the supernatant may be assayed for β-glucanase activity.

11. Cultivate the most active cell line(s) further as a suspension by weekly subculturing in the selection medium.

3.2. EGI in Transgenic Cultures

In order to examine chromosomal integration of the cDNA for EGI in the barley genome, the total DNA of cultured cells can be isolated according to the method presented by Dellaporta et al. *(16)*. Integration can be verified by Southern blot hybridization according to the method described in Maniatis et al. *(17)*, using plasmid pKAH21 as a probe as described by Aspegren et al. *(10)*.

3.2.1. Isolation and Purification of EGI

Because of the signal sequence encoded by EGI cDNA, the gene product is secreted into the growth medium. Isolation of EGI enzyme from the growth medium begins by removal of cells after 1 wk of growth by filtration through a 5 μm nylon net, and by clarifying the spent medium by centrifugation at 8350g for 25 min at 7°C. Clarified supernatant can be preserved for at least 2 mo at –20°C for further purification without significant loss of activity. Several batches are collected and pooled for further purification.

The first purification step is accomplished by hydrophobic chromatography.

1. Equilibrate the column with acetate buffer I. Adjust the pooled growth medium to pH 4.5 with acetic acid, and to a conductivity of 78 mS/cm with ammonium sulfate. Clarify the medium by centrifugation as earlier and apply it (i.e., 12 L) to the column at 7 mL/min (overnight) and continue with buffer I at 20 mL/min until no protein can be detected in the effluent. Detect protein on-line from the effluent by absorbance at 280 nm and collect fractions of, i.e., 30 mL.
2. Elute impurities from the column with buffers II (600 mL) and III (500 mL) at 20 mL/min. Build a decreasing salt gradient from 20 mM to 5 mM sodium acetate with buffers III (500 mL) and IV (500 mL), and elute the EGI enzyme from the column. Complete EGI elution with a further 600 mL of buffer IV.
3. Assay the fractions for β-glucanase activity and pool the active fractions. The EGI starts to elute during the wash with buffer III, resulting in approx 65 fractions, giving 1950 mL of EGI containing preparation.

The second step employs anion exchange chromatography on a DEAE column.

1. Desalt the pooled EGI sample by gel filtration on Sephadex G25. For this purpose, equilibrate a Sephadex column with acetate buffers V and VI, using approx 1.5 L of each. Apply the EGI sample to the column and elute it with buffer VI. Detect the protein in the effluent by absorption at 280 nm and collect the protein fractions (approx 3500 mL).
2. Apply the pooled fractions to a DEAE column after equilibration with acetate buffer VII. Wash proteinaceous impurities from the column with 800 mL of buffer VII and elute EGI with an increasing salt gradient formed with 450 mL of buffer VII and 450 mL of buffer VII containing 140 mM NaCl. Collect 25 mL fractions and assay them for β-glucanase activity. Pool active fractions (approx 25 fractions, giving 625 mL) and use them for further purification.

The third step involves cation exchange chromatography using CM-Sepharose FF.

1. Desalt the pooled EGI preparation from the previous step. Equilibrate the Sephadex G25 column with acetate buffers VIII and IX (approx 1.5 L each) and elute with buffer IX. Detect and collect the proteins in fractions, as previously. The pooled protein fraction is approx 850 mL.
2. Pre-equilibrate the CM Sepharose FF with acetate buffer X and apply the pooled enzyme solution (850 mL) to the column at 1 mL/min overnight. Elute the impu-

rities off with buffer X and, when no protein is detected in the effluent, elute EGI with an increasing salt (5–20 m*M* sodium acetate) and pH (pH 3.8–5.0) gradient, produced using equal volumes of buffers X and XI (approx 80 mL each).
3. Collect fractions of 5 mL and assay for β-glucanase activity. EGI is eluted as a sharp peak, and approx 20 mL of purified enzyme may be obtained. The enzyme should appear as a single protein band in SDS-PAGE.

3.2.2. Protein Assay

1. Protein content is determined according to the method of Bradford *(18)*. This microassay is suitable for concentrations below 25 μg/mL.
2. Prepare a standard curve with BSA by mixing 800 μL of BSA dilutions (0–20 μg/mL) with 200 μL of Bio-Rad protein assay dye reagent concentrate. After 5 min, read the absorbance at 595 nm.
3. Dilute the sample, as necessary, with water. Mix 800 μL of sample dilution with 200 μL of dye reagent concentrate. Measure the absorbance at 595 nm and read the concentration from the standard curve.

3.2.3. β-Glucanase Activity

EGI has high activity against barley (1–>3)(1–>4)-β-glucan, some activity against soluble cellulosic material, and appreciable xylanolytic activity. EGI activity may be assayed using barley β-glucan as substrate and determining the liberated reducing sugars with the DNS method *(15)*.

1. Remove the cells from the growth medium by filtration through a 5-μm nylon net and glass-fiber filter (*see* **Note 6**).
2. Remove reducing sugar in the culture filtrates by gel filtration through Sephadex-25 (Bio-Rad Biogel 10 DG). Apply a 3-mL sample to the column, which is first equilibrated with 50 mL of citrate buffer (I). Elute the enzyme with 4 mL of the same buffer. No pretreatment is needed for the samples of the purification steps.
3. The samples may be stored at –20°C for analysis.
4. The endo-β-glucanase (EGI) activity is measured according to the International Union of Pure and Applied Chemistry (IUPAC) standard instructions for the HEC assay *(19)*, but using 1% barley β-glucan as substrate. Prepare the substrate solution the day before by suspending 1% β-glucan in buffer (I), and by gradually heating to dissolve the substrate. Allow to cool slowly at 4°C overnight. The substrate solution keeps for 2 wk at 4°C.
5. Because of the thermotolerant nature of EGI enzyme, the assay is conducted at 50°C. Mix 0.1 mL of enzyme dilution with 0.9 mL of substrate. After 10 min incubation at 50°C, stop the reaction with 1.5 mL DNS.
6. Develop the color in a vigorously boiling water bath for 5 min, and cool the tubes rapidly in ice.
7. Measure the intensity of the developed color spectrophotometrically at 540 nm and read the result from the standard curve.

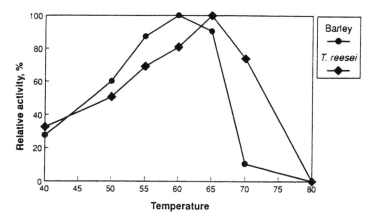

Fig. 1. Temperature optimum of fungal and heterologous barley-made EGI. Relative activity is expressed as percentage of maximum activity in optimum temperature.

8. Prepare a standard curve, with a series of glucose in citrate buffer, to range from 0 to 10 m*M*. Conduct **step 5** without sample (enzyme dilution). Add standard the dilution after stopping the reaction by DNS. Mix well and develop color as in **steps 7** and **8**.
9. If the dilution of the enzyme sample to be assayed is low, the possible background effect is corrected by an enzyme blank, which is obtained by treating the enzyme dilution as the sample, but adding it after the DNS reagent.

The activities are expressed in katals, i.e., moles of reducing sugars as glucose formed in 1 s. When screening cell cultures for EGI activity, the assay is also conducted with untransformed culture medium. However, at 50°C the background activity is almost nonexistent (*see* **Note 7**).

3.2.4. Enzyme Characteristics

Fungal EGI is a thermotolerant enzyme possessing maximum activity around 60°C and in slightly acidic conditions. In order to evaluate the barley-made enzyme, the conditions for maximum activity are determined by carrying out activity assays at different temperatures and pH values. For temperature optima, the assay is performed as described earlier (*see* **Subheading 3.2.3.**), except that incubation is carried out at temperatures varying from 40 to 80°C. It should be noted that the glucose standards are also treated accordingly at the different temperatures. **Figure 1** shows the behavior of both fungal and barley-made EGI in temperatures of 40–80°C. No activity can be measured at 80°C, and there is a slight deviation in optimum temperature, for the fungal enzyme at 65°C, and for the barley-made enzyme at 60°C. However, both enzymes can

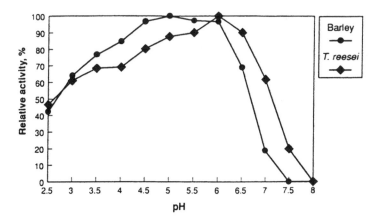

Fig. 2. pH optimum of fungal and heterologous barley-made EGI. Relative activity is expressed as percentage of maximum activity at pH optimum.

be regarded as thermotolerant and the difference in optima may be caused by differences in glycosylation.

For pH optima, the activity assay is carried out, as described in **Subheading 3.2.3.**, at 50°C. The pH is adjusted between 2.5 and 8.0 using McIlvaine's buffer. All dilutions are made in appropriate buffer, including enzyme samples, substrate, and glucose standards. The results are shown in **Fig. 2**. No activity is observed at pH 8; only in acidic conditions. The optimum of the fungal EGI is pH 6 and that of the barley-made enzyme at pH 5.0. However, the shape of the curve is flat in the range of pH 4.5–6.0. Here again some modifications of the protein structure may be the reason for the observed variation.

3.2.5. Molecular Weight

The proteins in the samples are separated in a 10% acrylamide running gel according to Laemmli *(20)*, using low molecular mass standards. The gel is analyzed for EGI by Western blotting, as described by Towbin et al. *(21)*, using EGI polyclonal antiserum for detection.

1. For SDS-PAGE, mix protein dilutions in water (20 μL) with SDS-Mix (10 μL) and boil them for 2 min in an Eppendorf tube. Spin the samples down by a short centrifugation and apply them in gel slots. The mol wt standard solution and the blank water control are treated similarly.
2. Run the gels at 10 mA for 30 min (samples in upper gel) and at 25 mA in the lower gel, until the blue dye front almost reaches the bottom of the gel.
3. For specific detection of EGI protein, antiserum is used in immunoblotting. After SDS-electrophoresis, the proteins are electrophoretically transferred to a nitro-cellulose (NC) filter, as described by Towbin et al. *(21)*. Before blotting, equili-

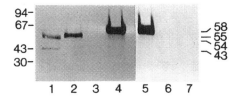

Fig. 3. SDS-PAGE and Western blotting of fungal and barley-made EGI, adapted from Mannonen *(11)*. Lane 1, growth medium of transgenic barley; lane 2, purified EGI from barley; lane 3, untransformed barley growth medium; lanes 4 and 5, fungal EGI; Lane 6, untransformed barley cell extract; lane 7, transgenic barley cell extracts. Lanes 1–3 contain 0.07 µg protein; lane 4, 0.03 µg; lane 5, 0.045 µg; and lanes 6–7, 0.1 µg.

brate the gel in blotting buffer for 30 min and fill the blotting chamber with the same buffer. Put the NC filter on top of the gel and assemble the sandwich between three layers of filter paper and one layer of superlon. Put the package between the electrode plates (gel on the cathode side). Transfer the protein to the NC filter at 400 mA for 2 h, or at 100 mA overnight.

4. Stain the NC filter after blotting with Ponceau S stain for 10 min, to visualize the molecular mass reference proteins. Remove the stain by washing with water.

5. Incubate the NC filter for 2 h at room temperature, or overnight at 4°C in blocking buffer, before the addition of primary antibody solution against EGI. Incubate with primary antibody for 30 min. Wash the filter three times with TBST buffer.

6. Incubate the NC filter in secondary antibody solution for 30 min. After three washes with TBST buffer, dry the filter on Whatman filter paper.

7. To visualize the alkaline phosphatase conjugated to the secondary antibody, incubate the filter in staining buffer until the bands appear. After staining, wash the filter with water. Allow to dry and photograph as necessary.

Figure 3 shows the result obtained by Western blotting. First, it can be observed that the EGI polyclonal antiserum reacts with barley-made EGI (lanes 1 and 2), but not with other barley proteins (intra- or extracellular, lanes 3 and 6). EGI is effectively secreted into the growth medium (lanes 1 and 7), and it possesses a slightly lower molecular weight (54–55 kDa) than the original fungal EGI (58 kDa). This difference may be caused by the different modes of glycosylation employed by fungi and plants, or proteolytic modification of the enzyme, which also explains the EGI active band at 43 kDa in lane 1.

3.2.6. Isoelectric Point

Analytical isoelectric focusing is performed with an LKB ampholine mixture for pH 3.5–5.0 (Pharmacia) according to the manufacturer's instructions, as described by Niku-Paavola *(22)*. After focusing, the pH along the gel is measured using a surface electrode. EGI activity is demonstrated with a substrate plate containing 0.1% β-glucan in agarose, as described by Kvesitadze et al. *(23)*.

1. Assemble the polyacrylamide gel in the focusing apparatus and switch the cooling on (4°C). Soak the filter paper electrode strips in suitable solutions, the anode in acetic acid and the cathode in NaOH, and put onto the gel 0.5 cm from the edges. Place the electrodes over the strips. Start the prerun at 25 mA, 1500 V, and 25 W. D a 20 min, during which time the pH gradient is formed.
2. Dilute samples in water to a protein concentration of 1.5 µg/30 µL and pipet them onto pieces of filter paper. Set the papers on the cathode side of the gel. Run the samples in the gel for 30 min using the same settings as in the prerun.
3. Take the filter paper pieces off with forceps and continue the run for 70 min.
4. After the run, separate a piece of the gel and measure the pH gradient with a surface electrode.
5. The EGI active bands are detected by incubating the gel on 0.1% barley glucan solidified with 1% agarose. Incubate the gel on the substrate plate for 10 min at 50°C. After incubation, wash the plate for 5–10 min in 0.1% Congo Red solution. The activity is observed as clear bands against the red background.

The purified fungal EGI preparation appeared to contain two isoenzymes or proteolytic products at pI 4.75 and 4.8 *(11)*. The purified barley-made EGI had a slightly lower pI of 4.45. No EGI activity was detected in the growth medium of untransformed barley. The decrease in pI may be caused by different glycosylation or proteolytic modification of the enzyme.

3.3. Preservation of Productive Cell Lines

Plant cells growing in an undifferentiated state are known to be genetically unstable, i.e., cell cultures producing secondary metabolites tend to lose their productivity during prolonged in vitro cultivation. Cryopreservation at ultra-low temperatures is the only method by which the whole metabolism can be halted, thus allowing the cells to retain their properties unchanged. For successful results, a specific freezing program must be developed for each cell type *(24)*. The P1 cells used in this paper are highly vacuolized cells approx 50 µm in diameter, growing in small aggregates, with a doubling time of 2.5 d.

1. When subculturing the cell culture for preservation, sieve the suspension through 1000 µm mesh in order to homogenize the aggregate size.
2. Before freezing, the cells need to be plasmolyzed in order to reduce their water content. Two to three days after subculture, at an early exponential growth phase, make the suspension 1*M* with respect to sorbitol by addition of an equal volume of 2*M* sorbitol in medium 211. Cultivate the culture as normal overnight.

3. Cool the suspension on ice and transfer the suspension to a graduated sterile vessel. Allow the cells to settle and concentrate the suspension to packed cell volume of 40%.
4. Make the suspension 5% (v/v) with DMSO by gradual addition over 30 min. Keep the suspension on ice for a further 30 min before pipeting into cryotubes (1 mL/1.2 mL tube).
5. Precool the freezing apparatus to 0°C before transfer of tubes to the freezing chamber. Start the freezing program with a holding period of 30 min at 0°C. Continue freezing at –0.5°C/min until –40°C, from which temperature the tubes are plunged into liquid nitrogen.
6. For recovery, thaw the samples rapidly in a 40°C water bath, in a container filled with sterile water (2 tubes/50 mL of water). Pipet or pour the suspension aseptically onto solid growth medium. The growth commences within 1 wk and, after 2–3 wk of cultivation, the culture can be transferred to liquid medium and cultivated normally (*see* **Note 8**).

4. Notes

1. A stable, homogenous, and fast-growing suspension culture is useful in the production of heterologous proteins in amounts needed for studies on their characteristics and function. Furthermore, an efficient gene transfer method, optimized for the cell type in question, is needed. A barley cell culture is used in our studies, because we want to introduce the same enzyme activity in barley plants.
2. In order to be able to harvest the gene product as easily as possible, it should be secreted. In our case, the *egl* cDNA contains a fungal signal sequence for secretion, which appeared to be functional in plants.
3. Instead of tungsten particles, gold can also be used. In our case, no marked difference has been observed between these two metals. The particle size is chosen with respect to the cell size.
4. The conditions for bombardment should be adjusted for the cell culture to be used. The penetrating force depends on the cell and aggregate size. Sometimes plasmolysis before bombardment is beneficial.
5. An efficient cloning method is to spread the suspension cells thinly on solid selection medium, pick up the growing cell clusters, and to suspend them again. The suspension may then be assayed for the product of the co-transferred gene. Repeating these steps finally results in stable cell clones.
6. Cellulosic Filter papers should not be used because glucanases tend to attach to their substrate, resulting in loss of the enzyme in the medium.
7. An alternative method to assay EGI activity is to use an artificial substrate, methylumbelliferyl-cellobiopyranoside (MUC), and to measure the cleaved methylumbelliferone fluorometrically (*10*). This is not fully quantitative, espe-

cially at high concentrations, but the method is sensitive and no pretreatment of samples is needed.

8. After a successful preservation, the percentage of viable cells should be 50–70%. A high percentage guarantees that there was no unintentional selection for cell type during freezing or preservation.

Acknowledgments

The skillful technical assistance of Taina Ala-Hakuni, Jaana Juvonen, and Tuuli Teikari is gratefully acknowledged. This work was financially supported by the Finnish malting and brewing industry, the Technology Development Centre Finland (TEKES), and the Academy of Finland.

References

1. Sanford, J. C., Klein, T. M., Wolff, E. D., and Allen, N. (1987) Delivery of substances into cells and tissues using a particle bombardment process. *Particulate Sci. Technol.* **5,** 3–16.
2. Gordon-Kamm, W. J., Spencer, T. M., Mangano, M. L., Adams, T. R., Daines, R. J., Start, W. G., O'Brien, J. V., Chambers, S. A., Adams, W. R., Willett, N. G., Rice, T. B., Mackey, C. J., Krueger, R. W., Kausch, A. P., and Lemaux, P. G. (1990) Transformation of maize cells and regeneration of fertile transgenic plants. *Plant Cell* **2,** 603–618.
3. Christou, P., Ford, T. L., and Kofron, M. (1991) Production of transgenic rice (*Oryza sativa* L.) plants from agronomically important indica and japonica varieties via electric discharge particle acceleration of exogenous DNA into immature zygotic embryos. *Bio/Technology* **9,** 957–962.
4. Somers, D. A., Rines, H. W., Weining, G., Kaepler, H. F., and Bushnell, W. R. (1992) Fertile, transgenic oat plants. *Bio/Technology* **10,** 1589–1594.
5. Vasil, V., Castillo, A. M., Fromm M. E., and Vasil, I. K. (1992) Herbicide resistant transgenic wheat plants obtained by microprojectile bombardment of regenerable embryogenic callus. *Bio/Technology* **10,** 667–674.
6. Ritala, A., Aspegren, K., Kurtén, U., Salmenkallio-Marttila, M., Mannonen, L., Hannus, R., Kauppinen, V., Teeri, T. H., and Enari, T. M. (1994) Fertile transgenic barley by particle bombardment if immature embryos. *Plant Mol. Biol.* **24,** 317–325.
7. Wan, Y. and Lemaux, P. G. (1994) Generation of large number of independently transformed fertile barley plants. *Plant Physiol.* **104,** 37–48.
8. Castillo, A. M., Vasil, V., and Vasil, I. K. (1994) Rapid production of fertile transgenic plants of rye (*Secale cereale* L.). *Bio/Technology* **12,** 491–506.
9. Salmenkallio-Marttila, M., Aspegren, K., Åkerman, S., Kurtén, U., Mannonen, L., Ritala, A., Teeri, T. H., and Kauppinen, V. (1995) Transgenic barley (*Hordeum vulgare* L.) by electroporation of protoplasts. *Plant Cell Rep.* **15,** 301–304.
10. Aspegren, K., Mannonen, L., Ritala, A., Puupponen-Pimiä, R., Kurtén, U., Salmenkallio-Marttila, M., Kauppinen, V., and Teeri, T. H. (1995) Secretion of a

heat-stable fungal β-glucanase from transgenic, suspension-cultured barley cells. *Mol. Breeding* **1,** 91–99.

11. Mannonen, L. (1993) *Barley as a Producer of Heterologous Protein.* VTT Publications 138, Espoo, Finland.

12. Séquin-Schwarz, G., Kott, L., and Kasha, K. J. (1984) Development of haploid cell lines from immature barley *Hordeum vulgare* embryos. *Plant Cell Rep.* **3,** 95–97.

13. Ritala, A., Mannonen, L., Salmenkallio-Marttila, M., Kurtén, U., Hannus, R., Aspegren, K., Teeri, T. H., and Kauppinen, V. (1993) Stable transformation of barley tissue culture by particle bombardment. *Plant Cell Rep.* **12,** 435–440.

14. Dawson, R. M. C., Elliott, D. C., Elliott, W. H., and Jones K. M. (1986) *Data for Biochemical Research,* 3rd ed., Oxford Science Publications, Clarendon, Oxford, UK.

15. Sumner, J. B. and Somers G. F. (1949) Dinitrosalisylic method for glucose. *Laboratory Experiments in Biological Chemistry.* 2nd ed. Academic, New York, pp. 38–39.

16. Dellaporta, S. L., Wood, J., and Hicks, J. B. (1983) A plant DNA minipreparation: version II. *Plant. Mol. Biol. Rep.* **1,** 19–21.

17. Maniatis, T., Fritsch, E. F., and Shambrook, J. (1982) *Molecular Cloning: A Laboratory Manual*, Cold Spring Harbor Laboratory, Cold Spring Harbor, NY.

18. Bradford, M. M. (1976) A rapid and sensitive method for the quantitation of microgram quantities of protein utilizing the principle of protein-dye binding. *Anal. Biochem.* **72,** 248–254.

19. IUPAC (International Union of Pure and Applied Chemistry) (1987) Measurement of cellulase activity. *Pure Appl. Chem.* **59,** 257–268.

20. Laemmli, U. K. (1970) Cleavage of structural proteins during the assembly of the head of bacteriophage T4. *Nature* **227,** 680–685.

21. Towbin, H., Staehelin, T., and Gordon, J. (1979) Electrophoretic transfer of proteins from polyacrylamide gels to nitrocellulose sheets: procedure and some applications. *Proc. Natl. Acad. Sci. USA* **76,** 4350–4354.

22. Niku-Paavola, M.-L. (1991) Isoelectric focusing electrophoresis of lignin. *Anal. Biochem.* **197,** 101–103.

23. Kvesitadze, E. G., Lomitashvili, T. B., Kvesitadze, G. I., and Niku-Paavola, M.-L. (1992) Thermostable endoglucanases of the thermophilic fungus *Allesheria terrestris. Biotechnol. Appl. Biochem.* **16,** 303–307.

24. Mannonen, L., Toivonen, L., and Kauppinen, V. (1990) Effects of long-term preservation on growth and productivity of *Panax ginseng* and *Catharanthus roseus* cell cultures. *Plant Cell Rep.* **9,** 173–177.

3

Introducing and Expressing Genes in Legumes

Adrian D. Bavage, Mark P. Robbins, Leif Skøt, and K. Judith Webb

1. Introduction

Legumes form an important element in many ecological and agricultural environments. In order to improve the range of phenotypes available for agriculture, plant breeding programs seek to produce improved varieties. Inevitably, these programs are limited by the available gene pool and restricted potential for hybridization between species. Molecular biology techniques offer an opportunity to modify the characteristics of legumes directly. The transgenic plants produced can then be used either in their own right or as a gene pool for inclusion in traditional breeding programs.

In this chapter, we will describe methods used for the genetic manipulation of three forage legumes: *Trifolium repens* (white clover), *Lotus corniculatus* (bird's foot trefoil), and *Lotus japonicus*.

Incorporating genetic material into plants requires a transformation system and methods for detecting the presence of functional transgenes as shown schematicaly in **Fig. 1**. Tannins and other polyphenolic compounds present in legumes may limit the efficacy of techniques used in the manipulation of other species. Such highly reactive compounds interfere with the recovery of nucleic acids, in particular. In our experience, adaptations of established methods are more successful than many of the commercially available kits; for example, DNA and RNA extraction kits are considerably less effective with legume tissues than with material from other species, such as tobacco.

From: *Methods in Biotechnology, Vol. 3:*
Recombinant Proteins from Plants: Production and Isolation of Clinically Useful Compounds
Edited by: C. Cunningham and A. J. R. Porter © Humana Press Inc., Totowa, NJ

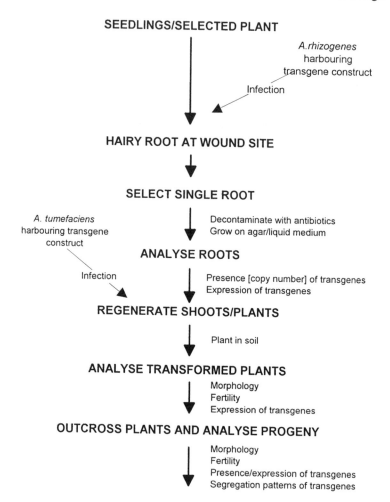

Fig. 1. Scheme for the production of transgenic legume plants via *Agrobacterium*.

2. Materials

2.1. Plant Transformation

2.1.1. Plant Material and Bacterial Strains

1. *Agrobacterium rhizogenes* strains LBA 9402, LBA 8196.
2. *Rhizobium loti* strain RCR3011 (Institute of Grassland and Environmental Research notation).
3. *Lotus corniculatus* cv Leo.

2.1.2. General Laboratory Equipment

1. Spectrofluorimeter (365 nm excitation, 455 nm emission, with slit width 10 nm) (Perkin Elmer, Bucks, UK).
2. Laminar flow/biohazard hood (Astecair, Astec, Weston-Super-Mare, UK).
3. Horizontal orbital shaker (in light and dark) at $2g$ (LH Fermentation, UK).
4. Growth rooms at 20°C and 25°C, with light intensities of 20 and 80 $\mu M/m^2/s$, and cold room/refrigerator at 2–4°C.
5. Plastic ware: 90-mm Petri dishes, 30-mL universal tubes (Sterilin, Stone Staffs, UK).
6. Scalpels, forceps, hypodermic needles, bacterial loops (Sarstedt, Leicestershire, UK), laboratory film (Nescofilm, Fisher, Leicestershire, UK).
7. Glassware: 250-mL flasks with cotton wool bungs and foil caps (Fisher).
8. Nitrocellulose filters 0.45 μm pore size, 25-mm diameter (Fisher).

2.1.3. Media

1. Yeast mannitol: 5% (w/v) K_2HPO_4, 2% (w/v) $MgSO_4$, 1% (w/v) (Merck, Dorset, UK), NaCl, 1% (w/v) mannitol, 0.04% (w/v) (Sigma, Dorset, UK), yeast extract, pH 7.2 (1.5% [w/v] Difco Bactoagar for semisolid media, Difco, Poole, Dorset, UK).
2. Nutrient broth No. 2 (Oxoid, Basingstoke, UK).
3. Root growth medium: Half strength B5 salts (Flow Labs, High Wycombe, Bucks, UK; **ref. *1***), 2% (w/v) sucrose (Merck), pH 5.6 (plus 1.5% [w/v] agar [Sigma] for semisolid medium). For decontamination medium, add ampicillin (Sigma) to a final concentration of 500 μg/mL.
4. Shoot growth medium: MS salts (Flow Labs; **ref. *2***), 3% (w/v) sucrose, 0.7% (w/v) agar, pH 5.6, in 30-mL Sterilin tubes.
5. Callus induction medium: MS salts (Flow Labs; **ref. *2***), 3% (w/v) sucrose, 0.7% (w/v) agar, 0.5 mg/L NAA, 0.5 mg/L BAP (Sigma), pH 5.6, **ref. *3***.
6. LB, TY, and nutrient broth media, as described in **ref. *4***.

2.1.4. Chemicals

1. Sodium hypochlorite (12% [v/v] available chlorine; BDH, Poole, Dorset, UK).
2. Antibiotics: 100 mg/L kanamycin (Sigma); 15 mg/L hygromycin (Melford Labs, Suffolk, UK); 15 mg/L G418 (Gibco BRL, Paisley, UK).
3. 2 mM 5-bromo-4-chloro-3-indolyl glucuronide (X-Gluc [Melford]) in 0.1M NaH_2PO_4 buffer, pH 7.0.
4. 4-Methylumbelliferyl glucuronide (MUG) [Sigma] in extraction buffer: 1 mM MUG, 50 mM NaH_2PO_4, pH 7.0, 10 mM ethylenediaminetetra-acetic acid disodium salt (EDTA [Merck]), 0.1% (v/v) 0.1% Triton X-100, 0.1% sarcosyl, 1 mM dithiothreitol (DTT [Sigma]).

2.2. CAT Assays

1. 50 mM Tris-HCl, pH 7.5, 1 mM EDTA, 400 mM NaCl, 2 mM phenyl-methylsulfonylflouride (PMSF).
2. 1.5-mL Micricentrifuge tubes and glass rods to fit.
3. 250 mM Tris-HCl, pH 7.5.
4. Acetylcoenzyme A (4 mg/mL stock solution).
5. ^{14}C Chloramphenicol (Amersham, Little Chalfont, Bucks, UK) 2 GBq/mmol, 925 kBq/mL.
6. Ethylacetate.
7. Precoated silicagel TLC plates (20 × 20 cm) (CamLab, Cambridge, UK).
8. X-ray film and cassettes.
9. Saran wrap.

2.3. Molecular Analysis

2.3.1. PCR Detection of Transgenes

1. Appropriate polymerase chain reaction kit or reagents (Amplitaq, Perkin-Elmer Cetus, Norwalk, CT).
2. Oligonucleotide primers specific for transgene or promoter/terminator combination of vector.
3. For small-scale genomic DNA isolation for PCR, all materials and solutions should be prepared to ensure they are DNA and DNase free.
    a. Fresh or frozen plant material.
    b. Extraction buffer: 15 mL 100 mM Tris-HCl, 10mM EDTA, pH 8.0, 1 mL 20% (w/v) sodium dodecylsulfate, 10 µL 2-mercaptoethanol.
    c. Liquid nitrogen.
    d. 5M Potassium acetate.
    e. T$_{10}$E$_1$: 10 mM Tris-HCl, 1 mM EDTA, pH 8.0.
    f. Phenol/chloroform (1:1 [v:v] prepared as described in **ref. 4**).
    g. 3M Sodium acetate, pH 5.2.
    h. RNase A: 50mg/mL in T$_{10}$E$_1$.
    i. 5M Sodium chloride.
    j. 80 and 100% ethanol.
    k. Mortars and pestles for each sample.
    l. Microcentrifuge (e.g., Eppendorf) and 1.5 mL microcentrifuge tubes.

3. Methods

3.1. Transformation of L. corniculatus with A. rhizogenes

We have successfully used *A. rhizogenes* and *A. tumefaciens* to transform both *L. corniculatus (5,6)* and *L. japonicus (7)*. The procedures with *A. tumefaciens* are described in full in the references. The procedure for *A. rhizogenes*-mediated transformation outlined below routinely yields transformed plants.

This method is based on procedures amended by Webb and co-workers *(8,9)*. *L. japonicus* and *T. repens* can also be transformed in this way, but the resulting hairy roots either regenerate shoots rarely (*L. japonicus*) or never (*T. repens*) *(6,7)*.

All procedures involving bacteria or plants must be performed under sterile conditions, unless stated otherwise.

3.1.1. Plant Material

1. Scarify seed with emery paper.
2. Surface-sterilize seeds with sodium hypochlorite (6.0% available chlorine) with 0.01% (v/v) Tween-20 for 30 min.
3. Wash six times in sterile tap water, leave to imbibe water at 20°C for 2 h.
4. Wash again to remove all traces of brown exudate in water.
5. Maintain selected genotypes (*see* **Notes 3** and **4**) on shoot growth media in 80 $\mu M/m^2/s$ light at 20°C.
6. Subculture by nodal cuttings at 1–2 mo intervals.

3.1.2. Bacteria

Transfer of plant transformation vectors to *Agrobacterium* is most often achieved by bacterial conjugation, although methods for direct transformation by electroporation are an alternative.

1. Grow the *Agrobacterium* recipient strain for 2–3 d at 28°C on agar plates or slants containing the appropriate antibiotic. Grow *A. tumefaciens* on LB agar and *A. rhizogenes* on TY agar.
2. Grow *Escherichia coli* donor and helper strains in individual 5 mL liquid LB, at 37°C for 1 d, with appropriate antibiotics.
3. Resuspend *Agrobacterium* cells in 2–3 mL of sterile distilled water (SDW). Spin down 1 mL *E. coli* cells in microcentrifuge for 15 s at 11,000*g* and resuspend in 200 mL sterile 0.8% (w/v) NaCl.
4. Mix 75 µL each of *Agrobacterium*, *E. coli* helper, and donor strains on a nitrocellulose filter (25 mm diameter) on a TY agar plate and incubate overnight at 28°C (*see* **Note 1**).
5. Suspend the mixture in 2–3 mL SDW and spread 100 µL of the appropriate dilution (0, 10^{-2}, and 10^{-4}) on selective agar plates (*see* **Note 2**). Single colonies are purified by repeatedly streaking individual colonies onto selective media. Verification of the presence of the gene construct in *Agrobacterium* can be obtained by Southern blot hybridization or PCR.
6. Grow the *Agrobacterium* on LB-agar plates with appropriate antibiotic(s) at 28°C for 2–3 d. Binary constructs in vectors pBin19 or pRok2 express kanamycin resistance in *Agrobacterium*.

3.1.3. Cocultivation

1. Inoculate in vitro shoots by cutting stems into 5-mm pieces with scalpel blade loaded with *Agrobacterium*.
2. Transfer stem pieces to sterile filter paper overlaying shoot-growth medium.
3. Leave for a maximum of 7 d in light (20 $\mu M/m^2/s$), then transfer explants to decontamination medium.
4. Transfer every 3–4 d to fresh decontamination media for 1 mo to prevent bacterial overgrowth, then reduce subculture to every 7–10 d.

3.1.4. Decontamination

1. Transfer single roots longer than 1 cm to decontamination medium, taking care not to damage the root tip. Those roots growing away from the medium or actually through the agar are cleanest.
2. Maintain on decontamination medium, adding the appropriate antibiotic for selection.

3.1.5. Selection

Select for noncontaminated roots that are growing well under appropriate selection. Test for bacterial contamination by squashing roots in nutrient broth and incubating for about 4 d at 35°C, with shaking; three consecutive clean tests indicate they are free of bacteria.

3.1.6. Maintenance

1. Grow selected roots on semisolid root-growth medium in darkness at 25°C, and multiply the stocks at each monthly transfer to fresh medium. Root cultures can be stored in darkness for 6 mo at 2–4°C.
2. Maintain root cultures in 50 mL liquid root-growth medium in 250-mL shaking flask, with transfers every 2 wk of 0.5 g of tissue (minimum 0.3 g).
3. Transfer to light (80 $\mu M/m^2/s$) for shoot initiation and do not subculture. Shoots appear after 6–8 wk.

3.1.7. Propagation of Shoots

Cut shoots of 2 cm and above from root and transfer to shoot-growth medium in light 80 $\mu M/m^2/s$ at 20°C. These can be propagated by nodal cuttings.

3.1.8. Establishment in Soil

1. Wash roots to remove agar containing sucrose and transfer plantlets to 50:50 soil:peat-based compost.
2. Harden plants off in a propagator, gradually increasing the airflow, until the cover is removed after 7–10 d.
3. Inoculate plants with *R. loti* if nodules required, otherwise, provide nitrogen (50 mM NH_2SO_4).

3.1.9. Seed Production

Lotus corniculatus is an outbreeder. Plants can be crossed by hand using the following method adapted from Williams *(10)*.

1. Clean hands with 100% ethanol.
2. Label flower head of female parent. Carefully remove the petals of the keel using forefinger and thumb to expose the stigma.
3. Collect pollen from intact flowers of male parent using a triangular piece of card (5 × 7 cm), folded longitudinally (roughen the tip of the card so that it will hold the pollen).
4. Place the pollen onto an exposed stigma.
5. Harvest after 5–6 wk, when the pods are brown and dry. High temperatures (≥39°C) may result in selfing of plants.

3.1.10. Analysis of Seedlings

1. Screen for expression of the transgenes in T_1 progeny, using a minimum of 20 T_1 seeds from each cross.
2. When transformed with the GUS reporter gene, test the first emerging root or the whole embryo for GUS activity by the histochemical method *(11)*.
3. Test antibiotic resistance of the seedlings by comparing growth of leaf and stem tissues after 3 wk on callus induction medium, both with and without antibiotic.

3.2. Selection/Confirmation

3.2.1. Antibiotic Resistance

1. Grow roots on 15 mg/L hygromycin, 50 mg/L kanamycin, or 15 mg/L G418, as appropriate.
2. Confirm transformation of tissues by growth on callus induction medium, with or without antibiotic.
3. Compare fresh weights after 21 d growth in darkness at 25°C.

3.2.2. Reporter Genes

3.2.2.1. ANALYSIS OF EXPRESSION OF β-GLUCURONIDASE

1. The fluorometric assay is based on that devised by Jefferson and co-workers *(11)*.
2. Collect samples (30 mg fresh weight) from young, fully expanded leaves of soil-grown plants. Grind samples in liquid nitrogen prior to addition of extraction buffer.
3. Centrifuge extracts for 10 min at 12,500g, 4°C, and assay for GUS activity using MUG as the substrate, using a spectrofluorimeter.
4. Relate GUS activity to the total concentration of soluble protein *(12)*.
5. To visualize distribution of GUS activity, incubate samples in X-Gluc at 37°C overnight and clear in ethanol.

3.2.2.2. ^{14}C CHLORAMPHENICOL ANALYSIS OF CAT ACTIVITY IN HAIRY ROOT CULTURES

This method is derived from that described by Stougaard et al. *(13)*. It can be used as both a qualitative and quantitative assay for the presence and activity of this enzyme in transgenic plant material.

1. Harvest 100 mg plant material and transfer to 1.5-mL Eppendorf tube.
2. Add 100 μL of extraction buffer to the tissue and homogenize with a glass rod.
3. Heat for 3 min at 65°C and cool on ice. Spin down in a microcentrifuge to remove precipitate.
4. Transfer 50 μL of the supernatant to a clean Eppendorf tube and add: 380 μL 250 mM Tris-HCl, pH 7.5; 50 μL acetylcoenzyme A (4 mg/mL stock); and 20 μL ^{14}C chloroamphenicol (2 Gbq/mmol, 925 kBq/mL).
5. Incubate at 37°C for 1 h.
6. Add 0.5 mL of ethylacetate, vortex, and centrifuge for 2–3 min.
7. Evaporate solvent in an air evaporator in a radiation handling area.
8. Resuspend in 10 μL ethylacetate and spot onto TLC plate.
9. Separate acetylated chloroamphenicol from chloroamphenicol by TLC, using chloroform:methanol (95:5, v:v) as the solvent.
10. After chromatography, air-dry the TLC plate and wrap it in Saranwrap before exposing to X-ray film in an autoradiography cassette, at room temperature, overnight.
11. For a quantitative assay, the regions of the TLC plate containing a positive signal can be scraped off and analyzed by scintillation counting.

3.3. Molecular Analysis

Transgenes can be monitored by probing blots of DNA or RNA isolated from transgenic material. However, we have found that the isolation of DNA and RNA is most effective using extensive phenol/chloroform extractions. We routinely use modified forms of the methods of Martin et al. *(14)* for DNA, or Ougham and Davies *(15)* for RNA. These methods are time-consuming and require relatively large amounts of tissue (≥3 g). Initial screening for the presence of transgenes can be more easily accomplished using the polymerase chain reaction (PCR). Only small quantities of DNA are needed and we have found the following method (derived from **ref. *16***) reliable for this purpose.

3.3.2. Isolation of DNA from Plant Tissues for PCR

1. Place 100 mg of tissue into a mortar precooled to –70°C. Cover with liquid nitrogen and grind tissue to a fine powder.
2. Pour powder into a microcentrifuge tube containing 1 mL freshly prepared extraction buffer at 65°C and mix well.
3. Incubate at 65°C for 10 min.

4. Divide sample equally between two microcentrifuge tubes, add 0.25 mL 5*M* potassium acetate to each, and incubate on ice for 20 min.
5. Centrifuge 11,000*g*, 4°C, for 20 min. Decant and discard the supernatants.
6. After air-drying the pellets for 10 min, resuspend them in a total volume of 750 μL $T_{10}E_1$, combining the two in the process.
7. Add 500 μL of phenol/chloroform, vortex vigorously, and then centrifuge at 11,000*g*, 4°C, for 5 min.
8. Transfer the supernatant to a fresh microcentrifuge tube and add 70 μL of 3*M* sodium acetate, pH 5.2, and 500 μL isopropanol. Incubate on ice for 20 min.
9. Centrifuge 11,000*g*, 4°C, for 10 min. Discard the supernatant and resuspend the pellet in 1 mL 80% ethanol. Centrifuge as before, discard supernatant, and air-dry the pellet until no ethanol remains.
10. Dissolve the pellet in 100 μL of $T_{10}E_1$ containing 50 mg/mL RNase A.
11. Add 5 μL 5*M* NaCl and 250 μL of 100% ethanol; incubate at –70°C for 20 min.
12. Centrifuge 11,000*g*, 4°C, for 20 min. Discard supernatant, air-dry, and then resuspend pellet in 40 μL sterile distilled water.

3.3.3. PCR Detection of Transgenes

Figure 2 shows the results of a PCR analysis of hairy root cultures transformed with a dihydroflavonol reductase cDNA construct. Genomic DNA was prepared from hairy-root cultures, as described in **Subheading 3.3.2.** Using degenerate oligonucleotide primers to a conserved region of the protein, discrete products were obtained from both the native gene and the introduced transgene. The transgene lacks any intron sequences and so produces a smaller fragment, as subsequently confirmed by Southern hybridization to the PCR products.

4. Notes
4.1. Plant Transformation

1. The helper strain provides the plasmid transfer genes, necessary to mobilize most plant transformation vectors, *in trans* on a plasmid such as pRK2013 *(17)*, which cannot be maintained in *Agrobacterium*. Certain *E. coli* strains, such as S17-1 *(18)*, have the transfer genes integrated into their genomes and will directly transfer plant transformation vectors to *Agrobacterium*. This may, however, increase the risk of accidental release.
2. Although selection can be achieved on complex media, we have found that minimal media, such as MS, are more effective for the counterselection against *E. coli*.
3. Seedlings offer alternative explants to in vitro-grown shoots *(8,9)*. Sow 10 surface-sterilized seeds in a line on water agar (2%) plates, seal, and place at an angle of 45 degrees in light (20 m*M*/m²/s). Stab hypocotyl of 7–10-d-old seedlings with a needle loaded with bacteria (just below cotyledons). Incubate in light (20 m*M*/m²/s); hairy roots should appear at the inoculation site after 10–14 d.

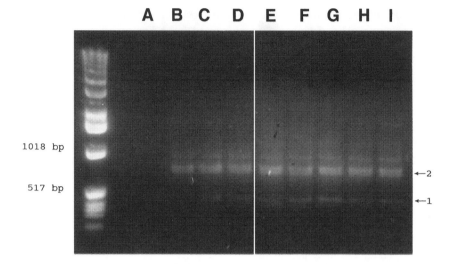

Fig. 2. Polymerase chain reaction amplification of dihydroflavonol reductase genes from Lotus corniculatus hairy root culture DNA: (A) *Antirrhinum majus* DFR cDNA clone, (B) untransformed *L. corniculatus* genomic DNA. (C–I) *L. corniculatus* transformed with *A. majus* DFR cDNA construct. (1) PCR product from *A. majus* cDNA, (2) PCR product from *L. corniculatus* genomic DNA.

4. Not all seedlings generate hairy roots and not all hairy roots produce shoots. Therefore, select for these characteristics.

4.2. A. rhizogenes Transformation

5. The advantages of this approach are that the generation of hairy-root cultures enables early analysis of the presence and expression of both marker gene and transgene, roots are probably derived from a single transformed cell, clones can be regenerated, nodulation and nitrogen fixation appear to be normal, and the maintenance and storage of cultured material is easier.
6. The disadvantages are that T-DNA left (TL) and right (TR) border sequences must be segregated away and may be present in multiple copies, depending on the *A. rhizogenes* strain used; also, female fertility is decreased in mature plants.

4.3. A. tumefaciens Transformation

7. The advantages of this system are: no interference from TL and TR and no phenotype of primary transformants to affect crossing, selfing, nitrogen fixation, or nodulation.
8. Major disadvantages include the greater possibility of chimeras, slower propagation, and storage has to be as shoots.

Acknowledgments

The authors wish to thank Alan Cookson for printing **Fig. 2**, and Steven Colliver for developing the genomic DNA protocol. Thanks also to Teri Evans, Ian Davies, Sue Mizen, and Emma Timms for technical assistance, and other members of the Cell Manipulation Group for useful discussions. We acknowledge financial support from Biotechnology and Biological Sciences Research Council (BBSRC) during the execution of these studies. Adrian Bavage is funded by a grant from the BBSRC Plant Molecular Biology Initiative (PG203/536).

References

1. Gamborg, O. L., Miller, R. A., and Ojima, K. (1968) Nutrient requirements of suspension cultures of soybean root cells. *Exp. Cell. Res.* **50,** 151–158.
2. Murashige, T. and Skoog F. (1962) A revised medium for rapid growth and bioassays with tobacco tissue cultures. *Physiol. Plant.* **15,** 473–497.
3. Bond, J. E. and Webb, K. J. (1989) Regeneration and analysis of plants from stolon segments of *Trifolium repens* (white clover). *Plant Sci.* **61,** 119–125.
4. Sambrook, J., Fritsch E. F., and Maniatis T. (1989), in *Molecular Cloning: A Laboratory Manual.* Second edition. Cold Spring Harbor Laboratory, Cold Spring Harbor, NY.
5. Gibbs, M. J. (1991) Genetic engineering of the forage legume *Lotus corniculatus* using *Agrobacterium*-mediated transformation systems. PhD thesis, University of Durham, UK.
6. Gibbs, M. J., Gatehouse, J. A., and Webb, K. J. (1990) *Lotus corniculatus*: a transformation system with *Agrobacterium tumefaciens*. 20th Meeting of FEBS, Budapest, 19–24 August 1990. Abstr. no. P-Th484.
7. Handberg, K. and Stougaard, J. (1992) *Lotus japonicus*, an autogamous, diploid legume species for classical and molecular genetics. *Plant J.* **2,** 487–496.
8. Webb, K. J., Jones, S., Robbins, M. P., and Minchin, F. R. (1990) Characterization of transgenic root cultures of *Trifolium repens*, *T. pratense* and *Lotus corniculatus* and transgenic plants of *L. corniculatus*. *Plant Sci.* **70,** 243–254.
9. Webb, K. J., Robbins, M. P., and Mizen, S. (1994) Expression of GUS in primary transformants and segregation patterns of GUS, T_L- and T_R-DNA in the T_1 generation of hairy root transformants of *Lotus corniculatus*. *Transgenic Res.* **3,** 232–240.
10. Williams, R. D. (1925) Studies concerning the pollination, fertilization and breeding of red clover. *Bull. Welsh Pl. Breed. Sta. Series H*, No. 4, p. 58.
11. Jefferson, R. A., Kavanagh, T. A., and Bevan, M. W. (1987) GUS fusions: b-glucuronidase as a sensitive and versatile gene fusion marker in higher plants. *EMBO J.* **6,** 3901–3907.
12. Bradford, M. (1976) A rapid and sensitive method for the quantitation of microgram quantities of protein utilizing the principal of protein-dye binding. *Anal. Biochem.* **72,** 248–254.

13. Stougaard, J., Marcker, K. A., Otten, L., and Schell, J. (1986) Nodule specific expression of a chimaeric soybean legheamoglobin gene in transgenic *Lotus corniculatus. Nature* **321,** 669–674

14. Martin, C., Carpenter, R., Sommer, H., Saedler, H., and Coen, E. S. (1985) Molecular analysis of instability in flower pigmentation of *Antirrhinum majus* following isolation of the pallida locus by transposon tagging. *EMBO J.* **4,** 1625–1630.

15. Ougham, H. J. and Davies, T. G. E. (1990) Leaf development in *Lolium temulentum*: gradients of DRA complement and plastid and non-plastid transcripts. *Physiol. Plant.* **79,** 331–338.

16. Robbins, M. P., Evans, T. E., Morris. P., and Carron. T. R. (1991) Some notes on the extraction of genomic DNA from transgenic *Lotus corniculatus. Lotus Newsletter* **22,**18–21.

17. Ditta, G, Stanfield, S. Corbin, D., and Helinski, D. R. (1980) Broad host range cloning system for Gram-negative bacteria: construction of a gene bank of *Rhizobium meliloti. Proc. Natl. Acad. Sci. USA* **77,** 7347–7351.

18. Simon, R. Priefer, U., and Puhler, A. (1983) A broad host range mobilisation system for *in vivo* genetic engineering: transposon mutagenesis in Gram-negative bacteria. *Biotechnology* **1,** 784–791.

4

Use of the GUS Reporter Gene

Joy Wilkinson and Keith Lindsey

1. Introduction

One of the most important considerations in the expression of heterologous proteins in plants is the choice of promoter. The study of promoter activity is simplified in the majority of cases by the use of a readily detectable reporter gene. Indeed, reporter genes can also be used to aid the isolation of promoters, through the transformation of plants with a promoterless reporter gene construct. Activation of the reporter gene will then only occur if the gene is inserted adjacent to a native promoter *(1)*.

Several reporter genes have been used for the analysis of regulatory sequences, including chloramphenicol acetyl transferase *(cat)* *(2,3)*, β-glucuronidase *(uid* A or *gus) (4)*, luciferase *(luc)*, and green fluorescent protein (GFP) *(6)*. The choice of which reporter gene to use depends on several factors, including available equipment, type of tissue to be assayed, and sensitivity required. For example, though the use of *cat* provides a very sensitive assay, it is a destructive method requiring that the tissue to be assayed is easily separated from surrounding tissue. Use of GFP, however, enables activity to be assayed in vivo. The majority of laboratories currently use *gus* and *luc*, but there is no doubt that once initial problems with GFP have been overcome, it too will become widely used.

The *gus* gene has many advantages, based mainly on the fact that it can be used for quantitative assays and also for histochemical staining of tissue sections to localize activity. The gene originates from *Escherichia coli* and is encoded by the *uid* A locus *(7)*. The encoded protein is a hydrolase that catalyzes the cleavage of a range of β-glucuronides, enabling its use in plant transformations in which background GUS activity levels are either very low or

From: *Methods in Biotechnology, Vol. 3:*
Recombinant Proteins from Plants: Production and Isolation of Clinically Useful Compounds
Edited by: C. Cunningham and A. J. R. Porter © Humana Press Inc., Totowa, NJ

nondetectable *(4,8)*. The *gus* gene has now been assayed in transgenic plants of several species, including soybean *(9)*, rice *(10,11)*, and melon *(12)*, as well as model species *(4)*, and is used by many as a control to assess successful transformation. GUS enzyme activity can easily be localized using the histochemical substrate 5-bromo-4-chloro-3-indolyl glucuronide (X-Gluc), which yields a blue dye following GUS-mediated hydrolysis, or can be quantitatively assayed using 4-methylumbelliferyl glucuronide (MUG), which yields a fluorescent product, 4-methylumbelliferone (4-MU), on hydrolysis, which can be measured using a fluorimeter.

Histochemical staining for GUS activity allows accurate localization of expression down to the single-cell level. The final product of the reaction with X-Gluc is insoluble and nondiffusable. This, therefore, allows accumulation of product and is particularly useful in cases in which activity is very low, since the tissue can be left to stain over a period of 2–3 d until enough product has accumulated to become visible. However, unlike the fluorimetric assay using MUG as a substrate, the histochemical stain is not quantitative. The advantages of the fluorimetric assay are that it provides a quantitative result and several samples can be processed simultaneously. By dissecting out test tissues, fluorimetric assays can be used to substantiate results obtained from histochemical staining.

Constructs available for the analysis of promoter activity using the *gus* reporter gene include pBI101 *(4)*, which contains a multiple cloning site upstream of the *gus* gene, enabling easy insertion of the promoter of interest. A *gus* reporter gene containing an intron has also been constructed *(13)*, which allows expression in plant cells but not in *Agrobacterium tumefaciens*, which is a commonly used vector in plant transformations. GUS activity, independent of that from potential bacterial contamination, can therefore be assessed easily. Commercial sources of constructs include companies such as Clontech.

In this chapter, we describe general methods for the fluorimetric assay using MUG, and histochemical staining for GUS activity using X-Gluc. We describe how some of the major problems, such as endogenous GUS activity found in some tissues, can be limited, and how methods can be altered to optimize conditions for particular tissues.

2. Materials
2.1. Plant Material

1. For the histochemical assay, it is desirable to use small amounts of plant material, in order to minimize the quantity of substrate required and to facilitate tissue infiltration by the substrate. Small organs, such as the flowers, leaves, and roots

of arabidopsis, and roots of tobacco, can be stained directly, once soil or agar are removed. However, for larger samples, sections can be hand-cut and placed immediately into histochemical staining buffer.

2. For the fluorimetric assay, woody material should be avoided, since it is difficult to isolate protein from such samples. The harvesting of plant material for either type of analysis may have to be modified, depending on the tissue. For example, pollen is easily harvested from tobacco by gently vortexing anthers in the appropriate buffer. Plant material for the fluorimetric assay can be frozen at $-80°C$ prior to use. It is, however, advisable not to freeze samples directly in GUS extraction buffer, since we have observed that this results in a significant loss of activity.

2.2. Special Equipment

1. Vortex mixer (Fisons, Loughborough, UK).
2. Microcentrifuge (Micro Centaur, Fisher Scientific, Loughborough, UK).
3. Water bath at 37°C (Grant, Cambridge, UK).
4. Incubator at 37°C (Gallenkamp, Fisher Scientific).
5. Fluorescence spectrometer (Perkin-Elmer LS-50, Warrington, UK).
6. Spectrophotometer (Perkin-Elmer Lambda 5 uv/vis).
7. 96-Well opaque microtiter plates (Perkin-Elmer).

2.3. Buffers and Other Solutions

Stock solutions of 4-MU (Sigma), MUG (Sigma), and X-Gluc (Sigma, Poole, Dorset, UK) are stored in one-use aliquots at $-20°C$ in the dark and can be kept for several months. The buffers are made up from concentrated, preautoclaved stocks, and are added to sterile water to give the required concentrations. Triton X-100 is not autoclaved and is added last to the required solutions. β-mercaptoethanol is added just before use.

1. 4-MU stock solution: 10 μM made up in 0.2M Na$_2$CO$_3$.
2. GUS extraction buffer: 50 mM NaPO$_4$ pH 7.0, 10 mM EDTA, 0.1% (v/v) Triton X-100, 10 mM β-mercaptoethanol.
3. MUG stock solution: 5 mM MUG made up in GUS extraction buffer.
4. GUS assay buffer: GUS extraction buffer containing 1 mM MUG diluted from MUG stock solution.
5. Stop solution: 0.2M Na$_2$CO$_3$.
6. Bovine serum albumin (BSA) stock solution: 0.4 mg/mL BSA made up in sterile distilled water.
7. X-gluc (20X stock solution): 20 mM X-Gluc made up in N-N-dimethyl formamide (DMF). 1.5 mL microfuge tubes should be used, since they are resistant to DMF, which degrades other types of plastics.
8. Histochemical staining buffer: 100 mM phosphate buffer, pH 7.0, 10 mM EDTA, 0.1% (v/v) Triton X-100.

9. Potassium ferri/ferrocyanide stock solution: 5 m*M* of each made up in histochemical staining buffer.
10. Ethanol solutions: 70% (v/v) and 95% (v/v) ethanol.

3. Methods

3.1. Fluorimetric Assay for GUS Activity

3.1.1. Preparation of Standard Curve

In order to quantify the fluorimetric assay results obtained, the fluorimeter should be calibrated using standards of the reaction product, 4-MU.

1. Prepare dilutions of 4-MU stock solutions in the range of 0–0.5 nmol in a final volume of 50 µL 0.2*M* Na$_2$CO$_3$. It may also be necessary to prepare other dilutions, depending on the sensitivity of the fluorimeter used. Five to 10 replicates of each concentration should be prepared, so that mean values can be calculated.
2. Add 4-MU dilutions to 150 µL stop solution (*see* **Note 1**) in a 96-well opaque microtiter plate.
3. Measure the fluorescence of each individual well using an excitation wavelength of 365 nm and an emission wavelength of 455 nm.
4. Use the results to plot nmols 4-MU against fluorescence units (**Fig. 1**). From this calibration curve, the relationship between fluorescence units and concentration of 4-MU can be calculated. Most computer graphics packages will perform this calculation. For example, the result shown in **Fig. 1** gives a curve in which one fluorescence unit is equal to 1.8837 pmol 4-MU. This result will differ depending on the type of fluorimeter used.

3.1.2. Preparation of Protein Extracts

1. Grind 10–100 mg tissue using a small pestle and mortar in 400 µL GUS extraction buffer (*see* **Note 2**).
2. Pour or pipet sample into a 1.5-mL microcentrifuge tube and place on ice until all samples have been prepared.
3. Centrifuge tubes at 13400*g* for 10 min to pellet cell debris, and remove supernatant to a fresh tube (*see* **Note 3**). Replace on ice until used in the assay, which should be carried out within 1 h.

3.1.3. GUS Enzyme Assay

The following method is designed for use with 96-well opaque microtiter plates, as described in **ref. 14**. Though the assay gives quantitative data, the information is relative between samples, and is easily affected, for example, by the tissue from which the protein was extracted (**Fig. 2**), since not all tissues are expected to have identical protein yields. Tissue from untransformed material should always be assayed at the same time, to enable background fluorescence to be measured and subtracted from figures obtained for transformed material.

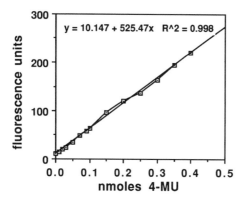

Fig. 1. Typical calibration curve of 4-MU. Fluorescence was measured using an excitation energy of 365 nm and reading the emission spectra at 455 nm.

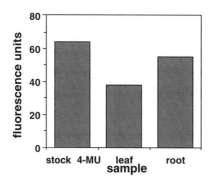

Fig. 2. The effect of tobacco leaf and root protein samples from an untransformed plant on the fluorescence readings obtained for a stock solution of 0.1 nmol 4-MU.

1. Prepare a microtiter plate by adding 180 µL stop solution to each well required (equal to the number of samples, including a blank, multiplied by the number of time-points to be taken) (*see* **Note 4**).
2. Place 180 µL GUS assay buffer in the required number of 1.5-mL microcentrifuge tubes (equal to the number of samples, plus one for a blank) and place in a water bath at 37°C to equilibrate.
3. Add 20 µL protein sample to the tubes containing GUS assay buffer (*see* **Note 5**). Vortex and immediately remove 20 µL into a well of the microtiter plate containing stop solution. Replace the tube in the 37°C water bath. This will be the zero time-point. Add 20 µL GUS extraction buffer to the blank.
4. At known intervals—for example, every 20 min—remove 20 µL from each sample into stop solution. Between taking samples, store the microtiter plate in the dark.

5. After taking the final time-point, measure the fluorescence as for the calibration curve (*see* **Note 6**).

3.1.4. Protein Concentration Measurements

The following method is designed for the use of 1-mL cuvets in a conventional spectrophotometer.

1. Add 5 µL of protein sample to a 1-mL cuvet containing 200 µL of Bradford's reagent (Bio-Rad, Hemel Hempstead, Hertfordshire, UK) and 795 µL of sterile distilled water. Mix by placing parafilm over the end and inverting the cuvet several times.
2. Incubate at room temperature for 5–10 min to allow the blue color to develop (*see* **Note 7**).
3. Read the absorbance at 595 nm against a nonprotein-containing blank.
4. Calculate protein concentrations against a calibration curve constructed using dilutions of BSA stock solution. Protein concentration measurements will then enable GUS activity to be expressed as: pmol 4-MU produced per minute per milligram of total protein. Bear in mind that the protein referred to in the equation is the amount of protein that ends up in the final stop solution.

3.2. Histochemical Staining for GUS Activity

One of the major problems with the use of X-Gluc as a histochemical stain is the potential for the soluble indoxyl reaction intermediate to diffuse to surrounding cells. This can be prevented by the addition of ferricyanide and ferrocyanide to the histochemical buffer *(15)*. However, this will cause some loss of GUS activity; a series of staining reactions will have to be set up containing varying quantities of ferri/ferrocyanide in order to determine the concentration that allows the GUS reaction to proceed without resulting in intermediate product diffusion.

The other major problem with the histochemical stain is the presence in some plant tissues of endogenous GUS-like activity, which has been found particularly in the seeds of some plants *(16)*. Kosugi et al. *(17)* claimed that endogenous activity could be eliminated by the addition of water-soluble organic solvents, particularly methanol, to the reaction mix. However, this has now been shown not to be the case *(18)*. Endogenous activity cannot at present be prevented, but it can be minimized. For example, we have noted that by staining wild-type tobacco pollen in 1.5-mL microcentrifuge tubes, it compacted at the bottom of the tube and the pollen appeared to stain blue. However, when equivalent pollen samples were stained at low densities in flat-bottomed microtiter plates, the samples did not stain. Also, Hu et al. *(17)* reported that endogenous staining patterns were more dispersed with weak intensities, but tissues transformed with *gus* showed more defined areas of staining with a high color intensity. We therefore suggest that all tissue staining be carried out with the appropriate wild-type control samples, to enable direct staining comparisons to be made.

1. Add X-Gluc from the stock solution to histochemical staining buffer to give a final concentration of 1 mM. Be careful not to allow the X-Gluc stock to touch any plastic susceptible to DMF before it has been diluted with the staining buffer. Diffusion of product intermediate can be prevented, if necessary, by the addition of potassium ferri/ferrocyanide stock solution, to give final concentrations in the range of 0.5–5 mM. Once the optimum concentration has been determined for a particular transgenic tissue, this can be used in all subsequent stainings.
2. Place tissue in a suitable container for staining. For example, pollen is easily stained in the wells of flat-bottomed, transparent microtiter plates; other samples can be placed in 1.5-mL microcentrifuge tubes.
3. Add enough histochemical staining buffer, containing X-Gluc and potassium ferri/ferrocyanide (if any), to cover the tissue, typically 200–400 µL (*see* **Note 8**). Ensure that sample containers are sealed to prevent drying out.
4. Incubate at 37°C for several hours or overnight, until a blue color develops (*see* **Note 9**).
5. Green tissue can be cleared following staining by being placed initially in 70% ethanol, followed by repeated changes in 95% ethanol. All tissues can be stored in 70–95% ethanol at 4°C for several months, since the blue precipitate is stable in ethanol.

4. Notes

1. The volume of stop solution used should always be at least three times the volume of sample added. Generally, a volume of 3–9 times stop solution is used. It must also be remembered that the capacity of a well of a microtiter plate is 200 µL.
2. The grinding procedure may have to be altered, depending on the tissue sample. A small pestle and mortar is convenient for relatively large quantities of tissue, and the addition of a small quantity of acid-washed sand can improve grinding. However, small tissue samples and pollen may be ground in 1.5-mL microcentrifuge tubes containing 200–400 µL GUS extraction buffer, using either a hand or mechanical grinder, such as those provided by Anachem, Luton, UK.
3. For the majority of tissue samples, one centrifugation step is enough to leave a clear supernatant. However, we have found that tissues such as pollen require a second centrifugation step in order to remove all cell debris.
4. Stop solution not only prevents further enzyme activity, but also maximizes the fluorescence of the product, 4-MU.
5. Ideally, similar quantities of protein should be used in each sample. In practice, protein concentrations are usually measured after the GUS assay has been performed, but if similar amounts of tissue are used for each sample, protein concentrations should not vary widely.

6. If very low activities are being measured, it is advisable to leave the microtiter plate for 1 h in the dark after taking the final sample, since this has been found to maximize fluorescence readings. For high activity levels, small differences in fluorescence readings will not be as important and this action need not be taken.

7. The protein samples can be left in Bradford's reagent for up to 1 h, but after this time the protein begins to precipitate.

8. At this point, some workers use vacuum infiltration to ensure that all of the tissue comes into contact with the staining solution. Place the tissue covered with staining solution in a vacuum dryer and apply a vacuum for approx 15 min.

9. Incubation of tissue can also be carried out at lower temperatures, such as at 25°C. This will slow the reaction and prevent some diffusion of soluble intermediate reaction product.

Acknowledgments

Funding from Biotechnology and Biological Sciences Research Council (BBSRC), Ministry of Agriculture, Fisheries and Food (MAFF), and European Community (EC), in support of our work on transgenic plants, is gratefully acknowledged.

References

1. Lindsey, K., Wei, W., Clarke, M. C., McArdle, H. F., Rooke, L. M., and Topping, J. F. (1993) Tagging genomic sequences that direct transgene expression by activation of a promoter trap in plants. *Transgenic Res.* **2,** 33–47.

2. Herrera-Estrella, L., Depicker, A., Van Montagu, M., and Schell, J. (1983) Expression of chimeric genes transferred into plant cells using a Ti-plasmid-derived vector. *Nature* **303,** 209–213.

3. An, G. (1986) Development of plant promoter expression vectors and their use for analysis of differential activity of nopaline synthase promoter in transformed tobacco cells. *Plant Physiol.* **81,** 86–91.

4. Jefferson, R. A., Kavanagh, T. A., and Bevan, M. W. (1987) GUS fusions: β-glucuronidase as a sensitive and versatile gene fusion marker in higher plants. *EMBO J.* **6,** 3901–3907.

5. Ow, D. W., Wood, K. V., Deluca, M., De Wet, J. R., Helsinki, D. R., and Howell, S. H. (1986) Transient and stable expression of the firefly luciferase gene in plant cells and transgenic plants. *Science* **234,** 856–859.

6. Niedz, R. P., Sussman, M. R., and Satterlee, J. S. (1995) Green fluorescent protein: an in vivo reporter of plant gene expression. *Plant Cell Rep.* **14,** 403–406.

7. Novel, G. and Novel, M. (1973) Mutants of *E. coli* K12 unable to grow on methyl-β-D-glucuronide: map of *uid* A locus of the structural gene of β-D-glucuronidase. *Mol. Gen. Genet.* **120,** 319–335.

8. Jefferson, R. A., Burgess, S. M., and Hirsh, D. (1986) β-glucuronidase from *E. coli* as a gene fusion marker. *Proc. Natl. Acad. Sci. USA* **83,** 8447–8451.

9. Yang, N-S. and Christou, P. (1990) Cell type specific expression of a CaMV 35S-GUS gene in transgenic soybean plants. *Dev. Genet.* **11,** 289–293.

10. Terada, R. and Shimamoto, K. (1990) Expression of CaMV 35S-GUS gene in transgenic rice plants. *Mol. Gen. Genet.* **220,** 389–392.

11. Battraw, M. J. and Hall, T. C. (1990) Histochemical analysis of CaMV 35S promoter-β-glucuronidase gene expression in transgenic rice plants. *Plant Mol. Biol.* **15,** 527–538.

12. Dong, J-Z., Yang, M-Z., Jia, S-R., and Chua, N-H. (1991) Transformation of melon (*Cucumis melo* L.) and expression from the Cauliflower Mosaic Virus 35S promoter in transgenic melon plants. *Bio/Technology* **9,** 858–863.

13. Ohta, S., Mita, S., Hattori, T., and Nakamura, K. (1990) Construction and expression in tobacco of a β-glucuronidase (GUS) reporter gene containing an intron within the coding sequence. *Plant Cell Physiol.* **31,** 805–813.

14. Rao, A. G. and Flynn, P. (1990) A quantitative assay for β-D-glucuronidase (GUS) using microtiter plates. *BioTechniques* **8,** 38–40.

15. Mascarenhas, J. P. and Hamilton, D. A. (1992) Artifacts in the localization of GUS activity in anthers of petunia transformed with a CaMV 35S-GUS construct. *Plant J.* **2,** 405–408.

16. Hu, C-Y., Chee, P. P., Chesney, R. H., Zhou, J. H., Miller, P. D, and O'Brien, W. T. (1990) Intrinsic GUS-like activities in seed plants. *Plant Cell Rep.* **9,** 1–5.

17. Kosugi, S., Ohashi, Y., Nakajima, K., and Arai, Y. (1990) An improved assay for β-glucuronidase in transformed cells: methanol almost completely suppresses a putative endogenous β-glucuronidase activity. *Plant Sci.* **70,** 133–140.

18. Wilkinson, J. E., Twell, D., and Lindsey, K. (1994) Methanol does not specifically inhibit endogenous β-glucuronidase (GUS) activity. *Plant Sci.* **97,** 61–67.

5

Expression of Recombinant Proteinase Inhibitors in Plants

Dominique Michaud and Thierry C. Vrain

1. Introduction

The importance of proteolytic enzymes in plant–pest and plant–pathogen interactions has recently been recognized, and control strategies based on their inhibition with protease inhibitors (PIs) have been developed or proposed to control herbivory insects *(1)*, plant parasitic fungi *(2,3)*, and nematodes *(4)*. The various roles of proteases in these organisms and the biochemical pathways affected by their interactions with PIs may differ, but their importance for normal growth and development is now evident. The repressive effects of PIs on insect growth and fecundity, notably, have been documented for several species *(1)*, and evidence for the implication of microbial proteases as phytopathogenic determinants has been reported in several cases *(5–13)*. Based on this information, transformation of plant genomes with PI cDNA clones appears to be an attractive approach for the control of plant pests and pathogens, and several economically important plants expressing exogenous plant PIs have been engineered during the past few years (**Table 1**). While allowing control of plant pests and pathogens, PIs expressed in transgenic plants may also serve as a source of active inhibitors for the study and the eventual control of some protease-related pathogenic processes in humans. Proteases are important not only in the intracellular regulation of peptides and proteins, but also in the development of several diseases, including tumor metastasis *(21)*, rheumatoid arthritis *(22,23)*, Alzheimer's disease *(24)*, emphysema *(25)*, pancreatitis *(26)*, and AIDS *(27,28)*. PIs thus appear interesting as therapeutic agents and strategies to control human pathogenic bacteria *(29)* and viruses *(30–33)* using these enzyme inhibitors have been proposed.

From: *Methods in Biotechnology, Vol. 3:*
Recombinant Proteins from Plants: Production and Isolation of Clinically Useful Compounds
Edited by: C. Cunningham and A. J. R. Porter © Humana Press Inc., Totowa, NJ

Table 1
PI-Expressing Transgenic Plants: Some Examples[a]

Plant	Inhibitor	Class	Ref.
Rapeseed	Oryzacystatin I	Cys	14[b]
Poplar	Oryzacystatin I	Cys	15[b]
Potato	Oryzacystatin I	Cys	16
Rice	Oryzacystatin I	Cys	17
Tobacco	Cowpea trypsin inhibitor	Ser	18[b]
	Potato proteinase inhibitor II	Ser	19[b]
	Tomato proteinase inhibitor I	Ser	19
	Tomato proteinase inhibitor II	Ser	19[b]
	Oryzacystatin I	Cys	20
Tomato	Oryzacystatin I	Cys	4[c]
	Oryzacystatin I (modified)	Cys	4[c]

[a]Active recombinant inhibitors *in planta* have been detected in each case.
[b]Adverse effects against growth, development, and/or fecundity of herbivory insects have been demonstrated.
[c]Adverse effects against growth, development, and/or fecundity of the plant parasitic nematode *Globodera pallida* have been demonstrated.

This chapter describes simple protocols for obtaining PI-expressing transgenic plants via the *Agrobacterium*-mediated transformation procedure, and for analyzing the activity of the recombinant inhibitors expressed in leaf cells. Several cDNA clones encoding plant PIs are available for plant transformation (**Table 2**); as an example, this chapter describes the transformation of potato with a cDNA clone of the rice cysteine PI, oryzacystatin I (OCI; **ref. 34**). Inhibitors of the cystatin superfamily *(35)* represent potentially useful molecules in planta for the control of herbivory insects *(14,15,36–39)* and plant parasitic nematodes *(4)*, and their use in medicine has been proposed, notably, for viral therapy *(31,40)*.

2. Materials

2.1. Plant Material

Axenically grown potato (*Solanum tuberosum* L. cv. Kennebec) plantlets are used for the experiments. The plantlets are cultivated onto a multiplication medium (*see* **Subheading 2.4.1.**), and maintained in a tissue-culture room (*see* **Subheading 2.3.**).

2.2. Agrobacterium *Strain and Growth Conditions*

The *Agrobacterium tumefaciens* strain LBA4404 carrying the plasmid pBInh-OCI *(16)* is used for inoculations. The plasmid pBInh-OCI is identical to the com-

Table 2
Some Plant PI-Encoding cDNA Clones Described in the Literature

Inhibitor	Accession numbers[a]
Aspartate PIs	
Tomato Asp-pin	X73986
Cysteine PIs	
Oryzacystatin I (rice)	J03469
Oryzacystatin II (rice)	J05595
Corn cystatin I	S49383
Potato cysteine proteinase inhibitor	X67844
Ambrosia artemisiifolia cysteine PI	L16624
Vigna cysteine PI	Z21954
Serine PIs	
Alfalfa Bowman-Birk inhibitor	X68704
Arabidopsis proteinase inhibitor II	X69139
Barley amylase/proteinase inhibitor	X05168
Potato proteinase inhibitor I	X67675
Potato proteinase inhibitor II	X03778
Soybean proteinase inhibitor CII	M20732
Soybean proteinase inhibitor IV	M20733
Tomato proteinase inhibitor I	K03290
Tomato proteinase inhibitor II	K03291

[a]Accession numbers from the GenBank nucleotide sequence database.

mercial plasmid pBI121 (Clontech, Palo Alto, CA), except that the β-glucoronidase gene has been replaced by an OCI cDNA clone isolated from the expression plasmid pOC26-5'-1 *(41)*. More precisely, pBInh-OCI includes a nopaline synthase (NOS) promoter expressing a NptII gene for kanamycin resistance and the cauliflower mosaic virus 35S promoter-driven OCI sequence *(16)*. Prior to inoculation, bacteria are grown for 36 h at 28°C in Luria-Bertani (LB) medium (*see* **Subheading 2.4.**) containing 50 mg/L kanamycin and 50 mg/L streptomycin.

2.3. Special Laboratory Tools and Materials

1. A tissue-culture room, maintained at 25°C under a light intensity of 60 μmol/m²/s and a photoperiod of 16 h/d provided by cool white fluorescent lights.
2. A controlled-environment chamber, maintained at 20°C (day) and 15°C (night) under a light intensity of 60 μmol/m²/s, a relative humidity of 75–80%, and a photoperiod of 16 h/d provided by cool white fluorescent lights.
3. General tissue culture and plant culture materials *(42)*.
4. Liquid nitrogen and a mortar and a pestle.
5. Miracloth (Calbiochem, La Jolla, CA).

2.4. Media, Buffers, and Other Solutions

All media, buffers, and other solutions are made up as aqueous preparations. The media for tissue culture are agar-solidified in 9-cm Petri dishes (25 mL/ Petri) after sterilization using standard procedures.

2.4.1. Tissue Culture Media (16)

1. Multiplication medium: MS salts (Sigma, St. Louis, MO) supplemented with 100 mg/L *myo*-inositol (Sigma), 0.4 mg/L thiamine-HCl (Sigma), 3% (w/v) sucrose, 0.8% (w/v) Bactoagar (Difco, Detroit, MI), and 0.075% (w/v) activated charcoal.
2. Callus induction medium: multiplication medium supplemented with 1 μ*M* naphtalene acetic acid (NAA), 10 μ*M* 6-benzylaminopurine (BA), and 25 μ*M* giberellic acid 3 (GA$_3$).
3. Selection medium: callus induction medium supplemented with 50 mg/L kanamycin (Sigma) and 500 mg/L carbenecillin (Sigma).
4. Regeneration medium: multiplication medium supplemented with 10 μ*M* BA, 25 μ*M* GA$_3$, and 500 mg/L carbenecillin.

2.4.2. Buffers

1. Leaf extraction buffer: 100 m*M* sodium phosphate, pH 6.5.
2. Papain proteolysis buffer: 100 m*M* sodium phosphate, pH 6.5, 1 m*M* EDTA, 5 m*M* L-cysteine (Sigma), 0.1% (v/v) Triton X-100 (Sigma).

2.4.3. Other Solutions

1. Papain solution: 20 μg papain (~0.2 units)/mL papain proteolysis buffer. Prepare 25 mL, distribute in 1-mL aliquots, and keep at –20°C until use. Stable for several months. Papain (E.C.3.4.22.2, from papaya latex) is purchased from Sigma.
2. Azocasein solution: 2% (w/v) azocasein (Sigma) diluted in papain proteolysis buffer. Prepare fresh as needed.
3. Trichloroacetic acid (TCA) solution: 100% (w/v) TCA (Sigma) in water. Prepare 100 mL, and keep at 4°C until use.

2.5. Gel System

All electrophoretic gels and solutions are made up as aqueous preparations. The protocol for gelatin-polyacrylamide gel electrophoresis (PAGE), derived from the widely used sodium dodecylsulfate (SDS)-PAGE procedure *(43)*, is not explicitly described here. A detailed, step-by-step protocol for SDS-PAGE is given in **ref. 44**.

1. A Mini-Protean II™ slab-gel unit (BioRad, Hercules, CA).
2. Gelatin-PAGE separation gel: 15% (w/v) acrylamide (ratio acrylamide-*bis*-acrylamide: 30:1.2), 0.1% (w/v) porcine type A gelatin (Sigma), 375 m*M* Tris, pH 8.8, 0.1% (w/v) SDS.

3. Gelatin-PAGE stacking gel: 5% (w/v) acrylamide (ratio acrylamide-*bis*-acrylamide: 30:1.2), 125 m*M* Tris-HCl, pH 6.8, 0.1% (w/v) SDS.
4. Gelatin-PAGE sample buffer 2X: 62.5 m*M* Tris-HCl, pH 6.8, 2% (w/v) SDS, 2% (w/v) sucrose, 0.01% (w/v) bromophenol blue.
5. Gelatin-PAGE running buffer: 25 m*M* Tris-HCl, 192 m*M* glycine, 0.1% (w/v) SDS.
6. Renaturation solution: 2.5% (v/v) Triton X-100.
7. Gel staining solution: 0.1% (w/v) Coomassie brilliant blue in 25% (v/v) isopropanol/10% (v/v) acetic acid. Dissolve Coomassie blue in isopropanol before adding water and acetic acid.
8. Gel destaining solution: 10% (v/v) isopropanol, 10% (v/v) acetic acid.

3. Methods

This section describes simple procedures and strategies for obtaining OCI-expressing transgenic potato plants, for assessing the activity of the recombinant inhibitor expressed in leaf tissues, and for achieving its efficient recovery from the plant tissues when planning its use as a therapeutic agent. Provided that efficient transformation procedures are available, the approaches presented to analyze and to recover recombinant OCI may be used for any plant species, without the need for major modifications. Transformation procedures with OCI constructs have been reported for several species, including rapeseed *(14)*, potato *(16)*, tobacco *(20)*, rice *(17)*, tomato *(4)*, and poplar *(15)* (*see* **Note 1**).

3.1. Transformation of Potato Leaf Sections

Potato leaf sections are transformed using *Agrobacterium*-mediated genetic transformation, essentially as described in **refs.** *16* and *45*. Growth conditions in the tissue-culture room are similar to those previously described for axenic plantlets (*see* **Subheading 2.3.**). For more details on *Agrobacterium*-mediated transformation procedures, see the detailed protocols of Draper and co-workers *(42)*.

3.1.1. Inoculation

1. Cut axenic potato leaves into small pieces (about 10 mm in width) using a sterile scalpel blade, and place the leaf pieces onto the callus induction medium (about 10–15 pieces per Petri dish) for 4 d in the tissue-culture room. This preliminary step is done to prevent subsequent excessive soaking of the leaf tissue by the *Agrobacterium* inoculum (**step 2**), by allowing callus to form on the cut edges of the explants before inoculation *(42)*.
2. Dip leaf pieces into a 1:10 dilution of an overnight culture of the *A. tumefaciens* strain LBA4404, harboring the plasmid pBInh-OCI (*see* **Subheading 2.2.**), for 10 min at room temperature. Shake off excess liquid and place the pieces on new plates containing the callus induction medium. Plates are left in the culture room for 4 d, or until visible bacterial colonies appear.

3.1.2. Regeneration

1. After coculture, transfer the calli onto the selection medium (10–15 sections/plate), and place the plates in the culture room for 2 wk.
2. After arrest of bacterial growth, place the calli onto the regeneration medium for 4 wk to allow regeneration of shoots.
3. Transfer the regenerated shoots onto the selection medium for root growth, shoot elongation, and multiplication of the transformants. For multiplication, cut the stems in nodal sections, each including an axillary bud. Trim off the leaf from each section, leaving a short petiole stump. Place the sections base down onto the selection medium. Subculture every 4 to 5 wk.

3.1.3. Establishment of Transgenic Plantlets Ex Vitro

1. Gently remove rooted plantlets from medium when they are 5–10 cm in height.
2. Quickly wash off agar sticking to the roots, place the plantlets in 20-cm diameter plastic pots filled with pasteurized greenhouse soil mix, and place the plants in the controlled environment chamber to protect the plants from desiccation.
3. Once they have formed a waxy cuticle, transfer the plants to normal culture conditions (e.g., in greenhouse). The healthy plants obtained serve as a source of OCI-expressing transgenic plants for further analyses (*see* **Note 2**).

3.2. Analysis of Recombinant OCI in Potato Transformants

As a general rule, the analysis of transgenic plants after integration of foreign DNA into their genome includes standard procedures aimed at confirming integration of the gene in the genome (Southern blotting), adequate expression of the cDNA clone (Northern blotting), actual accumulation of the recombinant protein in the transformed cells (Western blotting), and more specific procedures to assess the (putative) biological activity of the recombinant protein. The general Southern blotting, Northern blotting, and Western blotting procedures are described in detail in **ref. 46**, and are not discussed here. Rather, we emphasize the description of simple protocols useful in the analysis of recombinant PIs recovered from plant tissues.

3.2.1. Extraction of Recombinant OCI

Soluble proteins are recovered from either OCI-expressing and control potato leaves using an aqueous low ionic strength buffer (*see* **Note 3**). The resulting extracts are then used for both quantitative and qualitative analyses.

1. Grind 5 g of potato leaves (from either transgenic or control plants) to a fine powder in liquid nitrogen, using a mortar and a pestle.
2. Transfer the leaf powder to an ice-cold plastic tube, and add to the tube 10 mL of cold leaf-extraction buffer supplemented with 0.5% (w/v) polyvinylpyrrolidone (PVP; Sigma) (*see* **Note 4**).

3. After extraction of soluble proteins on ice for 30 min, pass the mixture through one layer of Miracloth to remove leaf-tissue debris. Centrifuge (15,000g) the aqueous solution for 15 min at 4°C, and discard the pellet.
4. Perform an assay of total protein in the supernatant fraction using the standard Bradford procedure *(48)* (*see* **Note 5**), with bovine serum albumin as a protein standard. This soluble extract serves as a source of recombinant OCI for further analyses. If not used immediately, it can be quick-frozen in liquid nitrogen and stored at –80°C for several weeks without loss of OCI inhibitory activity.

3.2.2. Quantitative Assay of Recombinant OCI Activity

The assay described involves the use of the protein substrate azocasein, a chromogenic derivative of casein. Detection of protease activity with this substrate depends upon the solubility of proteolysis-derived, low-mol wt, colored peptides present in the reaction mixture after precipitation of large fragments of the protein substrate in acid conditions *(50)*. The cysteine proteinase papain, a target proteinase of OCI *(34)*, is used as a test enzyme to assess PI activity in the OCI-expressing transgenic plants. Papain activity is assayed after its preincubation with various amounts of leaf extracts prepared from OCI-expressing transgenic plants (*see* **Subheading 3.2.1.**). For each clone, the amount of leaf-soluble protein causing the inhibition of 1 µg (~10 mU) of papain may be estimated, allowing confirmation that OCI is produced in an active form, and to compare the levels of active OCI in the various clones analyzed.

1. For each leaf extract to be analyzed, place 100 µL of papain solution (~2 µg papain) in 2 × 7 microcentrifuge (test) tubes, and 100 µL of leaf-extraction buffer in 2 × 7 (control) tubes.
2. While keeping the tubes standing on ice, add varying amounts of the potato leaf extract (0, 50, 100, 150, 200, 300, or 400 µL) to the test and control tubes, and add leaf-extraction buffer to 500 µL. Prepare each sample in duplicate.
3. Incubate the mixtures in a water bath for 10 min at 37°C to allow inhibition of papain by the recombinant OCI present in the leaf extract.
4. Add 500 µL of the azocasein solution to each tube, and place the tubes in the water bath for 180 min at 37°C.
5. After azocasein proteolysis, add 100 µL of the TCA solution, mix the tubes thoroughly, and allow them to stand on ice for 15 min to ensure complete precipitation of the remaining azocasein and azocasein large fragments.
6. Centrifuge (12,000g) the complete mixtures for 5 min at 4°C, and transfer 1 mL of the resulting supernatants to larger tubes containing 1 mL of 1N NaOH (*see* **Note 6**).
7. Determine the absorbance *(A)* of the solutions at 440 nm (A_{440}). For each amount of leaf extract tested, subtract the corresponding A_{440} value measured for the control samples (with no papain).
8. Express the A_{440} values on a relative basis, compared to the A_{440} values measured for papain in the absence of leaf extract (100% papain activity). Plot papain

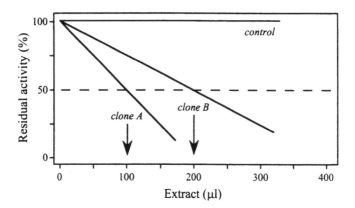

Fig. 1. Residual activity of papain after its incubation with various amounts of extracts prepared from a control plant and from two OCI-expressing transgenic potato clones, *clone A* and *clone B*. Assuming that the extracts of the two transgenic clones contain the same amount of total soluble protein, *clone A* produces twice the amount of active OCI produced by *clone B*. The arrows indicate the amount (μL) of extract needed to inhibit 50% of papain activity (~1 μg papain).

activity against the amount of leaf extract used (μL) and perform a simple (linear) regression with the plotted points. The amount of extract needed to inhibit 50% of papain activity roughly corresponds to the amount needed to inhibit 1 μg (~10 mU) of papain (*see* **Fig. 1**). The different OCI-expressing clones can be compared in their ability to inhibit papain by expressing the volume of extract needed on a protein content basis [μL needed × soluble protein content for a particular clone (μg protein/μL) = μg leaf-soluble protein needed for this clone].

3.2.3. Qualitative Assay of Recombinant OCI Activity

This procedure, adapted from **ref. 51**, allows the occurrence of active OCI in extracts of OCI-expressing transgenic plants to be visualized. After preincubation of transgenic and control leaf extracts with a small amount of papain, the enzyme/inhibitor mixtures are submitted to mildly denaturing gelatin-PAGE. Because OCI is a good inhibitor of papain ($K_i \sim 10^{-8}M$; **ref. 52**), the complexes formed between the enzyme and the recombinant inhibitor remain stable during electrophoresis, preventing restoration of papain activity during migration and allowing the identification of transgenic clones expressing OCI in an active form.

1. Incubate 20 μL of the transgenic leaf extract with 1 μL of 100 m*M* phenylmethyl-sulfonyl fluoride (PMSF; Sigma) for 10 min at room temperature in a microfuge tube. PMSF is added to inactivate endogenous proteinases present in the potato leaf extract *(16)*, which could interfere with papain activity (*see* **Note 7**).

2. Add 5 μL of the papain solution, diluted 10-fold in leaf-extraction buffer, to the extract, and incubate the enzyme/leaf-extract mixture in a water bath for 15 min at 37°C to allow inhibition of papain by recombinant OCI.
3. After inhibition of the papain, add 25 μL of gelatin-PAGE sample buffer 2X (*see* **Note 8**), and load 25 μL of the complete mixture onto a 0.75 mm-thick gelatin-containing polyacrylamide mini-gel. The gelatin-PAGE procedure is derived from the standard SDS-PAGE procedure (*43,44*).
4. Perform the electrophoretic migration at 4°C (150 V) until the bromophenol blue tracking dye reaches the bottom of the gel (*43,51*).
5. After migration, place the gel in the renaturation solution for 30 min at room temperature, with agitation. This allows Triton X-100 to remove the SDS from the gel and to renature papain.
6. Perform proteolysis by incubating the gel in the proteolysis buffer for 4 h at 37°C.
7. Stop the reaction by immersing the gel in the gel-staining solution for 1 h at 37°C and destain with the gel-destaining solution to visualize papain activity. Non-inhibited papain activities appear as clear (lysis) bands against the blue gelatin background. Previously, when mixed with an extract from a transgenic plant expressing an active recombinant OCI form, papain is "irreversibly" inactivated under the electrophoretic conditions used. No activity or only partial activity can be detected on the gel after staining (*see* **Fig. 2**, clone A).

3.3. Recovery of Recombinant OCI from Transgenic Leaf Tissue

In parallel with the increasing amount of information available about the potential of PIs for pest and pathogen control in agriculture and medicine, efficient strategies for their production in heterologous systems and their recovery in large quantities in a pure and active form are being developed (e.g., **refs.** *37* and *53* for the expression of OCI in *Escherichia coli*). However, no specific scheme has yet been proposed to efficiently recover recombinant PIs from transgenic plant tissues. The original recovery procedure reported for OCI, for instance, required multiple chromatographic steps and led to a final recovery yield of only 28% of the initial OCI activity present in the crude extract of rice seed (*34*). Ideally, a good recovery strategy should be designed to ensure stability of the inhibitor during extraction and to maximize its recovery using a simple, one-step purification scheme.

3.3.1. Ensuring Stability of the Protein Product

Whatever the procedure used to extract soluble plant proteins, care must be taken to ensure their stability after breakage of the tissue. Mature plant cells, especially those from green tissues, present special problems in the isolation of proteins, primarily because of the presence of proteases and phenolic compounds in their vacuoles (*47*). During the extraction process, these compounds are released from broken cells and may severely affect the integrity of the pro-

Fig. 2. Electrophoregram showing the gelatinolytic activity of papain after its incubation with an extract from an OCI-expressing transgenic potato clone. Lane 1: papain incubated with an extract from a control plant; lane 2, papain incubated with an extract from the OCI-expressing potato *clone A*; lane 3, noninhibited papain, incubated with leaf-extraction buffer. Although the enzyme remained fully active after its incubation with the extract from the control plant (lane 1 similar to the control lane 3), it was almost completely inactivated when placed in the presence of an extract from the transgenic *clone A* (lane 2).

teins present. A simple means of avoiding (or at least minimizing) protease- and phenolic-related modifications consists of adding appropriate protease inhibitors and phenolic-complexing compounds in the extraction buffer to minimize the extraction time, and performing all extraction steps at 4°C (*see* **ref. 47** for a discussion).

3.3.2 Single-Step Purification of the Inhibitor

Provided that the protein remains biologically active, the easiest way to a simple and efficient recovery of a given protein from a complex (crude) extract consists of fusing an affinity handle to one of its extremities using appropriate genetic constructs. Chen et al. *(37)*, for instance, fused a very basic peptidic tail to the N-terminal end of OCI, allowing purification of the inhibitor in an active form from *E. coli* crude extracts using a single anion-exchange chromatography step. Similarly, fusion of the 26 kDa-protein glutathione *S*-transferase to the N-terminal extremity of OCI allowed purification of the inhibitor in an active form, with a single affinity purification step using reduced glutathione-embedded agarose beads *(53)*. The stability of such abnormal chimeric proteins in the transformed cells, however, may require further analysis before planning their expression in plant cells (*see* **Chapter 14**).

4. Notes

1. The analytic approaches described are also suitable for the analysis of serine PI- or aspartate PI-expressing transgenic plants. In these cases, the serine proteinase trypsin (from bovine pancreas, type III; E.C. 3.4.21.4; Sigma) and the aspartate proteinase cathepsin D (from bovine spleen; E.C.3.4.23.5; Sigma) can be used instead of papain as a test enzyme.

2. PI assays must be carried out with plants acclimatized to greenhouse or field conditions, since the general metabolism of in vitro plantlets is modified compared to that of plants grown ex vitro. The nature and quantity of proteins in in vitro plants, in particular, is generally modified, resulting in significant differences when analyzing the biological activity of recombinant proteins extracted from either in vitro or acclimated plants. The activity of OCI accumulated in acclimated potato plants, for instance, is easily detected, but the activity measured in extracts from in vitro plantlets is very low (D. Michaud, unpublished observations).

3. Control plants are plants transformed with the construct pBInh-OCI (*see* **Subheading 2.2.**) lacking the OCI insert.

4. PVP inactivates phenols, which react with proteins by forming strong hydrogen bonds or, when oxidized to quinones, by condensing with the -SH and -NH_2 groups of proteins. When not controlled, these chemical interactions may cause the formation of proteins polymers crosslinked with polyphenols, thereby altering the nature of the proteins extracted. In combination with PVP, the inactivation of phenoloxidases by the addition of reducing agents, such as ascorbate, β-mercaptoethanol, or dithiothreitol, may be useful *(47)*. However, the use of antioxidant substances must be considered with care, since leaf cells often contain cysteine proteinases strongly activated under reduced conditions.

5. The Bradford procedure is based on the binding of the Coomassie brilliant blue dye to proteins *(48)*. Given the general occurrence of phenolic compounds in leaf cells, the well-known Lowry procedure for quantification of proteins *(49)*, which is based on the quantification of phenolics, is in many cases not suitable to properly quantify plant proteins extracted from green tissues. Because of less interference from phenolics, the Coomassie blue procedure is the method of choice for approximate but rapid and simple assay of plant leaf proteins *(47)*.

6. NaOH is added to favor color development by increasing the pH acidified by TCA. The volume used may be changed to satisfy technical constraints imposed by the available equipment.

7. For plants transformed with serine PI cDNA clones, PMSF must be hydrolyzed by heating the plant extract at 80 or 100°C for 2 min before adding trypsin, provided that the inhibitor remains active after heating. It might be necessary to

develop a strategy allowing inactivation of endogenous proteinases for each specific case. If no endogenous proteinase (gelatinase) co-migrates with trypsin, this inhibition step may be omitted without interfering with subsequent detection of trypsin activity.

8. No reducing agent, such as dithiothreitol or β-mercaptoethanol, is added to the sample buffer, since many enzymes are irreversibly inactivated in such conditions, preventing their subsequent detection using activity staining procedures *(47)*.

Acknowledgments

The authors thank Martine Korban and Line Cantin for critical reading of the manuscript. This work was partly supported by a fellowship from the Natural Science and Engineering Research Council of Canada to D. Michaud.

References

1. Hilder, V. A., Gatehouse, A. M. R., and Boulter, D. (1993) Proteinase inhibitor approach, in *Transgenic Plants: Engineering and Utilization*, vol. 1 (Kung, S.-D. and Wu, R., eds.), Academic, New York, pp. 317–338.
2. Lorito, M., Broadway, R. M., Hayes, C. K., Woo, S. L., Noviello, C., Williams, D. L., and Harman, G. E. (1994) Proteinase inhibitors from plants as a novel class of fungicides. *Mol. Plant-Microbe Interact.* **7,** 525–527.
3. Dunaevskii, Y. E., Pavlyukova, E. B., Belyakova, G. A., and Belozerskii, M. A. (1994) Anionic trypsin inhibitors from dry buckwheat seeds: isolation, specificity of action, and effect on growth of micromycetes. *Biochemistry (Moscow)* **59,** 739–743.
4. Urwin, P. E., Atkinson, H. J., Waller, D. A., and McPherson, M. J. (1995) Engineered oryzacystatin-I expressed in transgenic hairy roots confers resistance to *Globodera pallida. Plant J.* **8,** 121–131.
5. Keen, N. T., Williams, P. H., and Walker, J. C. (1967) Protease of *Pseudomonas lachrymans* in relation to cucumber angular leaf spot. *Phytopathology* **57,** 263–271.
6. Hislop, E. C., Paver, J. L., and Keon, J. P. R. (1982) An acid protease produced by *Monilinia fructigena in vitro* and in infected apple fruits, and its possible role in pathogenesis. *J. Gen. Microbiol.* **128,** 799–807.
7. Robertsen, B. (1984) An alkaline extracellular protease produced by *Cladosporium cucumerinum* and its possible importance in the development of scab disease of cucumber seedlings. *Physiol. Plant Pathol.* **24,** 83–92.
8. Bashan, Y., Okon, Y., and Henis, Y. (1986) A possible role for proteases and deaminases in the development of the symptoms of bacterial speck disease in tomato caused by *Pseudomonas syringae* pv. *tomato. Physiol. Mol. Plant Pathol.* **28,** 15–31.
9. Roby, D., Toppan, A., and Esquerré-Tugayé, M. T. (1987) Cell surfaces in plant microorganisms interactions. VIII. Increased proteinase inhibitor activity in melon plants in response to infection by *Colletotrichum lagenarium* or to treatment with an elicitor fraction from this fungus. *Physiol. Mol. Plant Pathol.* **30,** 453–460.

10. Tang, J. L., Gough, C. L., Barber, C. E., Dow, J. M., and Daniels M. J. (1987) Molecular cloning of protease gene(s) from *Xanthomonas campestris* pv. *campestris*: expression in *Escherichia coli* and role in pathogenicity. *Mol. Gen. Genet.* **210,** 443–448.

11. Ball, A. M., Ashby, A. M., Daniels, M. J., Ingram, D. S., and Johnstone, K. (1991) Evidence for the requirement of extracellular protease in the pathogenic interaction of *Pyrenopeziza brassicae* with oilseed rape. *Physiol. Mol. Plant Pathol.* **38,** 147–161.

12. Dow, J. M., Clarke, B. R., Milligan, D. E., Tang, J. L., and Daniels, M. J. (1990) Extracellular proteases from *Xanthomonas campestris* pv. *campestris*, the black rot pathogen. *Appl. Environ. Microbiol.* **56,** 2994–2998.

13. Dow, J. M., Fan, M. J., Newman, M.-A., and Daniels, M. J. (1993) Differential expression of conserved protease genes in crucifer-attacking pathovars of *Xanthomonas campestris*. *Appl. Environ. Microbiol.* **59,** 3996–4003.

14. Bonadé-Bottino, M. (1993) Défense du colza contre les insectes phytophages déprédateurs: étude d'une stratégie basée sur l'expression d'inhibiteurs de protéases dans la plante. Ph.D. Thesis, Université de Paris-Sud, Centre d'Orsay.

15. Leplé, J.-C., Bonadé-Bottino, M., Augustin, S., Delplanque, A., Dumanois, V., Pilate, G., Cornu, D., and Jouanin, L. (1995) Toxicity to *Chrysomela tremulae* (Coleoptera: Chrysomelidae) of transgenic poplars expressing a cysteine proteinase inhibitor. *Mol. Breed.*, **1,** 319–328.

16. Benchekroun, A., Michaud, D., Nguyen-Quoc, B., Overney, S., Desjardins, Y., and Yelle, S. (1995) Synthesis of active oryzacystatin I in transgenic potato plants. *Plant Cell Rep.* **14,** 585–588.

17. Hosoyama, H., Irie, K., Abe, K., and Arai, S. (1994) Oryzacystatin exogenously introduced into protoplasts and regeneration of transgenic rice. *Biosci. Biotechnol. Biochem.* **58,** 1500–1505.

18. Hilder, V. A., Gatehouse, A. M. R., Sheerman, S. E., Barker, R. F., and Boulter, D. (1987) A novel mechanism of insect resistance engineered into tobacco. *Nature* **330,** 160–163.

19. Johnson, R., Narvaez, J., An, G., and Ryan, C. A. (1989) Expression of proteinase inhibitors I and II in transgenic potato plants: effects on natural defense against *Manduca sexta* larvae. *Proc. Natl. Acad. Sci. USA* **86,** 9871–9875.

20. Masoud, S. A., Johnson, L. B., White, F. F., and Reeck, G. R. (1993) Expression of a cysteine proteinase inhibitor (oryzacystatin-I) in transgenic tobacco plants. *Plant Mol. Biol.* **21,** 655–663.

21. Sloane, B. F., Roxhin, J., Lah, T. T., Day, N. A., Buck, M., Ryan, R. E., Crissman, J. D., and Honn, K. V. (1988) Tumor cathepsin B and its endogenous inhibitors in metastasis. *Adv. Exp. Med. Biol.* **233P,** 259–268.

22. Mort, J. S., Recklies, A. D., and Poole, A. R. (1984) Extracellular presence of the lysosomal proteinase cathepsin B in rheumatoid synovium and its activity at neutral pH. *Arthritis Rheum.* **27,** 509–515.

23. Trabandt, A., Gay, R. E., Fassbender, H.-G., and Gay, S. (1991) Cathepsin B in synovial cells at the site of joint destruction in rheumatoid arthritis. *Arthritis Rheum.* **34,** 1444–1451.

24. Eriksson, S., Janciauskiene, S., and Lannfelt, L. (1995) α1-antichymotrypsin regulates Alzheimer β-amyloid peptide fibril formation. *Proc. Natl. Acad. Sci. U.S.A.* **92,** 2313–2317.

25. Chapman, H. A., Jr., and Stone, O. L. (1984) Comparison of live human neutrophil and alveolar macrophage elastolytic activity *in vitro*. Relative resistance of macrophage elastolytic activity to serum and alveolar proteinase inhibitors. *J. Clin. Invest.* **74,** 1693–1700.

26. Steer, M. L., Meldonlesi, J., and Figarella, C. (1984) Pancreatitis. The role of lysosomes. *Dig. Dis. Sci.* **29,** 934–938.

27. Kaplan, A. H., Zack, J. A., Knigge, M., Paul, D. A., Kempf, D. J., Norbeck, D. W., and Swanstrom, R. (1993) Partial inhibition of the human immunodeficiency virus type 1 results in aberrant virus assembly and the formation of noninfectious particles. *J. Virol.* **67,** 4050–4055.

28. Rosé, J. R., Babé, L. M., and Craik, C. S. (1995) Defining the level of human immunodeficiency virus Type I (HIV-1) protease activity required for HIV-1 particle maturation and infectivity. *J. Virol.* **69,** 2751–2758.

29. Björck, L., Akesson, P., Bohus, M., Trojnar, J., Abrahamson, M., Olafsson, I., and Grubb, A. (1989) Bacterial growth blocked by a synthetic peptide based on the structure of a human proteinase inhibitor. *Nature* **337,** 385–386.

30. Martin, L. N., Soike, K. F., Murphey-Corb, M., Bohm, R. P., Roberts, E. D., Kakuk, T. K., Thaisrivongs, S., Vidmar, T. J., Ruwart, M. J., Davio, S. R., and Tarpley, W. G. (1994) Effects of U-75875, a peptidomimetic inhibitor of retroviral proteases, on simian immunodeficiency virus infection in rhesus monkeys. *Antimicrob. Agents Chemother.* **38,** 1277–1283.

31. Korant, B. D., Towatari, T., Ivanoff, L., Petteway, S., Jr., Brzin, J., Lenarcic, B., and Turk, V. (1986) Viral therapy: prospects for protease inhibitors. *J. Cell. Biochem.* **32,** 91–95.

32. Wlodawer, A. and Erickson, J. (1993) Structure-based inhibitors of HIV-1 protease. *Annu. Rev. Biochem.* **62,** 543–585.

33. Richards, A. D., Roberts, R., Dunn, B. M., Graves, M. C., and Kay, J. (1989) Effective blocking of HIV-1 proteinase activity by characteristic inhibitors of aspartic proteinases. *FEBS Lett.* **247,** 113–117.

34. Abe, K., Hiroto, K., and Arai, S. (1987) Purification and characterization of a rice cysteine proteinase inhibitor. *Agric. Biol. Chem.* **51,** 2763–2768.

35. Barrett, A. J. (1987) The cystatins: a new class of peptidase inhibitors. *Trends Biochem. Sci.* **12,** 193–196.

36. Liang, C., Brookhart, G., Feng, G. H., Reeck, G. R., and Kramer, K. J. (1991) Inhibition of digestive proteinase of stored grain Coleoptera by oryzacystatin, a cysteine proteinase inhibitor from rice seeds. *FEBS Lett.* **278,** 139–142.

37. Chen, M.-S., Johnson, B., Wen, L., Muthukrishnan, S., Kramer, K. J., Morgan, T. D., and Reeck, G. R. (1992) Rice cystatin: bacterial expression, purification, cysteine proteinase inhibitory activity, and insect growth suppressing activity of a truncated form of the protein. *Protein Expres. Purif.* **3,** 41–49.

38. Michaud, D., Nguyen-Quoc, B., and Yelle, S. (1993) Selective inactivation of Colorado potato beetle cathepsin H by oryzacystatins I and II. *FEBS Lett.* **331,** 173–176.

39. Michaud, D., Bernier-Vadnais, N., Overney, S., and Yelle, S. (1995) Constitutive expression of digestive cysteine proteinase forms during development of the Colorado potato beetle, *Leptinotarsa decemlineata* Say (Coleoptera: Chrysomelidae). *Insect Biochem. Mol. Biol.*, **25,** 1041–1048.

40. Kondo, H., Ijiri, S., Abe, K., Maeda, H., and Arai, S. (1992) Inhibitory effect of oryzacystatins and a truncation mutant on the replication of poliovirus in infected Vero cells. *FEBS Lett.* **299,** 48–50.

41. Abe, K., Emori, Y., Kondo, H., Arai, S., and Suzuki, K. (1988) The NH_2-terminal 21 amino acid residues are not essential for the papain-inhibitory activity of oryzacystatin, a member of the cystatin superfamily. Expression of oryzacystatin cDNA and its truncated fragments in *Escherichia coli. J. Biol. Chem.* **263,** 7655–7659.

42. Draper, J., Scott, R., and Hamil, J. (1988) Transformation of dicotyledonous plant cells using the Ti plasmid of *Agrobacterium tumefaciens* and the Ri plasmid of *A. rhizogenes*, in *Plant Genetic Transformation and Gene Expression. A Laboratory Manual* (Draper, J., Scott, R., Armitage, P., and Walden, R., eds.), Blackwell Scientific, London, pp. 69–160.

43. Laemmli, U. K. (1970) Cleavage of structural proteins during the assembly of the head of bacteriophage T4. *Nature* **227,** 680–685.

44. Smith, B. J. (1984) SDS polyacrylamide gel electrophoresis of proteins, in *Methods in Molecular Biology*, vol. 1: *Proteins* (Walker, J. M., ed.), Humana, Clifton, NJ, pp. 41–55.

45. Wenzler, H., Mignery, G., May, G., and Park, W. (1989) A rapid and efficient transformation method for the production of large numbers of transgenic potato plants. *Plant Sci.* **63,** 79–85.

46. Sambrook, J., Fritsch, E. F., and Maniatis, T. (1989) *Molecular cloning: A Laboratory Manual*, 2nd ed. Cold Spring Harbor Laboratory, Cold Spring Harbor, NY.

47. Michaud, D. and Asselin, A. (1995) Review. Application to plant proteins of gel electrophoretic methods. *J. Chromatogr.* A **698,** 263–279.

48. Bradford, M. M. (1976) A rapid and sensitive method for the quantitation of microgram quantities of protein using the principle of protein-dye binding. *Anal. Biochem.* **72,** 248–254.

49. Lowry, O. H., Rosebrough, N. J., Farr, A. L., and Randall, R. J. (1951) Protein measurement with the Folin phenol reagent. *J. Biol. Chem.* **193,** 265–275.

50. Sarath, G., de la Motte, R. S., and Wagner, F. W. (1989) Protease assay methods, in *Proteolytic Enzymes: A Practical Approach* (Beynon, R. J. and Bond, J. S., eds.), IRL, New York, pp. 25–55.

51. Michaud, D., Faye, L., and Yelle, S. (1993) Electrophoretic analysis of plant cysteine and serine proteinases using gelatin-containing polyacrylamide gels and class-specific proteinase inhibitors. *Electrophoresis* **14,** 94–98.
52. Kondo, H., Abe, K., Nishimura, I., Watanabe, H., Emori, Y., and Arai, S. (1990) Two distinct cystatin species in rice seeds with different specificities against cysteine proteinases. Molecular cloning expression and biochemical studies on oryzacystatin II. *J. Biol. Chem.* **265,** 15832–15837.
53. Michaud, D., Nguyen-Quoc, B., and Yelle, S. (1994) Production of oryzacystatins I and II in *Escherichia coli* using the glutathione *S*-transferase gene fusion system. *Biotechnol. Prog.* **10,** 155–159.

6

Detection of Recombinant Viral Coat Protein in Transgenic Plants

Herta Steinkellner and Irina Korschineck

1. Introduction

The discovery that the expression of a viral coat protein in transgenic plants confers protection to infection by homologous and related viruses *(1–4)* revolutionized the field of plant breeding. The approach of "coat protein-mediated protection" (CPMP) became an extensively studied strategy for many researchers and companies. Coat protein-mediated protection has been demonstrated to be effective against members of more than 10 groups of RNA viruses *(3)*, but the molecular mechanism of the protection still remains unclear *(5,6)*. The phenotype of virus-derived resistance in transformed plants can vary, case-by-case, from a simple delay in normal symptom development, or partial inhibition of virus replication, to complete immunity to challenge virus or viral RNA inoculation.

Most examples of CPMP have involved viruses with genomes of positive sense ssRNA in a wide variety of dicot plants. So far no monocot-virus-CPMP system has been reported, presumably because of the technical difficulties with monocot transformation/regeneration. The availability of efficient gene vector systems and regeneration protocols is still a problem for many plant species. However, stable integration and expression of foreign genes in many dicot plants is now routine. The most successful existing method to transform dicot plants is via the Ti plasmid of *Agrobacterium tumefaciens* and, in this respect, one of the best-studied models for transformation/regeneration is *Nicotiana* sp.

Two different vector systems have been developed in order to place the recombinant gene construct between the border sequence of the T-DNA of the Ti plasmid of *Agrobacterium*, which becomes integrated into the plant genome: the binary system, in which the sequence of interest is directly cloned

From: *Methods in Biotechnology, Vol. 3:*
Recombinant Proteins from Plants: Production and Isolation of Clinically Useful Compounds
Edited by: C. Cunningham and A. J. R. Porter © Humana Press Inc., Totowa, NJ

between the T-DNA border repeats and the co-integrative vector system, in which the insertion of the recombinant gene depends on a single homologous recombination event between two sequences. Although in many cases the binary system is used *(7)*, the stability of these plasmids is not always optimal in *Agrobacterium* and it is possible that the transformation efficiency of some plant species is significantly lower than that obtained using co-integrative vectors.

So far, the most successful promoter that drives the transgene is the 35S-promoter from Cauliflower Mosaic virus, which is expressed constitutively in all plant cells *(8)*. Beside a strong promoter, an optimal expression of foreign genes in a transgenic plant depends on the translation efficiency and stability of the resultant mRNA. Although terminator signals ensure mRNA stability, the 5'-untranslated leader sequences can have a significant influence of stability and translation initiation *(9,10)*. Expression efficiency also depends very much on the site where the gene is inserted into the plant genome and how many copy numbers are integrated *(11)*.

The most common method for the transformation of cells of many plant species, because of its simplicity and efficiency, is the leaf-disks technique *(12)*. This technique involves the co-culture of leaf disks with a suspension of *Agrobacterium*, the elimination of the bacteria, and the regeneration of the plants on selection media suitable for transformed tissue.

Here we describe a protocol for transformation of *Nicotiana tabacum* with the coat protein gene of a plant viral coat protein and the subsequent regeneration of transformants. However, we will not discuss details about the construction of a suitable transformation vector (for details *see* **refs.** *13* and *14*).

The detection of recombinant viral coat protein should be done at both the nucleic acid and protein levels. Detection of the transgene can be done with Southern blot *(15)* or polymerase chain reaction (PCR) technique *(16)*; expression of the protein can be detected by immunological methods (ELISA, Western blot, immunosorbent electron microscope [ISEM]). In some cases it is advisable to analyze mRNA by either Northern blot or reverse PCR *(17)*.

Recently recombinant viral coat protein could be detected by ISEM in transgenic plants after infection with a related virus *(18–20)*. The heterologous coat protein was stably integrated into the particle of the challanged virus; therefore, a detection method for this heteroencapsidation will be descibed.

2. Materials

2.1. Transformation and Regeneration

1. 1.4% (w/v) Sodium hypochlorite containing 0.05% (w/v) Tween-80.
2. Sterile distilled water.

3. Steel mesh sieve.
4. Growth chamber (26°C in 16 h light per day (10–20 μE/m^2/s).
5. Sterile container (which allows good light transmission).
6. Sterile forceps.
7. *Nicotiana* seeds.
8. Sharp paper punch.
9. 3% (w/v) Glucose (sterile).
10. Bacterium culture shaker (28°C).
11. Flow hood.
12. MSO-medium: Murashige and Skoog basal salts solution with vitamins, containing no plant hormones: 1900 mg/L KNO_3, 1650 mg/L NH_4NO_3, 170 mg/L KH_2PO_4, 440 mg/L $CaCl_2·2H_2O$, 370 mg/L $MgSO_4·7H_2O$, 372 mg/L Na_2EDTA, 27.8 mg/L $FeSO_4·7H_2O$, 6.2 mg/L H_3BO_3, 22.3 mg/L $MnSO_4·H_2O$, 8.6 mg/L $ZnSO_4·4H_2O$, 0.83 mg/L KI, 0.25 mg/L $Na_2MoO_4·2H_2O$, 0.025 mg/L $CuSO_4·5H_2O$, 0.025 mg/L $CoCl_2·6H_2O$, 30 g/L sucrose, 0.5 mg/L nicotinic acid, 0.5 mg/L pyridoxin-HCl, 0.4 mg/L thiamine, 2 mg/L glycine, 100 mg/L myo-inositol. Stir to dissolve and adjust pH to 5.8 with KOH. Then add 7 g agar, adjust the volume to 1 L, and autoclave for 25 min.
13. Stock antibiotics:
    a. 200 mg kanamycin in 10 mL of distilled water.
    b. 500 mg carbenicillin in 10 mL of distilled water.
    The antibiotics are sterilized by passage through a 0.2 μm filter (store at –20°C).
14. Stock hormones: 100 mg benzylaminopurine (BAP, Gibco BRL, Gaithersburg, MD) and 10 mg napthalene acetic acid (NAA, Sigma, St. Louis, MO) are added to 50 mL distilled water. Storage is possible at room temperature for 1 mo.
15. MS with hormones: final concentration of 0.2 μg/mL NAA and 2.0 μg/mL BAP. 1 mL of stock hormones should be added to the MSO medium before the pH is adjusted to 5.8 and autoclaved.

2.2. DNA Extraction

1. Mortar and pestle.
2. Eppendorf tubes (1.5–2 mL).
3. Vortex.
4. Microcentrifuge.
5. Heating water bath.
6. Liquid N_2.
7. Exraction buffer: 0.1M Tris-HCl, pH 8, 1.4M NaCl, 0.05M EDTA.
8. Phenol: Distilled and equilibrated against Tris-HCl, pH 8.0, with 0.1% hydroxychinolin added.
9. Chloroform/isoamyl alcohol 1:24.
10. Isopropanol.
11. 70% Ethanol.

2.3. PCR Analysis

1. Thermocycler.
2. 10X PCR-buffer: 100 mM Tris-HCl, pH 8.3, 500 mM KCl, 15 mM MgCl$_2$.
3. 2 mM of each dNTP.
4. *Taq*-polymerase.
5. Mineral oil: Light white oil (Sigma).
6. Coat protein (cp) primers 1 and 2: 10 pmol/μL (cp primer 1 binds at the 5' end the cp gene; cp primer 2 binds at the 3'end of the cp gene).

2.4. Gel System

1. 1% (w/v) agarose-gel in 1X TAE.
2. 50X TAE stock: 242 g/L Tris, 57.1 mL/L acetic acid, 100 mL/L 0.5M EDTA.
3. Loading buffer (X5): 2.5 mg/mL bromephenol blue, 2.5 mg/mL xylencyanol, 33% (v/v) glycine.
4. Ethidium bromide: 10 mg/mL in distilled water.

2.5. ELISA

1. Microtiter plates.
2. Photometer.
3. Coating buffer: 0,1N NaHCO$_3$ buffer, pH 9.5.
4. Washing buffer: 1.15 g Na$_2$HPO$_4$, 0.2 g KH$_2$PO$_4$, 0.2 g KCl, 1 mL Tween-20 fill with water to 1l, pH 7.2–7.4.
5. Plant-extraction buffer: 0.05M Tris-HCl, pH 7.5, 2% (v/v) polyvinyl pyrrolidone (US Biochemical, Cleveland, OH), 1% (w/v) polyethylene glycol, 100 mM NaCl, 0.05% (v/v) Tween-20.
6. Dilution buffer: washing buffer + 1% (w/v) BSA.
7. Substrate solution for alkaline phosphatase: 1 mg *p*-nitrophenylphosphate in coating buffer.
8. Antibodies: anti-cp IgG (here we use antiserum against Arabis Mosaic Virus [ArMV]), anti-cp IgG conjugated to alkaline phosphatase (Sigma).

2.6. Virus Challange and ISEM

1. Celite.
2. Virus dilution buffer: PBS.
3. TPBS: PBS + 1% (v/v) Tween-20.
4. Copper grids: 400 mesh/Carbon film (LKB, Produkter, Stockholm).
5. 1% (v/v) uranyl acetate.
6. Antibodies (ab):
   a. Monoclonal PVY ab and PVY antimouse ab conjugated with 15 nm gold particles (Sigma).

    b. Polyclonal Plum Pox Virus (PPV, Sigma) antibody (Sanofi Phyto Diagnostics, France) and antirabbit ab conjugated with 5 nm gold particles.
7. Electron microscope.

3. Methods

3.1. Plant Transformation and Regeneration

1. To transform plant cells and obtain mature plants from them, it is necessary to start with sterile plant material. For the establishment of sterile plant culture from seed, sterilize the seeds by a 5 min treatment with sodium hypochlorite solution.
2. Wash seeds five times with sterile water, place them individually on MSO medium, and culture them in a growth chamber.
3. Place stem cuttings of the shoots every 4–5 wk onto fresh MS medium under the same conditions, and let sterile plantlets grow.
4. Alternatively, tobacco plants may be grown under greenhouse conditions and healthy young leaves sterilized immediately before use with 75% EtOH for 2 min, 1:10 dilution of bleach for 10 min and, after bleaching, rinse the leaves five times in sterile water.
5. For leaf-disk infection, cut sterile leaves into disks. (The transformation vector contains, beside the viral coat protein as a selection marker, a gene that confers kanamycin resistance.)
6. Place ten disks basal side up on an MS + hormones plate as a regeneration control and place about 10 disks on MS + hormones and 200 μg/mL kanamycin and 50 μg/mL carbenicillin plate as a control for leaf cell death.
7. Regeneration should begin within 2–3 wk if the hormone concentration is correct. Yellowing of the leaves should be observed within 2–3 wk, when the kanamycin is effective.
8. For infection of leaf disks with recombinant *Agrobacterium*, an overnight culture of the bacterium is used (28°C). Immerse the disks in the bacterium culture and place basal side up on an MS + hormones media plate (*see* **Note 1**).
9. Plates should be wrapped with parafilm and placed in a growth chamber for 24–48 h.
10. After incubation, wash the disks in sterile water three times for 5 min and blot dry on sterile paper.
11. Place the disks basal-side down in MS + hormones plates containing carbenicillin (500 mg/L), wrapped with parafilm, and incubate in a growth chamber for 6 d.
12. For selection, transfer the leaf disks to MS + hormones plates containing 200 μg/mL kanamycin and 50 μg/mL carbenicillin (*see* **Note 2**). Usually, callus formation appears after 2 wk, and shoots with internodes 10 d later (*see* **Notes 3** and **4**).
13. Remove each shoot as close as possible to the callus, but avoid taking any callus, and transfer the shoots to MSO with 200 μg/mL kanamycin and 50 μg/mL carbenicillin (*see* **Note 5**). In about 2 wk, roots should develop and the plants should be transferred to soil.

14. Gently remove the plantlet from any agar and place it in soil in 3 × 4 cm pots and cover with a clear plastic wrap. Shoots should have about four leaves before one of them is removed for any further analyses.

3.2. Detection of the Transgene

In order to estimate whether plantlets, regenerated on kanamycin containing MS media, have the transgene integrated, total DNA should be extracted from plantlets and the ArMV-cp gene amplified by PCR. Amplified fragments are visualized with ethidium bromide on an agarose gel.

3.2.1. Total DNA Extraction from Regenerated N. tabacum

1. Young leaves of putative transformed *N. tabacum* plants (ca. 100 mg) have to be powdered with liquid N_2 in a mortar and pestle. (It is very important to make a very fine powder.) Add 300–500 µL of plant extraction buffer and transfer suspension to a 1.5-mL Eppendorf tube.
2. Add an equal volumn of phenol, vortex, and incubate at 53–55°C for 5–10 min.
3. Add 400–500 µL chloroform, vortex, and incubate for a maximum of 5 min at 55°C.
4. Centrifuge for 15 min at 12,000g. Transfer supernatant to a new Eppendorf tube and repeat chloroform extraction until supernatant is clear.
5. Transfer supernatant to a new Eppendorf tube and add 1.5 vol isopropanol, mix (do not vortex), and incubate 5 min at room temperature.
6. Centrifuge for 20 min at 12,000g. Discard supernatant, add 1 mL 70% ethanol to the pellet, incubate for 5 min at room temperature, and centrifuge for 5 min at 12,000g.
7. Discard supernatant, air-dry the pellet at room temperature, and resuspend in 50 µL dH$_2$O for 1–2 h at 55°C.
8. Measure the DNA concentration in a fluorometer at 260 nm. (Absorbance at 260 nm in a photometer will give wrong concentration of DNA, because this purification procedure does not give pure DNA. If a fluorometer is not available, check DNA concentration on an 0.8% [w/v] agarose gel.) Store the DNA at +4°C (working solution) or –20°C (stock) (*see* **Note 6**).

3.2.2. PCR Amplification of the Transgene

When using plant DNA, it is advisable to use "hot start" PCR.

1. A typical PCR reaction is set up as follows (*see* **Note 7**):

10X PCR-buffer	10 µL
dNTP (2 m*M* each)	10 µL
Cp primer 1 (10 pmol)	2.5 µL
Cp primer 2 (10 pmol)	2.5 µL
50 ng template DNA	X µL
dH$_2$O	Y µL
Total volume	100 µL

All pipeting should be done on ice.

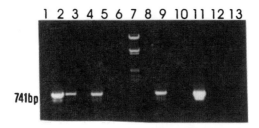

Fig. 1. 1% Agarose gel stained with ethidium bromide. PCR-amplified viral cp gene from putative transgenic plants. Lanes 1–11, different regenerated plants; lane 7, DNA marker; lane 12, positive control (purified virus); lane 13, negative control. It is clearly visible that plants in lanes 1, 2, 4, and 9 have the cp gene integrated. The different amount of amplified transgene is probably owing to the diffferent amount of copy numbers that are integrated into the genome. In lanes 3, 5, 6, 8, 10, and 11 no transgene was detected (*see* Notes section).

2. Cover reaction mix with mineral oil, incubate at 94°C for 5 min to allow plant DNA to denature completely, place on ice, and add 1 unit *Taq*-polymerase to the solution and start PCR reaction using the following parameters: 30 cycles of 92°C, 50 s; 50°C, 60 s; 72°C, 60 s (*see* **Note 8**).
3. Make a 1% (w/v) agarose gel in TAE to separate the amplified PCR fragment (**Fig. 1**).

3.3. Detection of the Heterologous Protein by ELISA

Plantlets that have the coat protein gene inserted into the genome have to be analyzed for expression of the heterologous protein. ArMV-cp expression in transgenic plants can be detected by direct ELISA analysis.

1. Anti-ArMV antibodies are diluted 1:100 in coating buffer (final concentration 2–10 µg/mL). 100 µL are added to each well and incubated overnight at 4°C. (Antibodies adsorb onto the wells, which are made of polystyrene, through hydrophobic forces.)
2. Wash the wells three times in washing buffer to remove unbound antibodies.
3. Add 100 µL plant sap (diluted 1:20 = 1 mg leaves + 20 µL dilution buffer) into each well and incubate for a minimum of 2 h at room temperature.
4. Wash the wells three times with washing buffer.
5. Add 100 µL diluted of anti-ArMV-AP-conjugate in dilution buffer to the wells and incubate for a minimum of 2 h at room temperature.
6. Wash the wells three times with washing buffer.
7. Add 100 µL of substrate solution. The color reaction depends on the enzyme, the substrate, and the temperature, and takes a few minutes to several hours.
8. Measure absorbance in a photometer at 402 nm (*see* **Note 10** and **Fig. 2**).

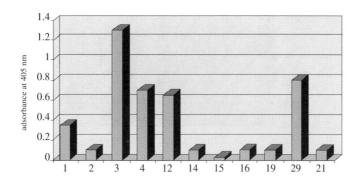

Fig. 2. Expression of transgenic ArMV coat protein detected by direct ELISA. Two antibodies specific to ArMV cp were used, the second one was labeled with alkaline phosphatase. Bars indicate the expression level of different transgenic plant lines 1–29. 21, negative control (plant sap from nontransformed plants). Only plant lines 1, 3, 4, 12, and 29 express the heterologous protein at a detectable level. Lines 3, 4, and 29 are high expressors and line 1 expresses very little amount of cp. Plant lines 2, 14, 15, 16, and 19 do not express cp at a detectable level (*see* Notes section).

3.4. Virus Challenge and ISEM

Transgenic tobacco plants that express the coat protein of PPV are challenged with Potato Virus Y (PVY).

1. Dilute PVY (purified virus or infected plant sap) in virus dilution buffer containing 1% Celite and mechanically inoculate transgenic plant leaves. After 1–2 wk, viral symptoms should be visible on plant leaves.
2. Grind plant sap in PBS (make several dilutions).
3. Place a drop of the virus solution on a hydrophobic surface of a copper grid and incubate for 5 min. Wash grid with 10–20 drops of TPBS.
4. Incubate grid in blocking buffer for 30 min.
5. Wash the grid as before.
6. Incubate grid in anti-PPV and anti-PVY antibodies diluted 1:100 in TPBS for 15–30 min.
7. Repeat washing and incubate grid for 1 h with gold-labeled conjugates diluted 1:100 in TPBS (antimouse ab will bind to the monoclonal anti-PVY-ab, antirabbit ab will bind to the polyclonal anti-PPV ab. With these two ab, both viral coat proteins can be detected).
8. Wash grid with 30 drops distilled water, stain for 30 s with 1% uranyl acetate, and air-dry.
9. Grids can now be analyzed in EM (**Fig. 3**).

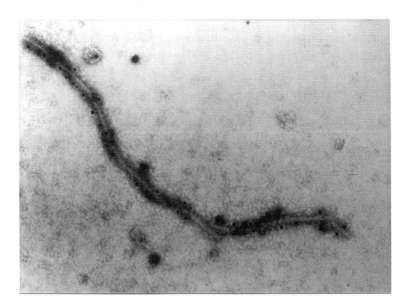

Fig. 3. ISEM: Negative stained potato virus Y rod-shaped particle. The virus particle is decotared with two gold-labeled antibodies: 5 nm gold particles indirectly detect transgenic PPV coat protein, 15 nm gold particles indirectly detect PVY coat protein. (Photograph provided by E. Maiss.)

4. Notes

1. Before transformation, the optical density at 600 nm (OD_{600}) of the recombinant bacteria should be about 0.6 and then diluted 1:50 in a sterile 3% (w/v) sucrose solution.
2. If leaves are still contaminated with *Agrobacterium* after carbenicillin incubation, wash leaf disks and subsequently incubate disks separately (carbenicillin 500μg/mL first, then kanamycin). Alternatively, other antibiotics can be used: cefotaxime 250–500 μg/mL, or carbenicillin, or cefotaxime in combination with vancomycin (250 μg/mL).
3. Regeneration of plantlets is more difficult in the presence of selection antibiotics. Therefore, in some cases regeneration may be initiated in the absence of antibiotic selection.
4. With some plant species shoots can be directly induced from disks (add 0.5 μg/mL BAP).
5. Any callus portion transferred with shoots can inhibit root formation in MSO + *kanamycin*. If the shoots do not form roots, cut remove ALL callus from the stems. If roots still do not develop, it is possible that the shoot is not a true transformant (escape mutant).

6. The developmental stage of a plant has a big influence of the quality of the purified DNA. The method described in this chapter gives DNA that is suitable for PCR reaction when young leaves are used. With many plant species, especially when they are in an older stage, many components, such as phenols, can disturb PCR reaction. Many people, therefore, use CTAP purification *(21)* which gives better quality of DNA.

7. Inhibition of PCR through the presence of plant compounds can be avoided through a better purification using Quiagen resin, or simply adding nonfat milk cocktails (1–5% *BLOTTO*) to the PCR mix.

8. The optimal conditions for PCR reaction have to be worked out individually in each lab. The reaction depends on many factors, such as *Taq* enzyme, nucleotides, primer sequence, quality of DNA, and so on. The parameters suggested in this paper are optimized for TRIO-Thermoblock BT1 from Biometra, Germany.

9. Plant lines 3, 5, 6, 8, 10, and 11 could regenerate on kanamycin containing media, but are not transformed. This phenomenon (escape mutants) happens frequently when *Nicotiana* is used for transformation. To avoid propagation of nontransformed plants, it is advisable to carry out the PCR reaction in an early stage of regeneration.

10. ELISA is very sensitive and an easy way to detect a specific protein in a protein mixture. However, the disadvantage of this system is that it does not give information about the size of the protein. If it is necessary to estimate the molecular weight of the protein, a Western blot should be made *(15)*.

11. Although all plant lines have the transgene integrated into their genome (proven by PCR; *see* **Fig. 1**), recombinant protein could not be detected by ELISA in all lines. However, it is worthwhile retaining these transgenic lines and checking them for viral tolerance, because, in some cases, these plants also show virus tolerance.

References

1. Beachy, R. N. B., Loesch-Fries, S., and Tumer, N. E. (1990) Coat protein-mediated resistance against virus. *Ann. Rev. Phytopathol.* **28,** 451–474.

2. Powell-Abel, P., Nelson, R. S., De, B., Hoffmann, N., Rogers, S. G., Fraley, R. T., and Beachey, R. N. B. (1986) Delay of disease development in transgenic plants that express tobacco mosaic virus coat protein. *Science* **232,** 738–747.

3. Hammond, J. and Kamo, K. K. (1994) Resistance to bean yellow mosaic virus (BYMV) and other potyvirus in transgenic plants expressing BYMV antisense RNA, coat protein, or chimeric coat proteins. *Proc. 5th Int. Symp. Biotechnol. Plant Protection*, College Park, MD.

4. Yie, Y., Wu, Z. X., Wang, S. Y., Zhao, S. Z., Zhanh, T. Q., Yoa, G. Y., and Tien, P. (1995) Rapid production and field testing of homozygous transgenic tobacco lines with virus resistance conferred by expression of satellite RNA and coat protein of cucumber mosaic virus. *Transgenic Res.* **4,** 256–263.

5. Wilson, T. M. A. (1990) Strategies to protect crop plants against viruses: pathogen-derived resistance blossoms. *Proc. Nat. Acad. Sci. USA* **90,** 3134–3141.

6. Farinelli, L. and Malnoe, P. (1993) Coat protein gene-mediated resistance to potato virus Y in tobacco: examination of the resistance mechanism—Is the transgenic coat protein required for protection? *Mol. Plant-Microbe. Interact.* **6,** 284–292.
7. Bevan, M. (1984) Binary *Agrobacterium* vectors for plant transformation. *Nucleic Acids Res.* **12,** 8711–8721.
8. Sanders, P. R., Winter, J. A., Barnason, A. R., Rogers, S. G., Fraley, R. T. (1987) Comparison of Cauliflower Mosaic virus 35S and nopaline synthase promoters in transgenic plants. *Nucleic Acids Res.* **15,** 1543–1558.
9. Gallie, D. R., Sleat, D. E., Watts, J. W., Turner, P. C., Wilson, T. M. A. (1987) The 5'-leader sequence of tobacco mosaic virus RNA enhances the expression of foreign gene transcripts in vivo and *in vitro. Nucleic Acids Res.* **15,** 3257–3273.
10. Jobling S. A. and Gehrke L. (1987) Enhanced translation of chimaeric massanger RNAs containing a plant viral untranslated leader sequence. *Nature* **325,** 622–625.
11. Wang, J., Lewis, M. E., Whallon, J. H., and Sink, K. C. (1995) Chromosomal mapping of T-DNA inserts in transgenic *Petunia* by *in situ* hybridization. *Transgenic Res.* **4,** 241–246.
12. Horsch, R. B., Fry, J. E., Hoffmann, N. L., Wallroth, M., Eichholtz, D., Rogers, S. G., and Fraley, R. T. (1985) A simple and general method for transferring cloned genes into plants. *Science* **227,** 1229–1231.
13. Steinkellner, H., Himmler, G., Mattanovich, D., and Katinger, H. (1990) Nucleotide sequence of AMV-capsid protein-gene. *Nucleic Acids Res.* **18,** 7182.
14. Steinkellner, H., Weinhäusl, A., Laimer, M., da Camara Machado, A., Cooper, I., and Katinger, H. (1992) Arabis Mosaic Virus: expression of the coat protein in transgenic plants. *Acta Horticulture* **308,** 37–42.
15. Sambrook, J., Fritsch, E. F., and Maniatis, T. (1989) *Molecular Cloning: A Laboratory Manual.* Cold Spring Harbor Laboratory, Cold Spring Harbor, NY.
16. McGarvey, P. and Kaper, J. M. (1991) A simple and rapid method for screening transgenic plants using polymerase chaim reaction. *BioTechniques* **11,** 428–432.
17. Korschineck, I., Himmler, G., Sagl, R., Steinkellner, H., and Katinger, H. (1991) A PCR membrane spot assay for the detection of plum pox virus RNA in bark of infected trees. *J. Virol. Meth.* **31,** 139–146.
18. Farinelli, L., Malnoe, P., and Collet, G. F. (1992) Heterologous encapsidation of Potato virus Y strain 0 (PVY°) with the transgenic coat protein of PVY strain N (PVY^N) in solanum tuberosum cv.bintje. *Bio/Technology* **10,** 1020–1025.
19. MacFarlane, S. A., Mathis, A., and Bol, J. F. (1994) Heterologous encapsidation of recombinant pea early browning virus. *J. Gen. Virol.* **75,** 1423–1429.
20. Maiss, E., Koenig, R., and Lesemann, D.-E. (1994) Heterologous encapsidation of viruses in transgenic plants in a mixed infections, in *Proceedings of the 3rd International Symposium on the Biosafety Results of Field Tests of Genetically Modified Plants and Microorganisms* (D. D. Jones, ed), University of California, Division of Agriculture and Natural Resources, Oakland, CA pp. 129–139.
21. Doyle, J. J. and Doyle, J. L. (1990) Isolation of plant DNA from fresh tissue. *Focus* **12,** 13–15.

7

Synthesis of Recombinant Human Cytokine GM-CSF in the Seeds of Transgenic Tobacco Plants

Ravinder K. Sardana, Peter R. Ganz, Anil Dudani, Eilleen S. Tackaberry, Xiongying Cheng, and Illimar Altosaar

1. Introduction

We are interested in studying plant systems as vehicles for the production of recombinant proteins of clinical relevance. There are a number of potential advantages to producing recombinant protein products in plants. Plants are more economical compared to fermentation or cell-culture facilities and their production scale can be easily increased. They offer the potential for very high levels of heterologous protein production and because many plant tissues are generally recognized as safe, the probability of health risks because of contamination with potential human pathogens and toxins in recombinant products derived from plants is minimized. Furthermore, plants have glycosylation machinery like other eukaryotic systems *(1)*.

In recent years, a number of valuable proteins from diverse origins have been produced in transgenic plants *(2–8)*. Plant seeds are the natural sites for the deposition of a variety of edible plant proteins. Seeds are easily stored at room temperature for long periods of time. Since the seeds of many plant species are edible, it is possible that the seed-derived recombinant products, in some instances, may not require further processing and purification, and so they offer the option of oral delivery *(9)*. Thus, seeds may be effective systems for producing valuable proteins *(5,10–13)*.

Plants represent an attractive expression system for recombinant blood proteins. Currently, many of these are produced in mammalian cells, with poor recoveries, thus reducing more widespread use, as in prophylactic treatment. Also, the costs for production in mammalian cells are prohibitive (**Table 1**).

From: *Methods in Biotechnology, Vol. 3:*
Recombinant Proteins from Plants: Production and Isolation of Clinically Useful Compounds
Edited by: C. Cunningham and A. J. R. Porter © Humana Press Inc., Totowa, NJ

Table 1
Proteins from Transgenic Sources:
Initial Capital Expenditures for Production

Transgenic Source	Cost (US$)[a]
Founder animal	$1.5 million
Mammalian cell line	$25–50 million
Stably transformed plant	$250–500 thousand

[a]Modified from **refs. *34*** and ***35***.

To study the feasibility of expression of blood proteins in plants, with benefits of increased production and lower cost, we have investigated the expression of the human cytokine, granulocyte macrophage colony-stimulating factor (GM-CSF). GM-CSF has many beneficial clinical uses in chemotherapy, bone marrow transplantation, and cancer and AIDS treatment *(14,15)*.

GM-CSF protein regulates the production and function of white blood cells (granulocytes and monocytes) *(14,15)*. The mature GM-CSF is a glycoprotein consisting of 127 amino acid residues *(16–18)*. In humans, the GM-CSF protein is produced by multiple cell types that are widely present throughout the body *(14)*. However, GM-CSF is actually present in very low amounts in human tissue *(14,15)*. Practical methods to produce it in sufficient quantities for clinical use have relied on recombinant DNA tools and techniques. For example, recombinant human GM-CSF has been produced in COS *(16)*, Namalwa *(19)*, and yeast cells *(20)*.

In this chapter, we describe the construction of a gene-fusion vector that is capable of directing the synthesis of recombinant GM-CSF in the seeds of transgenic tobacco plants. We also discuss the methods for the analysis of transgenic tobacco plants that produce recombinant human GM-CSF in the seeds. A method to test the biological activity of seed-derived GM-CSF is also outlined.

To construct a vector that is capable of directing gene expression in a specific tissue, a specific promoter is required. For example, to produce a desired protein in the seeds of transgenic plants, a seed-specific promoter is used. This promoter is then fused to a plant signal-peptide sequence that is in-frame with the coding sequence of the mature protein product of interest. The signal peptide is needed to direct the translated protein to the secretory pathway in the plant cell. Finally, a polyadenylation signal is needed to give some stability to the messenger RNA. This DNA is added to the 3' end of the coding sequence of interest.

To date, a number of seed-specific promoters have been isolated from different plant species. We are investigating the use of seed-specific promoters from oat and

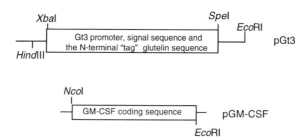

Fig. 1. The restriction maps of the Gt3 promoter and GM-CSF plasmids. The plasmid pGt3 contains the glutelin promoter (980 bp), glutelin signal sequence (72 bp), followed by the mature glutelin sequence. The *Spe*I site is located in the mature glutelin sequence and it is 18 bp downstream from the 5' end of the mature glutelin sequence. The pGM-CSF plasmid contains the cDNA coding for the mature GM-CSF protein. The size of the *Nco*I-*Eco*RI fragment is 402 bp.

rice. For instance, glutelin is a major rice seed storage protein *(21)*. In earlier studies, a rice glutelin promoter, Gt3, has been shown to direct the expression of bacterial genes in the seeds of transgenic tobacco plants *(22,23)*.

We have put the cDNA fragment encoding mature human GM-CSF under the control of a rice glutelin promoter Gt3. In our experiments, we have found that the Gt3 promoter is capable of directing the synthesis of biologically active recombinant GM-CSF protein in the seeds of transgenic tobacco plants *(24,25)*. The diagrams representing the restriction maps of the two plasmids used for the construction of the Gt3-GM-CSF fusion vector are shown in **Fig. 1**.

2. Materials

2.1. DNA Material

1. pGt3 plasmid.
2. pGM-CSF plasmid.
3. pBI221 plasmid (Clontech, Palo Alto, CA).

2.2. Reagents, Chemicals, and Commercial Kits

1. Restriction endonucleases (BRL, Bethesda, MD).
2. Klenow fragment of DNA polymerase (BRL).
3. dNTP solution: 1mM of each of dATP, dTTP, dCTP, and dGTP.
4. T4 DNA ligase (BRL).
5. T7 sequencing kit (Pharmacia, Uppsala, Sweden).
6. MicroSpin S-300 columns (Pharmacia).
7. Plasmid DNA purification kit (QIAGEN, Chatsworth, CA).
8. GeneClean kit (Bio 101, Vista, CA).
9. Agarose (Sigma, St. Louis, MO).

10. DH5α competent cells (BRL).
11. Buffer-saturated phenol (BRL).
12. Chloroform (BDH, Mississauga, Ontario).
13. Ethanol (BDH).
14. Ready To Go Kit (Pharmacia).
15. GM-CSF standard protein (R & D Systems, Minneapolis, MN).
16. Goat anti-GM-CSF antibody (R & D Systems).
17. Rabbit anti-goat IgG (R & D Systems).
18. NBT/BCIP phosphatase substrate (Kirkegaard & Perry, Gaithersburg, MD).
19. RPMI 1640 medium (BRL).
20. Phosphate-buffered saline (PBS) (BRL).

2.3. Buffers and Other Solutions

1. Agarose gel buffer: 1X: 0.04M Tris-acetate and 0.001M EDTA, pH 8.0.
2. Protein extraction buffer: 50 mM Tris, pH 7.5, 50 mM NaCl, 1 mM EDTA, 0.1% Tritron X-100, 1% ascorbic acid, 1% polyvinylpyrrolidone (PVP, insoluble; Sigma), 1 mM phenylmethylsulfonylfluoride (PMSF; Sigma) and 1% 2-mercaptoethanol. The latter two chemicals are added immediately prior to use.
3. LB broth base (BRL).
4. 10X TA buffer: 33 mM Tris-HCl acetate, pH 7.9, 66 mM potassium acetate, 10 mM magnesium acetate, 4 mM spermidine, 0.5 mM dithiothreitol (DTT; Sigma).
5. Hybridization buffer: 50% formamide (deionized), 5X Denhardt's, 5X SSC, 0.1% SDS, 100 µg/mL denatured salmon sperm DNA. This DNA is added last, at the time of prehybridization and hybridization.
6. TBST: 10 mM Tris-HCl, 150 mM NaCl, 0.05% Tween, pH to 8.0.
7. AP buffer: 100 mM Tris-HCl, 100 mM NaCl, 5 mM MgCl$_2$; pH to 8.5.

2.4. Plant Transformation Materials

1. *Agrobacterium* strain pGV3101/pMP90.
2. Plasmid pRD400 (National Research Council of Canada, Saskatoon).
3. Tobacco plants, cultivar Xanthi.
4. Appropriate tobacco tissue culture medium.

3. Methods
3.1. Construction of Gt3-GM-CSF Fusion Vector

The procedure, which was adapted from standard methods designed to test function and activity of endosperm-specific promoters driving marker enzymes, such as glucuronidase, involves fusing the rice promoter and part of the glutelin coding sequence in front of the human sequence coding for GM-CSF (*see* **Note 1**).

1. Digest 1–1.5 µg of pGM-CSF plasmid DNA with 6–8 units of *Nco*I, in a reaction volume of 20 µL. Incubate at 37°C for 3 h. After the digestion is complete, add

1 µL of 1 m*M* dNTPs and at least one unit of Klenow. Incubate the reaction for 20 min at room temperature.

2. To inactivate the enzymes present in the reaction, heat the sample at 75°C for 10 min. Purify the sample by passing it through a MicroSpin S-300 column (*see* **Note 2**).

3. Digest the cleaned DNA with 6–8 units of *Eco*RI, at 37°C for 3 h, in a final volume of 30 µL. This releases the fragment containing the GM-CSF coding sequence. Electrophorese the sample on a 1% agarose gel and cut out the GM-CSF DNA band. Clean it by using a GeneClean kit. Elute the DNA in a final volume of 15 µL of distilled water.

4. Digest 1–1.5 µg of the plasmid pGt3 with *Spe*I, blunt end it with Klenow, and digest again with *Eco*RI, as in **steps 1–3**. Purify by extraction with phenol:chloroform and collect DNA by ethanol precipitation. Resuspend in 15 µL of distilled water.

5. Set up an overnight ligation reaction at 16°C, in a total volume of 15 µL. Use 8 µL of the cleaned GM-CSF DNA (from **step 2**) and 3.5 µL of glutelin plasmid DNA (from **step 3**). Add 3 µL of 5X ligase buffer and 0.5 µL of DNA ligase.

6. Use 4–5 µL of the ligation mixture to transform DH5α competent cells, using the BRL protocol (*see* **Note 3**). Plate the transformed cells on the LB/agar plates containing ampicillin at a final concentration of 100 µg/mL. Incubate the plates overnight at 37°C, in an inverted position.

7. Pick up bacterial colonies using sterile toothpicks, grow them overnight in LB/ampicillin medium, and isolate the plasmid DNA. We use the QIAGEN plasmid kit for this purpose (*see* **Note 3**). Confirm the presence of the cloned fragment by *Eco*RI and *Hin*dIII restriction digests and agarose gel electrophoresis.

8. Sequence the selected plasmid clones. This is done to verify the in-frame fusion of the GM-CSF DNA to the glutelin sequence. We sequenced a number of plasmid clones. This was done by using a T7 sequencing kit (*see* **Note 3**). A plasmid clone was chosen that contained the Gt3 promoter, glutelin signal sequence, and the associated glutelin N-terminal tag sequence in-frame, with the GM-CSF coding sequence. This plasmid is called pGt3/GM-CSF1.

3.2. Addition of Polyadenylation Signal to the 3' End of the GM-CSF Coding Sequence

A DNA fragment containing the polyadenylation signal from the nopaline synthase gene (Nos/ter) was cloned from the plasmid pBI221. This was accomplished by replacing a *Hin*dIII-*Sst*I fragment (*Sst*I site is made blunt) in the plasmid pBI221 with the *Hin*dIII-*Eco*RI fragment (*Eco*RI site is blunt-ended) of the plasmid pGt3/GM-CSF1. This was done as follows:

1. Digest about 1–1.5 µg of pBI221 DNA in a final volume of 20 µL with *Sst*I at 37°C for 3 h. After the digestion is complete, add 1 µL of 1 m*M* dNTPs and at least one unit of Klenow. Incubate the reaction for 20 min at room temperature.

2. Heat the sample at 75°C for 10 min. Clean the sample by passing it through a MicroSpin S-300 column.

3. Redigest with 5–10 units of *Hin*dIII in a volume of 30 µL. Allow 3 h for completion of the digestion at 37°C. Run the sample on a 1% agarose gel and cut out the top plasmid DNA fragment. Clean it by using a GeneClean kit. Elute DNA with water in a final volume of 15 µL.

4. Digest about 1–1.5 µg of pGt3/GMCSF1 plasmid with *Eco*RI, in a final volume of 20 µL, at 37°C for 3 h. Add 1 µL of 1 m*M* dNTPs and one unit of Klenow. Incubate the reaction for 20 min at room temperature. Heat the sample at 75°C for 10 min. Clean the sample by passing it through a MicroSpin S-300 column.

5. Redigest the sample with 5–10 units of *Hin*dIII, in a reaction volume of 30 µL, at 37°C for 3 h. Run this sample on 1% agarose gel and cut out the Gt3-GM-CSF band and clean it by using the GeneClean kit. Elute DNA in 15 µL of water.

6. Set up a ligation reaction in a final volume of 15 µL. Use 4 µL of the eluted pBI221 DNA from **step 3**, 7 µL of the eluted DNA from **step 5**, 3 µL of 5X ligase buffer, and a unit of T4 DNA ligase. Mix well and leave it overnight at 16°C.

7. For bacterial transformation and DNA preparations from colonies, follow the instructions in **steps 6** and **7** of **Subheading 3.1.**

8. Confirm the cloned fragment by *Eco*RI and *Hin*dIII restriction digests and agarose gel electrophoresis.

We selected a plasmid clone and it is named pGt3/GM-CSF2. Its restriction map is shown in **Fig. 2**.

3.3. Cloning into Binary Agrobacterium Vector pRD400

The *Hin*dIII-*Eco*RI fragment from the plasmid pGt3/GM-CSF2 was cloned into a binary *Agrobacterium* vector pRD400 *(26)*. Standard cloning procedures were followed *(27)*. The engineered plasmid is called pGt3/GM-CSF3. This plasmid was used to transform the *Agrobacterium* strain pGV3101/pMP90, using a protocol from Pharmacia. These *Agrobacterium* cells were used to transform tobacco cv. Xanthi. The conventional transformation and regeneration procedures were used *(28,29)*. As the regenerated plants matured, they were transferred into the greenhouse and maintained.

3.4. Restriction Digestion of Tobacco Genomic DNA and Southern Analysis

The transgenic nature of tobacco plants was tested by Southern blots, as previously described in detail *(25,27)*. The tobacco chromosomal DNA was purified *(29)*. We outline some of the relevant steps of these procedures here.

1. Digest about 10 µg of tobacco chromosomal DNA with 8 µL each of *Eco*RI and *Hin*dIII (10 U/µL) in a final volume of 160 µL. Use 16 µL of 10X TA buffer. Incubate at 37°C overnight. To reduce the volume, concentrate the DNA by ethanol precipitation *(27)*.

Fig. 2. Restriction map of the plasmid pGt3/GMCSF2. The additional glutelin sequence shown here is 24 bp and it is in-frame with the GM-CSF sequence. The size of the Nos/ter fragment is 260 bp and it is derived from plasmid pPBI221.

2. Subject the restricted DNA to agarose gel electrophoresis, denature, and neutralize as described *(27)*. Use 0.8% agarose.
3. Transfer the separated DNA to a nylon membrane as instructed *(27)*.
4. Label the purified GM-CSF fragment. We use Ready To Go Kit for this purpose.
5. Prehybridize the nylon membrane for 2 h at 42°C. Hybridize with the denatured labeled probe at 42°C overnight in hybridization buffer containing 50% formamide.
6. Wash the membrane in 2X SSC, 0.1% SDS at room temperature for 10 min; 1X SSC, 0.1% SDS at 65°C for 15 min twice; and 0.4X SSC, 0.1% SDS at 65°C for 15 min.
7. Dry the membrane and use it for autoradiography.

3.5. Soluble Protein Extraction from Tobacco Seeds

To detect the recombinant GM-CSF protein in the transgenic tobacco seeds, extracts were made as follows (*see* **Note 4**).

1. Collect developing tobacco seeds (14 d after pollination in case of plants transformed with Gt3 promoter) and freeze immediately in liquid nitrogen. Store at −70°C until needed (*see* **Note 4**).
2. Homogenize 100–200 mg tobacco seeds in 100–200 μL of extraction buffer in a cold mortar with pestle.
3. Centrifuge (14,000g) seed extracts at 4°C and collect the supernatant.
4. Use this supernatant for further experiments, such as Western blots and biological assays.

3.6. Western Blot Analysis of Seed Protein Extracts from Transgenic Tobacco Plants

1. Estimate protein concentration of the seed extracts using the Bradford assay (Bio-Rad, Hercules, CA). Follow the supplier's instructions.
2. Subject aliquots of seed extracts (40–100 μg protein) to electrophoresis using denaturing SDS-polyacrylamide gel. Use 12% polyacrylamide resolving gel with a 4% stacking gel. We use Bio-Rad reagents and the Bio-Rad Mini-Gel System for electrophoresis and blotting experiments. The proteins on the gel are transferred onto Bio-Rad PVDF membrane as follows: 25mA for 45 min; 50 mA for 2 h; and 100mA for 15 min. For detailed discussion on making gels, buffers, and procedures for Western blots, follow the standard methods in the literature *(27)*.

3. After transfer onto PVDF membrane, wet it in TBST for 5 min.
4. Block with TBST containing 1% BSA for 1 h.
5. Replace blocking solution with goat anti-hGM-CSF antibody diluted 1:1000 in TBST. Leave it overnight on a shaker in the cold room.
6. Wash membrane with TBST, 3 × 5 min each to remove unbound antibody.
7. Incubate membrane in TBST containing the alkaline phosphatase-conjugated rabbit anti-goat IgG diluted 1:7500. Leave on a shaker in the cold room for 4 h.
8. Wash membrane with TBST, 3 × 5 min.
9. Visualize bands using buffer containing BCIP/NBT phosphatase substrate, according to the manufacturer's instructions.

3.7. Biological Activity Assay for GM-CSF Produced in Transgenic Tobacco Seeds

The biological activity assay utilizes TF-1 cells that are dependent for their growth on several cytokines, including GM-CSF *(30)*.

1. Grow TF-1 cells (ATCC, Bethesda, MD) as suspension cultures in RPMI 1640 medium with 1 ng/mL commercial GM-CSF and fetal bovine serum (10%). Determine cell viability by Trypan blue exclusion.
2. Wash twice with 1X PBS and resuspend cells at 2×10^5/mL in RPMI 1640 medium containing 10% fetal bovine serum.
3. Aliquot 0.5 mL resuspended cells (1×10^5 cells) in the wells of a 24-well tissue culture plate. To each of these wells, add 0.5 mL RPMI medium, with 10% fetal bovine serum containing only one of the following materials at a time: 1 ng/mL commercial GM-CSF, 1 ng/mL (10 µL) transgenic tobacco seed extract, 10 µL of seed extract from a nontransgenic tobacco plant, 10 µL of seed protein extraction buffer (excluding mercaptoethanol). Perform these experiments in triplicate and under sterile conditions.
4. Monitor cell growth every 48 h. Count live cells and add fresh medium containing appropriate extract.

4. Notes

1. The construct detailed here encompasses a strategy whereby the desired pharmaceutical protein is tagged at its N-terminal end with eight amino acids of a seed-specific protein, to help in its translational and posttranslational processing. The vector, consequently, was designed to contain not only the signal peptide coding sequence, but also 24 nucleotides of the mature glutelin coding sequence. This would result in the formation of a glutelin-GM-CSF fusion protein. However, we also have made another construct in which these mature glutelin tag sequences are absent. To ensure endosperm-specific deposition, other seed-specific promoters can also be used. For example, in a very recent study, a Gt1 glutelin promoter has been shown to direct the production of phaseolin (a plant protein) up to a level of 4% of the total soluble-seed proteins in transgenic rice plants *(31)*.

2. Digested samples can also be purified and cleaned by extraction with phenol:chloroform. In this case, DNA is precipitated with 2 vol of ethanol. The precipitated DNA is collected after centrifugation.
3. Commercial kits can be very expensive. For example, competent *Escherichia coli* cells can be easily prepared in the laboratory. Mini-preparations of plasmid DNA can also be performed using previously described methods and DNA sequencing may be done using standard protocols. For these methods and techniques, the reader is advised to follow protocols provided in the literature *(27)*.
4. Collecting seeds at the right stage of development (timing days after pollination or anthesis) is critical. This could vary with the use of different seed-specific promoters *(32)*. Furthermore, the choice of the buffer for the protein extraction can also be crucial. For example, a number of different compounds (Polyclar AT, albumins, powdered nylon) can also be used to remove phenolics from the plant homogenate material. The polyphenolics are readily oxidized and they bind and crosslink proteins. For detailed information, the reader is referred to a review article on protein purification from plants *(33)*.

Acknowledgements

The authors thank T. Okita (Washington State University) for providing the glutelin promoter plasmid pJH18. The excellent technical assistance of Angela Tyler and Connie Sauder is gratefully appreciated. This work was supported by a grant from the Canadian Red Cross Society Research & Development Fund and Natural Sciences and Engineering Research Council (NSERC) of Canada.

References

1. Chrispeels, M. J. (1991) Sorting of proteins in the secretory system. *Ann. Rev. Plant Physiol. Plant Mol. Biol.* **42,** 35–49.
2. Vandekerckhove, J., Van Damme, J., Van Lijsebettens, M., Botterman, J., De Block, M., Vandewiele, M., De Clercq, A., Leemans, J., Van Montagu, M., and Krebbers, E. (1989) Enkephalins produced in transgenic plants using modified 2S seed storage proteins. *Bio/Technology* **7,** 929–932.
3. Sijmons, P. C., Dekker, B. M. M., Schrammeijer, B., Verwoerd T. C., van den Elzen, P. J. M., and Hoekema, A. (1990) Production of correctly processed human serum albumin in transgenic plants. *Bio/Technology* **8,** 217–221.
4. Mason, H. S., Lam, D. M.-K., and Arntzen, C. J. (1992) Expression of hepatitis B surface antigen in transgenic plants. *Proc. Natl. Acad. Sci. USA* **89,** 11,745–11,749.
5. Pen, J., Molendijk, L., Quax, W. J., Sijmons, P. C., van Ooyen, A. J. J., van den Elzen, P. J. M., Reitweld, K., and Hoekema, A. (1992) Production of active *Bacillus licheniformis* alpha-amylase in tobacco and its application in starch liquefaction. *Bio/Technology* **10,** 292–296.

6. Herbers, K., Wilke, I., and Sonnewald, U. (1995) A thermostable xylanase from *Clostridium thermocellum* expressed at high levels in the apoplast of transgenic tobacco has no detrimental effects and is easily purified. *Bio/Technology* **13,** 63–66.

7. Turpen, T. H., Reinl, S. J., Charoenvit, Y., Hoffman, S. L., Fallarme, V., and Grill, L. K. (1995) Malarial epitopes expressed on the surface of recombinant tobacco mosaic virus. *Bio/Technology* **13,** 53–57.

8. Ma, J. K.-C., Hiatt, A., Hein, M., Vine, N. D., Wang, F., Stabila, P., van Dolleweerd, C., Mostov, K., and Lehner, T. (1995) Generation and assembly of secretory antibodies in plants. *Science* **268,** 716–719.

9. Mason, H. S. and Arntzen, C. J. (1995) Transgenic plants as vaccine production systems. *Trends Biotechnol.* **13,** 388–392.

10. Fiedler, U. and Conrad, U. (1995) High-level production and long-term storage of engineered antibodies in transgenic tobacco seeds. *Bio/Technology* **13,** 1090–1093.

11. Van Rooijen, G. J. H. and Moloney, M. M. (1995) Plant seed oil-bodies as carriers for foreign proteins. *Bio/Technology* **13,** 72–77.

12. Parmenter, D. L., Boothe, J. G., van Rooijen, G. J. H., Yeung, E. C., and Moloney, M. M. (1995) Production of biologically active hirudin in plant seeds using oleosin partitioning. *Plant Mol. Biol.* **29,** 1167–1180.

13. Pen, J., van Ooyen, A. J. J., van den Elzen, P. J. M., Quax, W. J., and Hoekema, A. (1993) Efficient production of active industrial enzymes in plants. *Ind. Crops Prod.* **1,** 241–250.

14. Metcalf, D. (1991) Control of granulocytes and macrophages: molecular, cellular, and clinical aspects. *Science* **254,** 529–533.

15. Metcalf, D. (1991) The Florey Lecture, 1991. The colony-stimulating factors: discovery to clinical use. *Phil. Trans. R. Soc. Lond.* **B 333,** 147–173.

16. Wong, G. G., Witek, J. S., Temple, P. A., Wilkens, K. M., Leary, A. C., Luxenberg, D. P., Jones, S. S., Brown, E. L., Kay, R. M., Orr, E. C., Shoemaker, C., Golde, D. W., Kaufman, R. J., Hewick, R. M., Wang, E. A., and Clark, S. C. (1985) Human GM-CSF: molecular cloning of the complementary DNA and purification of the natural and recombinant proteins. *Science* **228,** 810–815.

17. Cantrell, M. A., Anderson, D., Cerretti, D. P., Price, V., McKereghan, K., Tushinski, R. J., Mochizuki, D. Y., Larsen, A., Grabstein, K., Gillis, S., and Casman, D. (1985) Cloning, sequence, and expression of a human granulocyte/ macrophage colony-stimulating factor. *Proc. Natl. Acad. Sci. USA* **82,** 6250–6254.

18. Lee, F., Yokota, T., Otsuka, T., Gemmell, L., Larson, N., Luh, J., Arai, K., and Rennick, D. (1985) Isolation of cDNA for a human granulocyte-macrophage colony-stimulating factor by functional expression in mammalian cells. *Proc. Natl. Acad. Sci. USA* **82,** 4360–4364.

19. Okamoto, M., Nakayama, C., Nakai, M., and Yanagi, H. (1990) Amplification and high-level expression for human granulocyte-macrophage colony-stimulating factor in human lymphoblastoid Namalwa cells. *Bio/Technology* **8,** 550–553.

20. Ernst, J. F., Mermod, J.-J., DeLamarter, J. F., Mattaliano, R. J., and Moonen, P. (1987) O-glycosylation and novel processing events during secretion of α-factor/ GM-CSF fusions by *Saccharomyces cerevisiae*. *Bio/Technology* **5,** 831–834.

21. Okita, T. W., Hwang, Y. S., Hnilo, J., Kim, W. T., Aryan, A. P., Larsen, R., and Krishnan, H. B. (1989) Structure and expression of the rice glutelin multigene family. *J. Biol. Chem.* **264,** 12573–12581.
22. Leisy, D. J., Hnilo, J., Zhao, Y., and Okita, T. W. (1989) Expression of rice glutelin promoter in transgenic tobacco. *Plant Mol. Biol.* **14,** 41–50.
23. Zhao, Y., Leisy, D. J., and Okita, T. W. (1994) Tissue-specific expression and temporal regulation of the rice glutelin Gt3 gene are conferred by at least two spatially separated *cis*-regulatory elements. *Plant Mol. Biol.* **25,** 429–436.
24. Ganz, P. R., Dudani, A. K., Tackaberry, E. S., Sardana, R., Sauder, C., Cheng, X., and Altosaar, I. (1995) Plants as factories for producing human blood proteins. *Can. Soc. Transfusion Med. Bull.* **7,** 112–121.
25. Ganz, P. R., Dudani, A. K., Tackaberry, E. S., Sardana, R., Sauder, C., Cheng, X., and Altosaar, I. (1996) Expression of human blood proteins in transgenic tobacco plants: The cytokine GM-CSF as a model protein, in *Transgenic Plants: A Production System for Pharmaceutical Proteins* (Owen, M. R. L. and Pen, J., eds.), Wiley, New York, pp. 281–297.
26. Datla, R. S. S., Hammerlindl, J. K., Panchuk, B., Pelcher, L. E., and Keller, W. (1992) Modified binary plant transformation vectors with the wild-type gene encoding NPTII. *Gene* **122,** 383–384.
27. Sambrook, J., Fritsch, E. F., and Maniatis, T. (1989) *Molecular Cloning: A Laboratory Manual,* 2nd ed., Cold Spring Harbor Laboratory, Cold Spring Harbor, NY.
28. Horsch, R. B., Fry, J. E., Hoffmann, N. L., Eichholtz, D., Rogers, S. G., and Fraley, R. T. (1985) A simple and general method for transferring genes into plants. *Science* **227,** 1229–1231.
29. Robert, L. S., Thompson, R. D., and Flavell, R. B. (1989) Tissue-specific expression of a wheat high molecular weight glutenin gene in transgenic tobacco. *Plant Cell* **1,** 569–578.
30. Kitamura, T., Tange, T., Terasawa, T., Chiba, S., Kuwaki, T., Miyagawa, K., Piao, Y.-F., Miyazono, K., Urabe, A., and Takaku, F. (1989) Establishment and characterization of a unique human cell line that proliferates dependently on GM-CSF, IL-3, or erythropoietin. *J. Cellular Physiol.* **140,** 323–334.
31. Zheng, Z. W., Sumi, K., Tanaka, K., and Murai, N. (1995) The bean seed storage protein beta-phaseolin is synthesized, processed, and accumulated in the vacuolar type-II protein bodies of transgenic rice endosperm. *Plant Physiol.* **109,** 777–786.
32. Tanchak, M. A., Giband, M., Potier, B., Schernthaner, J. P., Dukiandjiev, S., and Altosaar, I. (1995) Genomic clones encoding 11S globulins in oats (*Avena sativa* L.). *Genome* **38,** 627–634.
33. Jervis, L. and Pierpoint, W. S. (1989) Purification technologies for plant proteins. *J. Biotechnology* **11,** 161–198.
34. Rudolph, N. S. (1995) Regulatory issues relating to protein production in transgenic systems. *Genet. Eng. News* **15,** 16–18.
35. Brower, V. (1996) PPL floats IPO as companies consider transgenic switch. *Nature Biotechnology* **14,** 692.

8

Tobamovirus Vectors for Expression of Recombinant Genes in Plants

Thomas H. Turpen and Stephen J. Reinl

1. Introduction

Plant and animal RNA viruses are increasingly used for gene delivery and expression in a variety of host cells. At a molecular level these agents can be viewed as cytosolic parasites of the ribosome, successfully competing for translation with the host messenger RNA (mRNA) population. In many virus-infected plants, modified plasmodesmata between adjacent cells enable movement of progeny virus throughout the whole organism. Because of this unique virus–host interaction, crop plants provide an inexpensive source of biomass available for rapid genetic manipulations.

Some of the more useful viral-gene transfer vectors for whole plants are based on diverse groups of RNA viruses (1). The RNA genomes of many plus-strand RNA viruses are directly infectious from inoculated leaf wounds (2). Here we describe the methods currently in use in our laboratory for the production and analysis of recombinant tobamoviruses. Tobacco mosaic tobamovirus, strain U1 (TMV), is the type member of the tobamovirus group (3). Members of the group encapsidate a plus-sense RNA genome of about 6.5 kb into rods 300 nm long and 18 nm in diameter, with a central hole of 4 nm in diameter. The structure of TMV has been refined at 2.9 Å from X-ray fiber diffraction studies (4). A right-handed helix is formed from about 2130 identical coat protein (CP) subunits of 17.5 kDa. By manipulating full-length cDNA clones of the virus and synthesizing infectious RNA transcripts in vitro, many genetically altered tobamoviruses have been described (5).

Virion RNA serves as mRNA for the 5' genes encoding the 126 kDa and overlapping 183-kDa subunits of the replicase complex (see **Fig. 1**). The

From: *Methods in Biotechnology, Vol. 3:*
Recombinant Proteins from Plants: Production and Isolation of Clinically Useful Compounds
Edited by: C. Cunningham and A. J. R. Porter © Humana Press Inc., Totowa, NJ

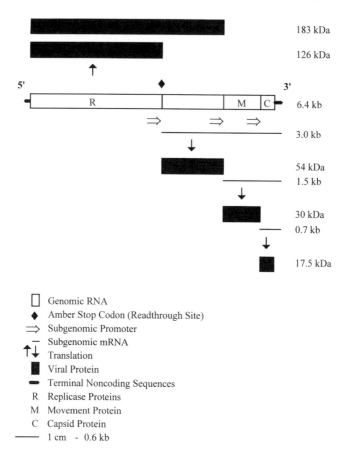

Fig. 1. Tobamovirus gene expression.

183-kDa protein is produced by readthrough *(6)* of an amber stop codon approx 5% of the time a ribosome passes this sequence. The requirements for leaky termination at the junction of the 126/183-kDa replicase reading frames of TMV have been defined in detail as the sequence CARYYA immediately following the termination codon *(7)*. The UAG stop codon is probably decoded as tyrosine in vivo by a naturally occurring plant tRNA^{Tyr} *(8)*. Expression of the internal genes is controlled by different promoters on the minus-sense RNA that direct synthesis of 3' co-terminal subgenomic mRNAs produced during replication. The predicted 54-kDa protein may be involved in the control of replication. The 30-kDa protein is required for the virus to move from cell to cell and is produced early and in relatively low amounts. The 17.5-kDa CP is produced late as the most abundant protein in infected cells. TMV RNA and CP aggregate spontaneously to form virions. A *cis*-acting sequence, the origin of assembly (OAS), is located in the movement protein gene and functions as a nucleation site to initiate the process of self-assembly. Heterologous sequences containing the OAS can be encapsidated, yielding rod-shaped virions whose lengths are proportional to the size of the RNA molecule. TMV particles are extremely stable; infectivity in plant sap survives heating to 90°C.

Previous results have demonstrated a remarkable genetic stability of foreign RNA sequences propagated in TMV-based vectors over brief time periods of replication in the absence of strong negative selection *(9)*. Viral promoters are used for the synthesis of new messenger RNAs, or peptide coding regions are fused directly to the CP gene. Using these two transient vector expression designs, we have established that heterologous mRNA produced by viral RNA amplification in whole plants is fully functional *(10,11)*. Both types of vectors have been tested in outdoor field trials (*see* **Note 1**).

1.1. Peptides Fused to the TMV Coat Protein

The abundance of the TMVCP in plants, and the purification advantages presented by the unique physical properties of the TMV virion structure, prompted us to focus on fusing peptides directly to the viral surface. The core of a single TMVCP subunit contains a bundle of four antiparallel α-helices projecting away from the RNA binding site. The region at high radius contains the N- and C-termini as well as a loop connecting the so-called right-slewed (RS) and right-radial (RR) α-helices. This four-helix bundle domain is one of the more common and simple protein folds. The stability of the structure is based on the packing of hydrophobic side chains on the interior and the orientation of hydrophilic groups on the surface. Recently we demonstrated, using malarial epitope sequences, that the TMV virion structure will accommodate peptide fusions in at least two regions, the C-terminus and the surface loop *(11)*. Native recombinant virions were bound specifically by protective monoclonal antibodies, indicating that the respective epitopes were on the surface of intact particles. Furthermore, these recombinants were not significantly reduced in their ability to assemble into stable particles and systemically infect whole tobacco plants. These initial results and parallel experiments from additional laboratories *(12,13)* are encouraging, but the limitations and utility of this system must be empirically determined. For example, it is possible that the particular peptides used to date do not significantly disrupt the TMVCP fold or the processes of assembly, disassembly, and long-distance movement, because they are of a similar hydrophilic nature to the surface upon which they are attached. It is also unclear to what extent the plant defense systems may recognize and respond to foreign peptides. In at least one case, a recombinant virus having the sequence MYGGFL directly fused to the C-terminus induced a lesion response on a host cultivar, Samsun, that is systemically infected by the parental virus *(14)*.

Restriction endonuclease cleavage sites are introduced into desired regions of the TMVCP gene by standard methods of site-directed mutagenesis. Peptide coding regions are most conveniently inserted as double-stranded synthetic DNA fragments into a suitable TMVCP subclone. Several different peptide fusions and substitutions at the TMVCP C-terminal region result in recombinants that fail to accumulate virions in upper leaves of inoculated plants (Turpen and Reinl, unpublished, *14,15*). However, the readthrough stop codon signal from the replicase reading frame may be placed upstream of the fusion peptide coding region, resulting in synthesis of wtTMVCP and fusion TMVCP in transfected plants at a ratio of <20:1. ELISA measurements of the peptide content in virions confirmed that for a 15-amino acid malarial epitope peptide, there is no significant exclusion of the TMVCP containing the C-terminal fusion peptide during virion assembly *(11)*.

If synthetic DNA fragments are used, one must design nucleotide sequences to encode the desired peptide. We believe this is an important but somewhat arbitrary decision. We generally try to avoid possible mRNA destabilizing sequences (AUUUA) *(16)*, avoid duplication of adjacent codons, and match the sequence to the most frequent codons used in the TMV genome. There are five codons used rarely (<10% frequency) in the TMVU1 genome: ARG (CGG, CGT, CGC) and LEU (CTA, CTC). Finally, any chance direct or inverted repeat sequences are removed to the extent possible.

1.2. Recombinant Messenger RNA from Additional Subgenomic Promoters

Coat protein subgenomic promoters from divergent members of the tobamovirus group can be functionally added to the TMVU1 genome. For example, the vectors TB2 and TTO1 contain the CP and subgenomic promoter from odontoglossum ringspot and tomato mosaic tobamoviruses, respectively *(17,18)*. For the synthesis of recombinant proteins in plants, a *Xho*I site adjacent to the TMVU1 CP subgenomic promoter is available for insertion of the desired coding region *(19)*. In some cases, increased stability of the foreign gene in the viral population is observed upon removal of noncoding regions of the cDNA. The stability of expression over multiple passages is highly dependent on individual gene sequences.

2. Materials
2.1. Plant Culture

Seeds are germinated in bedding flats at 90% humidity, at a temperature of 28°C, in a growth chamber before transplanting. Plants are grown in commercial potting soil (Sunshine Mix #1, Sungro Horticulture, Bellevue, WA) at a temperature of 25–29°C, with a controlled-release fertilizer (Osmocote,14-14-

14), using natural light supplemented with fluorescent light (GroLux, Sylvania, Danvers, MA) on a 16 h/d length, in an indoor greenhouse. Plants and soil to be discarded are autoclaved at the termination of each experiment. Fiberglass trays placed beneath the pots are treated with hypochlorite (10% Clorox) to disinfect any possible contaminant virus.

2.2. Laboratory Equipment

1. Thin-layer chromotagraphy (TLC) reagent sprayer (Alltech, Deerfield, IL).
2. Caframo stirrer (VWR, South Plainfield, NJ).
3. Kimble disposable pellet pestle (VWR).
4. DNA thermal cycler (Perkin-Elmer, Foster City, CA).
5. Waring blender (VWR).

2.3. Plasmid DNA

Plasmid DNA is purified from bacterial cultures (JM109) grown under appropriate antibiotic selection, using the alkaline lysis method *(20,21)*. Plasmid preparations should be treated with RNase A prior to phenol extraction. Cesium chloride gradient purification is not required.

2.4. Enzymes and Reagents

1. RNase inhibitor (Perkin-Elmer).
2. MuLV reverse transcriptase (Perkin-Elmer).
3. *Taq* DNA polymerase (Perkin-Elmer).
4. SP6 RNA polymerase (Gibco BRL, Gaithersburg, MD).
5. T7 RNA polymerase (Gibco BRL).
6. NTPs and dNTPs (ultrapure; Pharmacia Biotech, Piscataway, NJ).
7. RNA Cap structure analog G(5')ppp(5')G (New England Biolabs, Beverley, MA).
8. Bovine serum albumin (BSA), RNase- and DNase-free (Pharmacia Biotech).
9. Proteinase K (molecular biology grade; Sigma, St. Louis, MO).
10. Mineral oil (molecular biology grade; Sigma).
11. Silicon carbide, −400 mesh (Aldrich, Milwaukee, WI).
12. 8000 Da polyethylene glycol (PEG; Sigma).

2.5. Buffers and Solutions

1. DEPC-treated water: Bring water to 0.1% diethylpyrocarbonate (DEPC), mix well, and autoclave. **Caution: DEPC is a suspected carcinogen. Wear gloves and use a fume hood when handling.**
2. Proteinase K: 20 mg/mL in DEPC water, self-digested 10 min, 37°C, prior to use.
3. 5X SP6 buffer: $0.2M$ Tris-HCl, pH 7.9, 30 mM $MgCl_2$, 10 mM spermidine-$(HCl)_3$.
4. 5X T7 buffer: $0.2M$ Tris-HCl, pH 8.0, $0.125M$ NaCl, 40 mM $MgCl_2$, 10 mM spermidine-$(HCl)_3$.

5. 10X PCR buffer: 0.1*M* Tris-HCl, pH 8.3, 0.5*M* KCl.
6. dNTP mix: 10 m*M* each of dATP, dCTP, dGTP, and dTTP.
7. Phosphate buffer: Titrate 0.5*M* Na$_2$HPO$_4$ to pH 7.2 with 0.5*M* KH$_2$PO$_4$.
8. Extraction buffer: Phosphate buffer with 1% 2-mercaptoethanol.

3. Methods

Nicotiana tabacum cvs. Xanthi, Xanthi/nc are well-known tobacco cultivars for laboratory use *(22)*. Xanthi/nc contains the N gene, conferring hypersensitivity to tobamoviruses. The lesion response of this cultivar is useful for estimates of viral titer and rapid assessment of biological activity. MD609 is a commercial tobacco variety we have used for increased biomass *(23)*. Xanthi and MD609 are systemic hosts for TMV U1 and derived peptide/CP fusion vectors. *N. benthamiana* is particularly susceptible to many viruses and recombinants and is the species of choice for expression of recombinant cDNA from vectors carrying additional subgenomic promoters (TB2, TTO1).

It is preferable to increase virus from plants inoculated with RNA transcripts and use one passage for protein production. Recombinant virus replicates in transcript-inoculated plants for 1–2 wk to increase inoculum for subsequent passaging. After 1–4 wk of viral replication in virion-inoculated plants, accumulated product is extracted. This production cycle can be scaled as necessary from benchtop, growth chamber, or greenhouse to field-grown quantities of biomass.

RNA transcripts synthesized in vitro from full-length cDNA clones are most conveniently used to create recombinant tobamoviruses. The protocols described below are based on bacteriophage SP6 or T7 promoters. Viral cDNA or bacterial cultures of *Agrobacterium tumefaciens* containing viral cDNA can also be used to directly infect plants from abraded leaf surfaces. These may be useful alternative sources of inoculum for some experiments (*see* **Note 2**). In this case, infectious RNA is synthesized in vivo from host RNA polymerase II. The cDNA must contain appropriate initiation and termination signals. A self-cleaving ribozyme can be used to generate a functional 3' terminus *(24,25)*.

3.1. In Vitro Transcription

1. Linearize 10–20 µg of plasmid with an appropriate restriction endonuclease. A minimal number of additional nucleotides (<6) should remain in the plasmid template at the 3' terminus of the viral genome.
2. Remove contaminating nucleases from linearized plasmid by digesting with 50 µg/mL proteinase K for 30 min at 37°C.
3. Extract digest twice with an equal volume of phenol:chloroform:isoamyl alcohol (25:24:1) to remove proteinase K.
4. Precipitate DNA with 0.33 vol 10*M* NH$_4$OAc and 2 vol 100% ethanol. Pellet the DNA and wash with 70% ethanol:DEPC water. Resuspend the pellet in DEPC water at a concentration of 1 µg/mL.

5. Set up either SP6 or T7 transcription reactions:
   a. SP6 transcription reactions contain 1X SP6 buffer, 1 m*M* dithiothreitol, 40 ng/μL BSA, 0.5 m*M* each of ATP, CTP, and UTP, 0.25 m*M* GTP, 0.5 m*M* RNA cap analog, 2 U/μL RNase inhibitor, 0.05 μg/mL linearized DNA, and 0.3 U/μL SP6 RNA polymerase in DEPC water. Incubate reactions at 37°C for 60 min.
   b. T7 transcription reactions contain 1X T7 buffer, 10 m*M* dithiothreitol, 40 ng/μL BSA, 2 m*M* each ATP, CTP, and UTP, 0.2 m*M* GTP, 0.5 m*M* RNA cap analog, 2 U/μL RNase inhibitor, 0.05 μg/μL linearized DNA, and 2 U/μL T7 RNA polymerase in DEPC water. Incubate 5 min at 37°C, add GTP to 2 m*M*, and incubate for an additional 60 min at 37°C.

3.2. Inoculation

1. Dust young plants (<10 cm tall) with silicon carbide as an abrasive so that a thin layer is deposited on each fully expanded leaf to be inoculated. A reagent sprayer designed for thin-layer chromatography is convenient for this purpose. Optionally, susceptibility may be increased by placing plants in the dark the day preceding inoculations.
2. Inoculate each plant with 15–25 μL of transcription reaction immediately following the 37°C incubation. Include a Xanthi/nc cultivar and a plant mock-inoculated with buffer only (*see* **Note 3**).
3. Apply small droplets of the transcription reaction on dusted leaves and lightly rub with a gloved hand. Only light abrasion is required to cause wounding of trichomes at the leaf surface sufficient for infection. A heavy touch may cause severe damage to the inoculated leaf, reducing the efficiency of the inoculation. Dark necrotic lesions form on Xanthi/nc in 2–3 d; lighter colored abrasion damage is apparent the following day. Systemic symptoms appear 5–15 d postinoculation and vary according to the properties of individual recombinants and the size and growth rate of the plant.

3.3. Virus Purification

Virus is purified from systemically infected upper leaves of transcript-inoculated plants. Typically, the earliest emerging leaves with visible signs of infection are used as starting material for initial purification of a recombinant virus. The method is a scaled-down version of the procedure originally described by Gooding and Hebert *(26)*, and typically yields over 1 mg virus per g tissue extracted (*see* **Note 4**).

1. Remove 50–200 mg of tissue and place in an Eppendorf tube, being careful not to compress the tissue.
2. Freeze with liquid nitrogen and grind, using a Caframo stirrer and disposable pestle, until powdered.
3. Add 3–4 vol extraction buffer, 0.4 vol *n*-butanol, and 0.4 vol chloroform to the tissue.
4. Homogenize for 1 min, followed by vortexing once every few minutes for 15 min. At this stage, samples may be held on ice until processing is completed.

5. Centrifuge at 10,000*g* for 15 min at 4°C and transfer the clear aqueous phase to a clean tube.
6. Add 40% PEG 8000 to the aqueous phase to a final concentration of 4% and vortex once every few minutes for 15 min. Pellet precipitated virus at 10,000*g* for 15 min at 4°C.
7. Aspirate all of the supernatant with a pipet. Resuspend viral pellet in 10 m*M* phosphate buffer, pH 7.2, at a concentration of 1 mg/mL. Once pellet is completely resuspended, clarify with a 5-min spin at 10,000*g*, and transfer supernatant to a clean tube. A portion of this sample is placed at −20°C for long-term storage. The remainder is stored at 4°C as a viral stock. Sodium azide (0.02%) may be added as a preservative.

3.4. Titer

An extinction coefficient ($E^{0.1\%,\ 1\ cm}$) of 2.9 at 260 nm, uncorrected for light scattering, is commonly used for TMV. The $A_{260/280}$ is 1.2. Virus is bioassayed on the lesion host Xanthi/nc by inoculations at a concentration of 1–50 µg/mL. There is good proportionality between 10–100 lesions per leaf.

3.5. PCR Analysis of a Recombinant Gene in a Viral Population

The viral genome is analyzed to confirm the integrity of the recombinant sequences. RT-PCR *(27,28)* is a fast and informative method for analyzing viral genomes. Primer design depends on the sequence of the recombinant virus, but typically a primer pair that anneals external to the foreign sequence is desired over a pair that anneals internal to the foreign sequence. This allows the use of standard primers that only require initial optimization, and will not require synthesis of a new pair for every insert. In addition, smaller inserts may produce very small PCR products when amplified with internal primers. For larger inserts, an internal pair or a mixed pair may provide additional information and more accurate gel sizing.

1. Release viral RNA in a 20 µL reaction containing 100 ng/µL virus and 2 mg/mL proteinase K in a disposable PCR reaction tube. Layer mineral oil on the sample and run 1 cycle each of 30 min at 37°C, 10 min at 99°C, and 5 min at 4°C in a thermal cycler.
2. Set up a 20 µL cDNA synthesis reaction containing 0.5 ng/µL viral RNA (approx equivalent to 200 ng virus), 1X PCR buffer, 5 m*M* MgCl$_2$, 1 m*M* dNTPs, 0.5 m*M* 3' primer, 1 U/µL RNase inhibitor, and 2.5 U/µL reverse transcriptase in DEPC water. Layer mineral oil on the sample and run 1 cycle of 5 min for each 200 bp of synthesis at 42°C, 5 min at 99°C, and 5 min at 4°C.
3. To the 20 µL cDNA reaction, add 80 µL of 1X PCR buffer, 1.25 m*M* MgCl$_2$, 0.125 µ*M* 5' primer, and 0.03 U/µL *Taq*-polymerase. Run 25–30 cycles of 1 min at 94°C, 1 min at an appropriate annealing temperature, and 1 min for each 1 kb of synthesis at 72°C, followed by a final 5 min extension at 72°C. The expected

fragment is amplified from plasmid template as a control and provides an accurate apparent mol wt standard for gel electrophoresis. This reaction contains 1X PCR buffer, 2 mM MgCl$_2$, 0.2 mM dNTPs, 0.1 µM 3' primer, 0.1 µM 5' primer, 0.1 ng/µL plasmid, and 0.025 U/µL *Taq*-polymerase.

PCR products are directly sized on an appropriate gel system. For finer resolution, phenol extract PCR products, precipitate, and digest with a diagnostic restriction enzyme prior to gel electrophoresis. Control RNA template for RT-PCR may be prepared from the in vitro transcription reaction after treatment with RNase-free DNase. This is particularly useful if one suspects strong stops in the cDNA synthesis reaction as a source of template for the synthesis of unpredicted PCR reaction products (*see* **Note 5**).

3.6. Viral Passage and Large Scale Purification

1. Inoculate the desired number of plants with a 10–50 µg/mL virus solution. For peptide production, recombinant virions are frequently needed in multigram quantities. One can expect an average lab tobacco plant to yield <100 g of leaf tissue in a month of growth postinoculation. Commercial field cultivars produce a substantially higher leaf weight.
2. Homogenize frozen systemically infected leaves in a large Waring blender, using 1–1.25 mL of extraction buffer per gram of tissue.
3. Add 8 mL of n-butanol per 100 mL of extract and stir 15 min.
4. Centrifuge at 10,000g for 30 min.
5. Decant supernatant and add 4.0 g of PEG per 100 mL while stirring.
6. After PEG has completely dissolved, centrifuge at 10,000g for 15 min and discard supernatant.
7. Resuspend pellets in 20 mL of 10 mM phosphate buffer, pH 7.2, per 100 mL of initial extract.
8. Clarify by centrifugation at 10,000g for 15 min.
9. Repeat a second PEG precipitation step as follows: While stirring, add 0.4 g of NaCl and 0.4 g of PEG per 10 mL of virus suspension.
10. Centrifuge at 10,000g for 15 min and discard supernatant.
11. Resuspend pellets in 2 mL of 10 mM phosphate buffer, pH 7.2, per 100 mL of initial extract.
12. Centrifuge at 10,000g for 5 min to clarify and discard pellet. This solution is normally a viral suspension of remarkably high purity.

3.7. Protein Expression

The characterization of individual proteins and peptides for specific applications is described in Chapters 5, 7, 9, 10, and 11. Many issues, such as protease sensitivities during either biosynthesis or purification and quantifying expression results, must be determined on a case-by-case basis. Few generali-

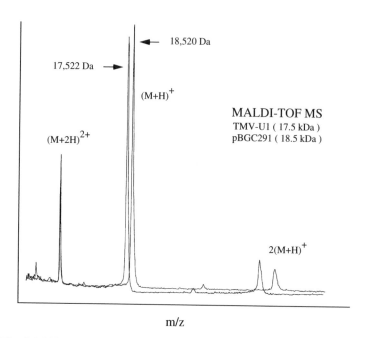

17,522 Da →

← 18,520 Da

(M+H)⁺

(M+2H)²⁺

2(M+H)⁺

MALDI-TOF MS
TMV-U1 (17.5 kDa)
pBGC291 (18.5 kDa)

m/z

Fig. 2. MALDI-TOF mass spectrum of TMVCP. Virus was isolated from plants inoculated with TMV U1 or with RNA transcripts from the plasmid of pBGC291 as described in
the text. Samples were prepared in 10% acetic acid (1–10 µg/mL), diluted 10-fold and
mixed 1:1 with sinapinic acid matrix solution before laser desorption with a HP-2025 A
MALDI-TOF System. Calibrants included a Protein Standard Solution for a mass range of
12,000–67,000 Da and the TMVCP of the U1 strain as an internal standard. The predicted
molecular weight of the malarial epitope recombinant (291) is 18,538 Da.

zations are meaningful and future success will depend as much on innovations
in downstream processing of plant extracts as in the development of new
expression vectors and methods. However, it is clear that plants are a new and
renewable source of an ever-increasing number of high-value products, including human pharmaceuticals. For example, the recombinant tobamovirus
BGC291 is candidate subunit vaccine for malaria *(11)*. BGC291 virus carries a
12-amino acid insertion of a protective *Plasmodium* epitope in the surface loop
of the TMVU1CP, and was conveniently produced in gram quantities necessary for animal testing. The molecular mass and sample purity was confirmed
by matrix assisted laser desorption ionization time-of-flight mass spectrometry
(MALDI-TOF/MS, *see* **Fig. 2**).

4. Notes

1. These vectors are also useful for indirectly manipulating endogenous protein levels via RNA-mediated responses *(18)*.
2. Methods for the inoculation and analysis of viral recombinants in transfected plant protoplasts were recently described by Kearney et al. *(29)*.
3. Transcription products may be analyzed for premature termination or other problems by electrophoresis in formaldehyde or glyoxal gels *(21)*.
4. The physical properties, yields, and symptoms induced by each new CP fusion recombinant may be unpredictable.
5. Helpful discussions regarding PCR and primer design can be found in Innis et al. *(30)*.

Acknowledgments

We thank our colleagues at Biosource Technologies, and William O. Dawson (University of Florida, Lake Alfred, FL) for their support and Earl L. White (R. J. Reynolds Tobacco Company, Winston-Salem, NC) for the MALDI-TOF/MS analysis.

References

1. Zaccomer, B., Haenni, A.-L., and Macaya, G. (1995) The remarkable variety of plant RNA virus genomes. *J. Gen. Virol.* **76,** 231–247.
2. Boyer, J.-C. and Haenni, A.-L. (1994) Infectious transcripts and cDNA clones of RNA viruses. *Virology* **198,** 415–421.
3. Goelet, P., Lomonossoff, G. P., Butler, P. J. G., Akam, M. E., Gait, M. J., and Karn, J. (1982) Nucleotide sequence of tobacco mosaic virus RNA. *Proc. Natl. Acad. Sci. USA* **79,** 5818–5822.
4. Namba, K., Pattanayek, R., and Stubbs, G. (1989) Visualization of protein-nucleic acid interactions in a virus. Refined structure of intact tobacco mosaic virus at 2.9Å resolution by X-ray fiber diffraction. *J. Mol. Biol.* **208,** 307–325.
5. Dawson, W. O. (1992) Tobamovirus-plant interactions. *Virology* **186,** 359–367.
6. Pelham, H. R. B. (1978) Leaky UAG termination codon in tobacco mosaic virus RNA. *Nature* **272,** 469–471.
7. Skuzeski, J. M., Nichols, L. M., Gesteland, R. F., and Atkins, J. F. (1991) The signal for a leaky UAG stop codon in several plant viruses includes the two downstream codons. *J. Mol. Biol.* **218,** 365–373.
8. Zerfass, K. and Beier. H. (1992) Pseudouridine in the anticodon GYA of plant cytoplasmic tRNA^{Tyr} is required for UAG and UAA suppression in the TMV-specific context. *Nucleic Acids Res.* **20,** 5911–5918.
9. Kearney, C. M., Donson, J., Jones, G. E., and Dawson, W. O. (1993) Low level of genetic drift in foreign sequences replicating in an RNA virus in plants. *Virology* **192,** 11–17.

10. Kumagai, M. H., Turpen, T. H., Weinzettl, N., della-Cioppa, G., Turpen, A. M., Donson, J., Hilf, M. E., Grantham, G. L., Dawson, W. O., Chow, T. P., Piatak, M., Jr., and Grill, L. K. (1993) Rapid, high-level expression of biologically active α-trichosanthin in transfected plants by an RNA viral vector. *Proc. Natl. Acad. Sci. USA* **90,** 427–430.

11. Turpen, T. H., Reinl, S. J., Charoenvit, Y., Hoffman, S. L., Fallarme, V., and Grill, L. K. (1995) Malarial epitopes expressed on the surface of recombinant tobacco mosaic virus. *Bio/Technology* **13,** 53–57.

12. Hamamoto, H., Sugiyama, Y., Nakagawa, N., Hashida, E., Matsunaga, Y., Takemoto, S., Watanabe, Y., and Okada, Y. (1993) A new tobacco mosaic virus vector and its use for the systemic production of angiotensin-I-converting enzyme inhibitor in transgenic tobacco and tomato. *Bio/Technology* **11,** 930–932.

13. Sugiyama, Y., Hamamoto, H., Takemoto, S., Watanabe, Y., and Okada, Y. (1995) Systemic production of foreign peptides on the particle surface of tobacco mosaic virus. *FEBS Lett.* **359,** 247–250.

14. Takamatsu, N., Watanabe, Y., Yanagi, H., Meshi, T., Shiba, T., and Okada, Y. (1990) Production of enkephalin in tobacco protoplasts using tobacco mosaic virus RNA vector. *FEBS Lett.* **269,** 73–76.

15. Saito, T., Yamanaka, K., and Okada, Y. (1990) Long-distance movement and viral assembly of tobacco mosaic virus mutants. *Virology* **176,** 329–336.

16. Green, P. J. (1993) Control of mRNA stability in higher plants. *Plant Physiol.* **102,** 1065.

17. Donson, J., Kearney, C. M., Hilf, M. E., and Dawson, W. O. (1991) Systemic expression of a bacterial gene by a tobacco mosaic virus-based vector. *Proc. Natl. Acad. Sci. USA* **88,** 7204–7208.

18. Kumagai, M. H., Donson, J., della-Cioppa, G., Harvey, D., Hanley, K., and Grill, L. K. (1995) Cytoplasmic inhibition of carotenoid biosynthesis with virus-derived RNA. *Proc. Natl. Acad. Sci. USA* **92,** 1679–1683.

19. Dawson, W. O., Bubrick, P., and Grantham, G. L. (1988) Modifications of the tobacco mosaic virus coat protein gene affecting replication, movement and symptomatology. *Phytopathology* **78,** 783–789.

20. Yanisch-Perron, C., Vieira, J., and Messing, J. (1985) Improved M13 cloning vectors and host strains: nucleotide sequences of the M13mp18 and pUC19 vectors. *Gene* **33,** 103–119.

21. Sambrook, J., Fritsch, E. F., and Maniatis, T. (1989) *Molecular Cloning: A Laboratory Manual,* 2nd ed., Cold Spring Harbor Laboratory, Cold Spring Harbor, NY.

22. Holmes, F. O. (1938) Inheritance of resistance to tobacco mosaic disease in tobacco. *Phytopathology* **28,** 553–561.

23. Morgan, O. D. and McKee, C. (1965) Two new Maryland tobacco varieties resistant to Black Shank. *University of Maryland Extension Service Fact Sheet* **173.**

24. Weber, H., Haeckel, P., and Pfitzner, A. J. P. (1992) A cDNA clone of tomato mosaic virus is infectious in plants. *J. Virol.* **66,** 3909.

25. Turpen, T. H., Turpen, A. M., Weinzettl, N., Kumagai, M. H., and Dawson, W. O. (1993) Transfection of whole plants from wounds inoculated with *Agrobacterium*

tumefaciens containing cDNA of tobacco mosaic virus. *J. Virol. Meth.* **42,** 227–240.

26. Gooding, G. V., Jr., and Hebert, T. T. (1967) A simple technique for purification of tobacco mosaic virus in large quantities. *Phytopathology* **57,** 1285.

27. Saiki, R. K., Scharf, S., Faloona, F. A., Mullis, K. B., Horn, G. T., Erlich, H. A., and Arnheim, N. (1985) Enzymatic amplification of β-globin genomic sequences and restriction site analysis for diagnosis of sickle cell anemia. *Science* **230,** 1350–1354.

28. Veres, G., Gibbs, R. A., Scherer, S. E., and Caskey, C. J. (1987) The molecular basis of the sparse fur mouse mutation. *Science* **237,** 415–417.

29. Kearney, C. M., Chapman, S., Turpen, T. H., and Dawson, W. O. (1997) High levels of gene expression in plants using RNA viruses as transient expression vectors, *Plant Molecular Biology Manual* 2nd ed. (Gelvin, S. B. and Schilperoort, R. A., eds.), Kluwer, Dordrecht, The Netherlands.

30. Innis, M. A., Gelfand, D. H., Sninsky, J. J., and White, T. J. (1990) *PCR Protocols: A Guide to Methods and Applications*, Academic, San Diego, CA.

9

Single-Chain Fv Antibodies Expressed in Plants

Udo Conrad, Ulrike Fiedler, Olga Artsaenko, and Julian Phillips

1. Introduction

Methods of gene cloning and genetic engineering of immunoglobulins and transgenic plant techniques have given rise to new approaches in plant biotechnology *(1)*. The high affinity and specificity of antibodies to different structures can be used to block regulation factors in plant cells by expression of specifically designed antibody constructs *(2)*. Antibodies expressed in plants can also be used to influence and block plant pathogen development and action *(3,4)*. Furthermore, plant cells, especially plant storage organs, can serve as factories for the production of antibodies of biotechnological interest *(5)*.

Several types of engineered antibodies have been developed in past years and some have already been expressed in plant cells *(1)*. In this chapter, we focus on the expression of single-chain Fv (scFv) antibodies **(Fig. 1)**. A single-chain Fv molecule is a synthetic antibody derivative consisting of the heavy- and light-chain variable domains of an immunoglobulin (VH and VL) joined together by a flexible peptide **(Fig. 2)**. The scFv is the smallest antibody fragment that can be made containing the complete antigen binding site, and is formed by the interaction of the VH and VL domains alone. In most cases investigated, the affinities of the scFv proteins were almost identical to the affinities of the complete parental antibody *(6–8)*. The use of such small antibody fragments offers several advantages. First, after PCR (polymerase chain reaction)-fusion the VH and VL sequences are encoded by one gene and synthesized as a single polypeptide. Both chains are therefore expressed in equimolar concentrations and covalent linking of the two chains facilitates the association of the VH and VL domains after folding. Second, cloned scFv libraries can be displayed on the surface of filamentous bacteriophages *(9)*,

From: *Methods in Biotechnology, Vol. 3:*
Recombinant Proteins from Plants: Production and Isolation of Clinically Useful Compounds
Edited by: C. Cunningham and A. J. R. Porter © Humana Press Inc., Totowa, NJ

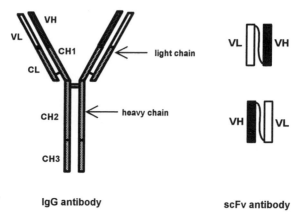

Fig. 1. General structures of complete IgG antibody and the scFv fragment. VH = variable domain of the immunoglobulin heavy chain; VL = variable domain of the immunoglobulin light chain; CH1–CH3 = constant domains of the immunoglobulin heavy chain; CL = constant domain of the immunoglobulin light chain.

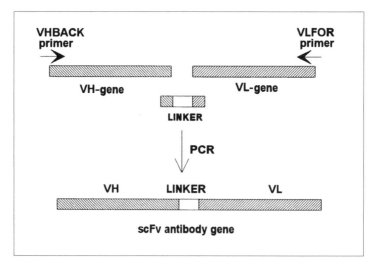

Fig. 2. Construction of a scFv antibody gene by fusion PCR. LINKER = $(Gly_4Ser)_3$ peptide coding sequence.

which provide a powerful affinity selection system. Third, bifunctional proteins can be produced by fusing scFv genes to sequences providing other functions. Fourth, scFv proteins, because of their small size, may access some intracellular compartments more easily and, since they lack a glycosylation site, are expected to be less immunogenic (**6**). These properties of the scFv proteins are of great importance for therapeutic purposes and intracellular immunization studies *(10,11)*.

1.1. Expression and Characterization of scFvs in Bacteria and Phage

ScFv genes can be constructed by PCR assembly of the rearranged V-genes of hybridomas. The VH and VL genes are separately amplified using degenerate primers and cDNA from the hybridoma as a template. Reamplification of the VH and VL genes with the linker sequence overlapping them both generates one chain in which the V-gene sequences are joined in-frame for expression *(12,13*; **Fig. 2)**.

The most attractive system for the engineering and characterization of the antibody fragments is *Escherichia coli*. Cloning, expression, and detection of the proteins are quite simple and well established. Many different types of cloning, sequencing, and expression vectors are available. Two main strategies for the expression of antibody fragments in *E. coli* have been developed. The first strategy is the production of the antibody fragments as intracellular inclusion bodies. Polypeptides expressed in this way are mostly denatured and require in vitro refolding *(14)*. The second approach employs the secretion of the antibody fragments into the periplasmic space that leads to the correct folding and assembly of both domains of the scFv proteins.

To enable rapid and sensitive detection, purification, and quantification of the secreted scFv antibodies by immunological methods, different tag sequences are used. These are short peptides fused to the antibody fragments and recognized by specific monoclonal antibodies. The *c-myc* tag has been used in many experiments *(4,5,9,15)*. This peptide contains 11 amino acids fused to the C-terminus of the scFv and is recognized by the 9E10 antibody *(16)*. Other tag sequences utilized for analytical studies are an 11-amino acid peptide from the herpes simplex virus glycoprotein *(8)*, the His_6 tag *(17)*, and the FLAG tag *(14)*.

Along with the Western blot analyses, use of the tag sequences also allows the design of enzyme-linked immunoassays (ELISA) for testing the affinity and specificity of the scFv antibodies. The scheme of a direct scFv antibody ELISA is shown in **Fig. 3A**. The microtiter plate is coated with antigen conjugated to bovine serum albumin (BSA). The scFv proteins bound to the antigen are detected by anti-c-myc tag antibodies, followed by the antimouse immunoglobulin-alkaline phosphatase (ALP) conjugate. Binding of the scFv antibodies to the free antigen can be detected by competition ELISA (**Fig. 3B**). In this test, binding of the scFv antibodies to the immobilized antigen-BSA conjugate is inhibited by the addition of free antigen. Therefore, the amount of scFv molecules bound to the conjugate is inversely proportional to the concentration of free antigen.

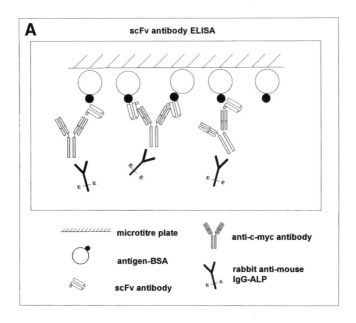

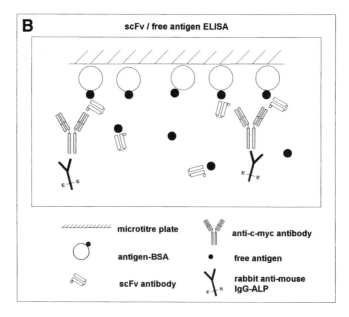

Fig. 3. Schematic presentation of ELISA techniques. (**A**) Direct ELISA for the detection of active scFv antibodies. (**B**) ELISA for the detection of scFv antibodies binding to free antigen in solution.

An alternative phage display system has been developed for the construction and expression of scFv antibodies *(9,18,19)*. ScFv polypeptides fused to the minor coat protein of the bacteriophage are displayed on the viral surface at approximately five molecules per phage particle and retain their functional properties. The advantage of this expression system is that the protein of interest is linked to the phage particle-containing DNA that encodes this protein. Such a system provides a powerful means of affinity selection of the antibody fragments and their maturation in vivo.

1.2. High-Level Expression of scFv in Different Organs of Transgenic Tobacco

As a result of the interest in using antibodies and single-chain Fvs for therapeutic purposes, agricultural production offers the possibility of obtaining large quantities of antibodies at relatively low costs.

1.2.1. Plant Expression Cassettes

After verifying the functionality of the scFv in the bacterial system, the scFv gene has to be cloned into a plant expression cassette. A primary consideration that must be addressed prior to vector construction is whether the antibody fragment should be expressed ubiquitously or in specific organs only. For the overall expression of the scFv in plants, the CaMV 35S promoter *(20)* is most widely used. This has been well characterized, but expression in the whole plant has potential disadvantages. A plant expressing a foreign gene in all its tissues may be less productive under field conditions, because of the energy required for formation of engineered antibodies or the interference of expressed proteins with metabolic processes *(21)*. Correct harvesting and storage of the plant biomass containing the scFv is very important. Normally, the harvested green material has to be stored frozen or at least refrigerated. However, seed-specific expression seems to be advantageous because of the stable deposition of proteins in ripe seeds for a long period of time without the necessity of cooling *(22,23)*. The scFv protein can be extracted as a functionally active scFv at any convenient time *(5)*. So far, for the seed specific expression of scFv in transgenic tobacco, the LeB4 promoter from *Vicia faba* has been used *(24)*. If the foreign protein is produced only in the parts of the plants that will be harvested (e.g., seeds), regulatory guidelines associated with transgenic plants expressing chimeric genes may be eased so that the remaining portion of the plant can be disposed of in the usual way *(21)*.

The choice of promoter appears to influence the expression yield and the targeted expression of the scFv in different plant cell compartments. Expression levels of scFv in plants vary from 0.06 to 0.1% (**Table 1**) of total soluble protein (TSP)

Table 1
Expression of Single-Chain Fv Antibodies in Transgenic Tobacco Plants

scFv	Construction	Signal sequence (SS)	Expression level
Antiphytochrome scFv *(25)*	CaMV 35S-VKLVH	None	0.06–0.1% of TSP
Antiartichoke mottled crinkle virus scFv *(4)*	CaMV 35S-VHLVK	None	0.1% of TSP
Antiphytochrome scFv *(27)*	CaMV 35S-SS-VKLVH	PR1a	0.5% of TSP
Antiabscisic acid scFv *(2)*	CaMV 35S-SS-VHLVK-KDEL	LeB4	0.05–4.8% of TSP
Antioxazolone scFv *(5)*	LeB4-SS-VHLVK	LeB4	0.02–0.67% of TSP

(4,25), when the constitutive CaMV 35S promoter is used to drive cytoplasmic expression. Up to 0.5% TSP was detected if a signal peptide sequence *(26)* was included in the scFv constructs *(27)*. Fused to a signal peptide, the scFv protein enters the secretory pathway and may be protected from the cytoplasmic proteases. ScFv protein could only be detected in seeds of transgenic tobacco when constructs including the legumin LeB4 signal peptide, were used **(Table 1)**. Ripe tobacco seeds contained scFv up to 0.67% of TSP *(5)*.

Targeting proteins to special compartments like the endoplasmic reticulum (using an ER-retention signal) could also increase protein expression levels *(28)*. The expression of a scFv recognizing the phytohormone abscisic acid reached up to 4.8% of TSP when the KDEL-encoding sequence was attached to the 3' end *(2)*. **Figure 4** shows a general scheme of possible plant expression cassettes.

1.2.2. Agrobacterium tumefaciens *Mediated Plant Transformation*

Among the various methods of gene transfer into plants, the most common and convenient is to employ *Agrobacterium tumefaciens* as a vehicle for the introduction of recombinant vectors to the plant cell nucleus *(29)*. Therefore, the plant expression cassette has to be cloned into a plant transformation vector. Suitable binary vectors include pGA 482 *(30)*, pBin 19 *(31,32)*, and pGSGLUC1 *(33)*. The cassette is cloned between the left and the right border sequences of the T-DNA. Transfer of the T-DNA from *Agrobacterium* into the plant cell nucleus is followed by heritable genomic integration of the T-DNA.

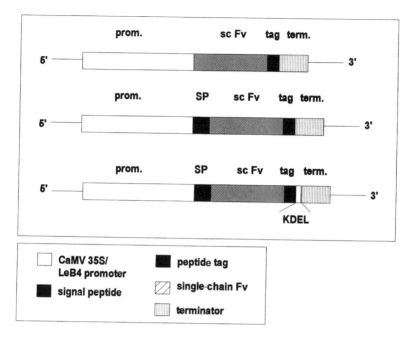

Fig. 4. Schematic representation of cassettes for the expression of scFv genes in transgenic plants. KDEL, tetrapeptide for ER targeting.

Vectors have been developed that contain a plant-expressed kanamycin resistance gene (NPT II), allowing the selection of transformed plants. *Agrobacterium* can be transformed by electroporation, and positive clones are selected by isolation of plasmid DNA and by subsequent Southern analysis.

The leaf disk method is the most efficient and preferable method for transformation of tobacco. Transformed tobacco plants can be regenerated and selected in the presence of kanamycin.

1.2.3. Investigation of Transgenic Tobacco Plants

Plants can be investigated by Southern analysis for the presence of the transgene in the plant genome (**Fig. 5**). Plants transformed with constructs for ubiquitous expression can be directly tested by Western analysis for the expression of the scFv. For detection of the scFv, antibodies recognizing the tagged peptide (*c-myc*), or the scFv itself, can be used. A typical Western blot, detecting the expression of the scFv in ripe tobacco seeds, is shown in **Fig. 6**. The level of expression can be determined by comparison between the signal strength of known amounts of scFv (affinity-purified from bacterial or even

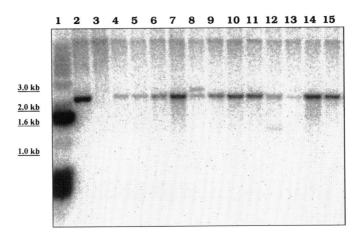

Fig. 5. Southern analysis of transgenic tobacco plants. 20 μg genomic DNA isolated from each plant were digested with *Hin*dIII, run on an agarose gel, blotted, and hybridized with a ^{32}P (α-dATP labeled scFv fragment. Lane 1: 1 kb DNA size marker; lane 2: purified scFv-*Hin*dIII-fragment related to five copies; lane 3: *Hin*dIII cleaved genomic DNA from a nontransformed plant; lanes 4–15: *Hin*dIII cleaved genomic DNA from different transgenic plants.

plant extracts) and the signal from plant extracts to be tested. Expressed scFv can be purified from crude plant extracts by immunoaffinity chromatography **(Fig. 7)**, in which an antibody recognizing either the tagged peptide or the scFv itself is coupled to the column.

1.3. Manipulation of Plant Physiological Functions by Expression of scFvs

ScFvs that bind to endogenous plant regulatory molecules can be used to create phenotypic mutants **(2)**. A promising area of investigation using this technology is offered by scFvs that bind phytohormones. These are present in levels sufficiently low to allow a reduction in activity by binding to the scFv expressed at levels that can be obtained in transgenic plants. Reduction of activity (immunomodulation) of a phytohormone, and the study of the changes in growth, development, and physiology, can lead to a deeper understanding of its regulatory role in plants.

Abscisic acid (ABA) is involved in many processes in plants, such as stomatal movements, stress responses, and seed development and dormancy. ABA functions have been studied using ABA-deficient and ABA-insensitive mutants, or by application of ABA or ABA-synthesis inhibitors to plant cells.

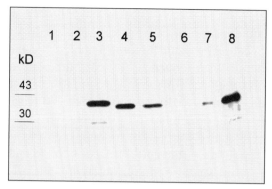

Fig. 6. Detection of scFv proteins in mature seeds of transgenic tobacco plants by Western blotting. Molecular mass of marker proteins are indicated on the left. Lane 1: 40 μg seed extract of a nonexpressing plant; lane 2: 40 μg seed extract of a non-expressing plant; lane 3: 40 μg seed extract, expression level 0.67% of TSP; lane 4: 40 μg seed extract, expression level 0.63% of TSP; lane 5: 40 μg seed extract, expression level 0.43% of TSP; lane 6: 40 μg seed extract, wild type plant; lane 7: 50 ng scFv; lane 8: 200 ng scFv.

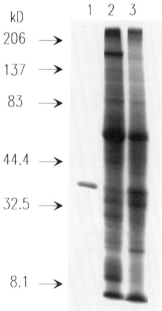

Fig. 7. SDS-PAGE of affinity-purified scFv and leaf proteins extracted from transgenic plants. Lane 1: 5 μg of anti-ABA–scFv antibody purified from leaves of a transgenic plant; lane 2: 80 μg protein from leaf extracts of wild-type tobacco plants; lane 3: 80 μg protein from leaf extracts of a transgenic tobacco plant expressing 4.5% scFv of total soluble protein.

The knowledge gained from such experiments allowed prediction of the possible effects of expression of an anti-ABA-engineered antibody in transgenic plants. Another prerequisite for this approach is the availability of a suitable scFv-encoding DNA sequence. A well-characterized, highly specific monoclonal antibody, with high affinity for ABA, allowed the production of a scFv that binds to free ABA *(15)*. In order to bind its endogenous target, a scFv must be present in the same cellular compartment. ABA moves freely across membranes in its protonated form and can therefore be bound wherever the scFv is localized. Targeting of the anti-ABA scFv to the endoplasmic reticulum (ER) by a legumin signal peptide, together with a KDEL sequence, allowed accumulation of active scFv up to 4.8% of total soluble protein in transgenic plants. These plants have a wilty phenotype **(Fig. 8)**, which is typical for ABA deficient mutants in other species, e.g., *Nicotiana plumbaginifolia, Arabidopsis thaliana,* and tomato *(34–36)*, and is caused by their inability to close stomata. Plants expressing a scFv that binds to the hapten, oxazolone, have a phenotype indistinguishable from that of wild-type plants, demonstrating the specific nature of the effects caused by the anti-ABA scFv. Measurement of ABA levels in leaves showed that anti-ABA scFv transgenic plants have elevated levels compared with wild-type plants. It is expected that most of this ABA is bound to scFv molecules and thus is trapped in the ER as the plants nevertheless show symptoms of ABA-deficiency. Hence, using this scFv, it was demonstrated, for the first time, that phytohormone action can be manipulated in plants by expression of an engineered antibody.

This approach has several advantages over the use of mutants to study ABA action. Like antisense technology, which involves knocking out genes that have previously been cloned and characterized, it is applicable for any species that can be transformed. It should be possible, with either strategy, to block the function of molecules in particular cells, tissues, or organs of the plant by directing scFv expression. Furthermore, utilizing antisense mRNA transcription technology, a series of transformed plants with different expression levels could be produced, allowing the severity of the effect to be studied in more detail. Unlike antisense technology, the scFv approach does not require the plant gene of interest to be cloned and characterized. However, the work required for the isolation and characterization of a scFv with the necessary specificity and affinity for its target in order to obtain unambiguous results should not be underestimated.

Potentially, the approach described here can be used to directly influence the action of any molecule for which a scFv can be produced. It is expected that this new technology will increasingly be used to study plant processes.

Fig. 8. Wilting of tobacco plants expressing anti-ABA scFv antibodies in leaves. Transgenic (left and right) and control plants (middle) were grown in the greenhouse under plastic bags to maintain high humidity. Removal of bags led to rapid wilting of the plants, expressing anti-ABA scFv antibodies.

2. Materials
2.1. DNA Materials

1. Primers *(13)*:   VH1BACK 5'-AGG TSM ARC TGC AGS AGT CWG G-3'
   VK4FOR is an equimolar mix of:
   5'-CCG TTT GAT TTC CAG CTT GGT GCC-3'
   5'-CCG TTT TAT TTC CAG CTT GGT CCC-3'
   5'-CCG TTT TAT TTC CAA CTT TGT CCC-3'
   5'-CCG TTT CAG CTC CAG CTT GGT CCC-3'

> VH1FOR 5'GGA GAC GGT GAC CGT GGT CCC TTG GCC CC3'
>
> VK2BACK 5'GAC ATT GAG CTC ACC CAG TCT CCA3'
>
> Degeneracy code: S = C or G; M = A or C; R = A or G; W = A or T

2. Linker sequence *(9,13)*: 5'-GGG ACC ACG GTC ACC GTC TCC TCA GGT GGA GGC GGT TCA GGC GGA GGT GGC TCT GGC GGT GGC GGA TCG GAC ATT GAG CTC ACC CAG TCT CCA-3'

3. Herring sperm DNA (Boehringer/Mannheim GmbH, Mannheim, Germany).

2.2. Agrobacterium Strains

1. C58C1Rf [r] (pGV 2260 in C58C1) *(37)*.
2. EHA 101 (pEHA 101 in C58) *(38)*.

2.3. Plant Material

Nicotiana tabacum cv. Samsun NN plants were used for leaf disk transformation.

2.4. Special Laboratory Tools and Materials

1. Gene-Pulser™ (Bio-Rad, Hercules, USA).
2. Transfer electrophoresis unit (Hoefer, San Francisco, CA).
3. ELISA reader (Dynatech MR 7000, Denkendorf, Germany).
4. Centriprep-30 concentrator (Amicon, Witten, Germany).
5. Nitrocellulose membranes (Cellulosenitrate membranes BA 85, Schleicher and Schuell, Dassel, Germany).
6. Nylon membranes (Hybond-N, Amersham, Amersham, UK).
7. Blotting paper GB 002 (Schleicher and Schuell).
8. Microtiter plates (Falcon 3915, Falcon, Los Angeles, CA).
9. Millipore filters Type VS 0.025 µm (Millipore, Bedford, MA).

2.5. Reagents and Column Materials

1. ECL Western blotting analysis system (Amersham).
2. Megaprime DNA labeling system (Amersham).
3. Affi-10 Gel Active Ester Agarose (Bio-Rad).
4. *Taq* DNA polymerase (Boehringer/Mannheim GmbH).
5. Bio-Rad Protein assay (Bio-Rad).
6. Domestos Sanitär-Reiniger (chlorine-free, Lever GmbH, Hamburg, Germany).
7. Marvel dried skimmed milk fat free (Premier Beverages, Stafford, UK).
8. Acrylamide/*Bis*-acrylamide Stock Solution Protogel (National Diagnostics, AGS, Heidelberg, Germany).
9. Antibodies: Antimouse Ig, horseradish peroxidase linked whole antibody (Amersham); Anti-mouse IgG alkaline phosphatase conjugate (Sigma, Deisenhofen); anti-c-myc antibody 9E10 *(16)*.
10. Redivue (α ^{32}P) dATP (Amersham).

11. Hybridization solution Roti-Hybri-Quick (Roth, Karlsruhe, Germany).
12. Mineral oil (Sigma).

2.6. Buffers and Other Solutions

1. Carbonate buffer: 50 mM Na$_2$CO$_3$, pH 9.6.
2. Phosphate-buffered saline (PBS): 8 mM Na$_2$HPO$_4$, 2 mM KH$_2$PO$_4$, 0.15M NaCl, pH 7.5.
3. Lysis buffer: 10 mM Tris-HCl, pH 8.0, 10 mM EDTA, 0.1M NaCl.
4. TE buffer: 10 mM Tris-HCl, pH 8.0, 1 mM EDTA.
5. Gel-loading buffer: 50% (v/v) glycerol, 10 mM Tris-HCl, pH 8.0, 0.05% (w/v) SDS, two crystals bromophenol blue per 100 mL.
6. 1X TBE buffer: 89 mM Tris-borate, 89 mM boric acid, 2 mM EDTA.
7. Glycine-NaCl, pH 2.2: 0.1M glycine, 0.1M NaCl.
8. 20X SSC: 3M NaCl, 0.3M sodium citrate.
9. Protein extraction buffer: 50 mM Tris-HCl, pH 8.0, 200 mM NaCl, 5 mM EDTA, 0.1% (v/v) Tween-20.
10. SDS-gel loading buffer: 10 mL containing 1 mL 10% (v/v) glycerol, 1.4 mL separating buffer, 2 mL 10% (w/v) SDS, 0.5 mL β-mercaptoethanol, 5.1 mL water, 2 crystals of bromophenol blue.
11. Separating buffer: 0.8% (w/v) SDS, 1.5M Tris-HCl, pH 8.8.
12. Stacking buffer: 0.8% (w/v) SDS, 0.5M Tris-HCl, pH 6.8.
13. SDS-PAGE running buffer: 25 mM Tris, 0.25M glycine, pH 8.3, 0.1% (w/v) SDS.
14. Transfer buffer: 400 mL 10X SDS-PAGE running buffer, 800 mL methanol in 4 L.
15. Marvel buffer: 20 mM Tris-HCl, pH 8.0, 180 mM NaCl.
16. CBY medium: 1 g/L yeast extract (Difco, Detroit, MI), 5 g/L tryptone (Difco), 5 g/L sucrose, pH 7.2; sterilize by autoclaving for 15 min, 121°C. 2 mM MgSO$_4$·7H$_2$O is added after autoclaving from a 200 mM filter-sterilized (0.2 μm pore size) stock solution. For solidified medium, add 20 g/L agar (Difco) prior to autoclaving.
17. SOC: 5 g/L yeast extract (Difco), 20 g/L tryptone (Difco), 0.58 g/L NaCl, 0.186 g/L KCl. Sterilize by autoclaving for 15 min, 121°C; when cool add 1 mL/L filter-sterilized 2M glucose.
18. Murashige and Skoog (MS) liquid medium: 4.49 g/L MS medium basal salt mixture (Duchefa, Haarlem, The Netherlands), 1 mL/L vitamin stock solution (100 mg/L myo-inositol, 2.0 mg/L glycine, 0.5 mg/L nicotinic acid, 0.5 mg/L pyridoxine-HCl, 0.1 mg/L thiamine-HCl), 30 g/L sucrose. Adjust to pH 5.8 with 1N KOH. Sterilize by autoclaving for 15 min, 121°C.
19. MS/NAA/BA: Add agar to MS medium to 0.8–1.0% for solidified medium. After autoclaving for 15 min, 121°C, leave to cool to around 50°C and add 0.1 mL/L NAA (α-naphthaleneacetic acid) and 1.0 mL/L BA (6-benzyl-amino-purine) from 1 mg/mL stock solutions. To make stock solutions, dissolve 100 mg powder in a small volume (1–2 mL) of 1N NaOH. Make up to 100 mL with distilled water, and filter-sterilize (0.2 μm pore size). Store at 4°C.
20. Kanamycin and cefotaxime: dissolved each in water, sterilized by filtration (0.2 μm pore size), and added after media have been autoclaved.

21. dNTP solutions: 2.5 m*M* each dATP, dCTP, dGTP, and dTTP (Boehringer Mannheim).
22. 10X PCR buffer (Boehringer Mannheim).
23. RNase: (20 µg/mL).
24. Phenol solution (Roth).

3. Methods

3.1. PCR Assembly of the VH and VK Genes in scFv Configuration

This method was developed by Clackson et al. *(13)*.

1. Make the following mixture in a 0.5 mL Eppendorf tube: 47 µL water, 10 µL 10X PCR buffer, 10 µL dNTPs, 5 µL linker DNA fragment (100 ng), 5 µL VH1BACK primer (10 µ*M*), 5 µL VK4FOR primer (10 µ*M*), 8 µL VH fragment amplified before with VH1FOR2 and VH1BACK or VK4FOR and VK2BACK primers (300 ng), and 8 µL VK fragment (300 ng).
2. Overlay with mineral oil and heat to 94°C for 4 min.
3. Add 2 µL *Taq*-polymerase (1–5 units/µL) under the oil.
4. Start temperature cycling: each cycle consists of 94°C, 1.5 min; 72°C, 2.5 min run for 30 cycles.
5. Analyze the PCR-assembly on a 2% (w/v) agarose gel (*see* **Note 1**).

3.2. Transformation of Agrobacterium by Electroporation

The transformation of *Agrobacterium* by electroporation is a quick and simple method for the introduction of plasmid DNA into competent cells. It has the advantage that the cells may be stored for several months and transformed as soon as vectors have been cloned in *E. coli*. It is wise to check transformant colonies for presence of the plasmid introduced, for example, by Southern analysis of isolated DNA.

1. Inoculate 10 mL CBY medium containing appropriate antibiotics with a single colony from a plate of *A. tumefaciens*. Incubate in a shaker (150–200 rpm) at 28°C for 2 d.
2. Use 0.5 mL of the culture to inoculate 100 mL CBY medium containing appropriate antibiotics, and incubate as above.
3. Harvest the cells by centrifugation, for example, in a Sorvall SS34 rotor, 4300*g*, 15 min, 4°C.
4. Decant the medium and resuspend the pellet in 100 mL sterile, distilled water. Centrifuge as before.
5. Wash the cells twice more with 50 mL water and finally in 10 mL 10% (v/v) glycerol.
6. Centrifuge again and resuspend in 500 µL 10% (v/v) glycerol. Transfer 45 µL aliquots to sterile microcentrifuge tubes.

7. Freeze in liquid nitrogen and store at –80°C (*see* **Note 2**).
8. Drop dialyze plasmid DNA (isolated from *E. coli* by conventional methods) by pipeting 10 µL onto a piece of millipore filter, floating, glossy side up, in a Petri dish full of distilled water. Leave at room temperature for 1 h.
9. Thaw an aliquot of cells and put on ice. Add 100–1000 ng dialyzed *E. coli* plasmid DNA, mix, and transfer to an ice-cold electroporation cuvet. Electroporate, for example, with a Bio-Rad Gene Pulser™ 25 µF, 2.5 V, 200 Ω (*see* **Note 3**).
10. Rinse out the cuvet with 1 mL SOC and transfer to a sterile microcentrifuge tube (*see* **Note 4**). Incubate, shaking gently at 28°C for 2–4 h.
11. Spread 100 µL aliquots, as well as 1:10 and 1:100 dilutions, onto selective CBY plates. Incubate at 28°C until colonies are visible. This usually takes 2–3 d, depending on the strain of *Agrobacterium* used.

3.3. Agrobacterium-*Mediated Transformation of Tobacco Leaf Explants*

The transformation of tobacco leaf disks was first described by Horsch et al. *(39)*. This method is very efficient and can be used to generate large numbers of transgenic plants in a short time. After surface sterilization, all operations should be performed in a laminar flow cabinet using aseptic techniques appropriate for plant tissue culture work.

1. Grow the desired variety of tobacco *(N. tabacum)*, for example, Samsun NN, until the leaves are about 10–15 cm long, preferably in a controlled-environment growth room, or similar (*see* **Note 5**).
2. Remove several leaves, wash with tap water and surface-sterilize by immersion in 30% (v/v) Domestos for 15 min in a large sterile beaker.
3. Rinse six times with sterile water and cut leaves into about 1-cm squares, avoiding the midrib. Place the squares in 9-cm Petri dishes containing MS 0.1 NAA 1.0 BA medium, 10 explants per plate. Seal dishes with Parafilm and incubate for 2 d in dim light at 25°C.
4. Inoculate 5–10 mL CBY medium containing appropriate antibiotics with a single colony of *Agrobacterium*. Incubate 2 d, shaking at 28°C.
5. Dilute *Agrobacterium* culture 1:50 in liquid MS medium in a Petri dish. Control explants can be dipped into MS medium without bacteria.
6. Dip leaf explants briefly into the bacterial suspension and return to the original plates. Co-cultivate for 2 d in dim light at 25°C.
7. Transfer leaf explants to MS/NAA/BA plates containing 100 mg/L kanamycin and 250 mg/L cefotaxime (*see* **Note 6**).
8. Transfer to fresh selective plates after 4 wk. Callus will start to appear at the edge of the explants prior to shoot formation.
9. After a further 2–4 wk, shoots appear. Excise shoots, including apices and 2–3 leaves, and push the cut base into MS medium without plant growth regulators containing 100 mg/L kanamycin and 250 mg/L cefotaxime, and incubate in the light at 25°C.

10. After 2–3 wk, roots will grow from the transgenic shoots. Transfer plantlets to standard potting compost in small plant pots. Grow in humid conditions for the first 1–2 wk, for example, in a propagator with a plastic cover. Wedge the cover slightly open for a few days to allow the plants to acclimatize to ambient conditions. Remove the cover completely and repot plants as necessary (*see* **Note 7**).

3.4. Southern Analysis

To determine whether the transformed tobacco plants are transgenic for the scFv gene, genomic DNA is isolated from tobacco leaves. After digestion with a suitable restriction enzyme, DNA fragments are separated by electrophoresis through an agarose gel, denatured, transferred to a nylon filter and immobilized. The DNA attached to the filter is then hybridized to a ^{32}P-labeled DNA fragment containing the scFv gene. Plants giving a positive signal of the expected size in autoradiography are chosen for further experiments.

3.4.1. Isolation of Genomic DNA from Transformed Tobacco Plants

1. Grind 200–300 mg of a young tobacco leaf in liquid nitrogen to a fine powder.
2. Collect the frozen powder into a 15-mL disposable plastic centrifuge tube containing 20 µL proteinase K (25 mg/mL).
3. Add 3 mL lysis buffer and 0.8 mL 10% SDS and shake well.
4. Incubate at 56°C for at least 2 h.
5. Cool to room temperature, add 1 mL saturated NaCl, and shake for 15 s.
6. Centrifuge for 15 min at room temperature at 2800*g*.
7. Pipet supernatant into a beaker containing 10 mL cold ethanol.
8. Collect precipitated DNA by stirring with a glass rod.
9. Put the DNA into a 1.5 mL tube, centrifuge, and wash the pellet with 70% ethanol at room temperature.
10. Dry the DNA pellet for 10 min at room temperature and resuspend in 200 µL RNase in TE and leave at 4°C overnight.
11. After adding 400 µL TE to the dissolved DNA, remove proteins by extraction once with 500 µL phenol, once with 500 µL of a 1:1 mixture of phenol/chloroform, and once with 500 µL chloroform. Always mix the contents of the tubes until an emulsion forms and centrifuge for 5 min at full speed in an Eppendorf centrifuge at room temperature. Use the upper, aqueous phase for the next extraction.
12. Precipitate DNA by adding 0.1 vol 3*M* sodium acetate, pH 5.2, and 1 mL ethanol.
13. After leaving for a few minutes on ice, spin the precipitated DNA down, and wash the pellet with 70% ethanol.
14. Resuspend the air-dried DNA pellet in 100–200 µL TE.

3.4.2. Digestion of Isolated Genomic DNA

1. Estimate the amount of isolated DNA by measuring the absorbance at 260 nm (recommended dilution, 1:100). An OD of 1 corresponds to approx 50 µg/mL double-stranded DNA.

2. Mix water with DNA solution containing 20 μg genomic DNA, and add the appropriate 10X digestion buffer and 75 units of the restriction enzyme, to a final volume of 30–40 μL. Incubate for at least 4 h at the recommended temperature.
3. Add 6–8 μL gel-loading buffer to the restricted DNA.

3.4.3. Separation of Digested DNA in an Agarose Gel

1. Make a 1% (w/v) agarose gel. Add 1 g powdered agarose to 100 mL 1X TBE electrophoresis buffer. Melt the slurry in a microwave oven and add 10 μL ethidium bromide (10 mg/mL) to the dissolved agarose.
2. Load the samples into the slots of the 1X TBE submerged gel and electrophorese the digested DNA at 6 V/cm for 6 h.
3. Visualize the DNA under UV light. Place a ruler beside the gel and photograph so that the distance that any given band of DNA has migrated can be read directly from the photographic image.

3.4.4. Transfer of DNA to a Nylon Membrane

1. Hydrolyze the DNA by acid depurination in $0.25M$ HCl for 15 min at room temperature with constant shaking. This acid-induced cleavage aids in the transfer of large DNA fragments.
2. Denature the DNA by soaking the gel in $1.5M$ NaCl, $0.5M$ NaOH for 30 min with constant shaking.
3. Neutralize the gel by soaking in a solution of $1M$ Tris-HCl, pH 8.0, and $1.5M$ NaCl for 45 min at room temperature with constant shaking.
4. Fill a dish with 20X SSC and place a rack or stack of glass plates in the dish.
5. Place a piece of blotting paper on top that is in contact with the SSC solution.
6. Place a piece of wet blotting paper on top of the support and smooth out all air bubbles with a glass rod.
7. Place the gel on the damp blotting paper. Make sure there are no air bubbles between blotting paper and gel.
8. Cut a piece of nylon membrane the same size as the gel. Use gloves to handle the membrane.
9. Wet the nylon membrane in SSC solution and place it on top of the gel so that one edge is just at the line of the wells. Remove all air bubbles.
10. Place two wet pieces of blotting paper on top of the nylon filter. Again remove all air bubbles.
11. Place a stack of paper towels (5–8 cm high) on the blotting paper, put a glass plate on top and weigh it down with a 500 g weight.
12. Allow transfer of the DNA to proceed for about 12 h.
13. After transfer, remove towels, dry the filter, and crosslink the DNA on the filter by 15 s of UV-light exposure. Alternatively, the filter can be baked for 2 h at 80°C.

3.4.5. Hybridization of Southern filters

1. Immerse the baked filter in 6X SSC.
2. Put the wet filter in hybridization solution. Prehybridize the filter for 4 h at 65°C.

3. Place the filter in hybridization solution containing 100 μg/mL denatured herring sperm DNA and ^{32}P-labeled denatured probe DNA, and incubate at 65°C overnight.
4. Remove the filter from the hybridization solution and wash in a tray containing 2X SSC and 0.5% SDS at room temperature.
5. Wash the filters at 65°C for at least 3 h, with gentle agitation, in 1X SSC and 0.1% SDS.
6. Wash one additional time in 0.2X SSC, 0.1% SDS, at 65°C, for a further 30 min.
7. Wrap the filter in Saranwrap and apply to X-ray film to obtain an autoradiographic image.

3.5. Western Blot Analysis

Total soluble proteins are extracted from leaves and ripe or developing seeds. After separation by SDS-PAGE, the proteins are transferred onto nitrocellulose. *C-myc* peptide-tagged scFv can be detected by incubation with a mouse monoclonal antibody recognizing this tag (9E10), followed by incubation with a secondary antibody (antimouse 9E10 conjugated to horseradish peroxidase). The ECL detection system can be used for developing the enzymatic reaction.

3.5.1. Preparation of Protein Extracts from Seeds

1. Grind 50–150 mg ripe seeds from transgenic and control plants with liquid nitrogen and acid-washed sand.
2. Add 0.5–0.8 mL of protein-extraction buffer to the frozen powder.
3. Transfer the homogenate to an Eppendorf tube and centrifuge for 10 min at 4°C, 14000g.
4. Collect the supernatant and determine the protein concentration using, for example, the Bio-Rad Protein Assay.
5. For Western blotting, use 40 μg of total soluble protein from each plant for SDS-PAGE.

3.5.2. Preparation of Protein Extracts from Leaves

1. Homogenize a piece of a leaf (100 mg) with an electric drill attached to a pestle that fits into a microcentrifuge tube, and add 200 μL SDS-gel loading buffer.
2. Boil the homogenate for 10 min.
3. Centrifuge at 14,000g, room temperature, for 15 min.
4. Use the supernatant for determination of protein content using, for example, the Bio-Rad Protein Assay.
5. For Western blot, separate 40 μg of total soluble protein by SDS-PAGE per lane (*see* **Note 8**).

3.5.3. Polyacrylamide Gel Electrophoresis (PAGE)

1. Clean and assemble the glass plates and plug the bottom with agarose (1–2% w/v).

2. Make up the following 12.5% SDS-PAGE separating-gel mix (50 mL): 20.8 mL Protogel, 12.5 mL separating buffer, 16 mL water, and 81 μL TEMED. Just before pouring, add 0.46 mL 10% (w/v) ammonium persulfate.
3. Pour the gel, leaving enough space at the top for the stacking gel. Pour a few mL of water on top of the separating gel. Polymerize for at least 30 min.
4. Drain water and remove remaining liquid with filter paper.
5. Make up the following 6% stacking gel mix (15 mL): 3 mL Protogel, 3.75 mL stacking buffer, 8.5 mL water, and 30 μL TEMED. Just before pouring, add 75 μL 10% (w/v) ammonium persulfate.
6. Pour the stacking gel and position the comb. Leave to polymerize for 15–30 min.
7. Clamp the gel into a vertical electrophoresis apparatus; fill upper and lower reservoirs with SDS-PAGE running buffer.
8. To the samples, add an equal volume of SDS-gel loading buffer (*see* **Note 9**).
9. Incubate samples for 5 min at 94°C.
10. Load the samples (40 μg total soluble protein) and run the gel at 5 V/cm.

3.5.4. Transfer onto Nitrocellulose

1. Cut a piece of nitrocellulose and eight pieces of blotting paper to a size slightly larger than the gel.
2. Wet the nitrocellulose and blotting paper in transfer buffer.
3. Make a sandwich in the electroblotting device as follows:
   a. Negative electrode.
   b. Two sheets blotting paper.
   c. Foam rubber.
   d. Two sheets blotting paper.
   e. Gel (remove stacking gel).
   f. Nitrocellulose (check no air bubbles between this and gel).
   g. Two sheets blotting paper.
   h. Foam rubber.
   i. Two sheets blotting paper.
   j. Positive electrode.
4. Electrophorese at 200 mA, 18 V, for 8–12 h (length of time depends on protein size).
5. Remove nitrocellulose from the apparatus. Transfer was successful if the colored marker proteins from the gel are visible on the nitrocellulose filter. Do not dry the nitrocellulose filter before proceeding and during the following steps.

3.5.5. Reaction and Detection with Antisera

1. Incubate nitrocellulose in 5% (w/v) Marvel in Marvel buffer for 2 h, or in 0.4% (w/v) Marvel at 4°C overnight on a gentle rocking plate.
2. Incubate in a suitable dilution of primary antibody (with 9E10 supernatant from hybridoma cell culture, about a dilution of 1:50 or 1:500) in 0.4% (w/v) Marvel in Marvel buffer for 2 h at room temperature with gentle agitation.

3. Wash the filter three times for 10 min in 0.4% (w/v) Marvel (about 100 mL per wash).

4. Incubate in a suitable dilution of second antibody conjugate in 0.4% (w/v) Marvel for 1–2 h at room temperature (antimouse Ig, horseradish peroxidase-linked whole antibody 1:2000).

5. Wash once for 20 min, followed by four washes for 5 min with 0.05% (v/v) Tween-20 in PBS.

6. Develop with the ECL detection system. Pour substrate solution onto the blot, and allow to react for 1 min. Wrap the filter in Saranwrap and apply the filter for 1 min to a X-ray film.

7. Expression levels can be quantified by comparing signal strength of known amounts of scFv with tested plants.

3.6. Direct ELISA Using the 9E10 Antibody for Detection of the scFv

1. Coat a microtiter plate with 100 µL per well of antigen conjugated to BSA and diluted in carbonate buffer. Leave overnight at 4°C. Optimal antigen-coating concentration should be determined experimentally for each conjugate–antigen system.

2. Rinse wells three times with distilled water and block with 100 µL per well of 2% (w/v) BSA in PBS containing 0.05% (v/v) Tween-20 (PBST-BSA) for 1 h at room temperature.

3. Discard PBST-BSA solution and add 100 µL per well of the scFv-containing solution. Incubate 3 h at room temperature (RT).

4. Rinse wells three times with distilled water and pipet 100 µL per well of the affinity-purified 9E10 antibody diluted in PBST-BSA at a concentration of approx 4 µg/mL (crude ascitic liquid may be also used). Leave for 1 h at RT.

5. Discard the 9E10 antibody solution, rinse wells three times with distilled water and add 100 µL per well of antimouse immunoglobulin-alkaline phosphatase conjugate diluted 1:2000 in PBST-BSA. Incubate for 1 h at RT.

6. Discard the solution and rinse wells three times with distilled water. Add 100 µL per well of 1 g/L *p*-nitrophenyl phosphate in carbonate buffer. Incubate for 1 h at 37°C.

7. Measure absorbance at 405 nm on the ELISA plate reader.

3.7. Competition scFv ELISA-Detecting Free Antigen

1. Coat a microtiter plate with 100 µL per well of the antigen conjugated to BSA and diluted in carbonate buffer. Leave overnight at 4°C.

2. Rinse wells three times with distilled water and block with 100 µL per well of PBST-BSA for 1 h at RT.

3. Pipet 50 µL per well of different dilutions of free antigen in PBST. To each well add 50 µL of the affinity-purified scFv antibody diluted in PBST, mix, and incubate overnight at 4°C. The concentration of the scFv antibodies should be determined by constructing a dilution curve for the direct ELISA described above and estimating 50% binding to the antigen–BSA conjugate.

4. Rinse wells three times with distilled water. Add 9E10 antibody diluted in PBST-BSA and incubate for 1 h at RT.
5. Discard 9E10 antibody solution, rinse wells three times with distilled water and add 100 μL per well of antimouse immunoglobulin-alkaline phosphatase conjugate diluted 1:2000 in PBST-BSA. Incubate for 1 h at RT.
6. Discard the solution and rinse wells three times with distilled water. Add 100 μL per well of 1g/L *p*-nitrophenyl phosphate in carbonate buffer. Incubate for 1 h at 37°C.
7. Measure absorbance at 405 nm on the ELISA plate reader.

3.8. Purification of scFv by Affinity Chromatography

The techniques used for affinity purification depend on the C- or N-terminal tag. For protein fusions with the penta-histidine tag, the expressed scFv can be purified by metal affinity purification. Purification of scFv tagged with the c-myc peptide can be done by coupling the antibody 9E10 recognizing the c-myc peptide tag to activated agarose. Affi-Gel 10 is an activated immuno-affinity support that offers rapid, high-efficiency coupling for all ligands with a primary amino group, including proteins. Affi-Gel 10 support is a N-hydroxysuccinimide ester of a derivatized crosslinked agarose gel bead support. It couples to ligands spontaneously in aqueous or nonaqueous solution.

3.8.1. Preparation of Protein Extracts

1. Grind leaves (200–300 mg) from transgenic plants expressing scFv with liquid nitrogen to a fine powder.
2. Suspend powder in 400 μL PBS/1% (v/v) Triton X-100 as extraction buffer (*see* **Note 10**).
3. Centrifuge for 10 min at 4°C, 14,000*g*. The supernatant contains soluble proteins, including scFv, and can be applied to the column.

3.8.2. Preparation of a 9E10 Column

1. Mix 22.3 mL 0.1*M* MOPS, pH 7.5, and 2.5 mL ascites fluid from the 9E10 hybridoma. Concentrate to 2.5 mL by Centriprep-30 concentration.
2. Mix the concentrated solution again with 22.3 mL 0.1*M* MOPS and concentrate to a final volume of 1.5 mL. There should be 25–50 mg protein in 1.5 mL concentrated solution.
3. Shake the Affi-gel 10 agarose vial to get a uniform suspension.
4. Wash 2 mL of the slurry three times with three bed volumes of ice-cold millipore water. Centrifuge at low speed to separate agarose and water. Care should be taken not to allow the gel bed to become dry.
5. Transfer the moist gel cake to a test tube and add cold ligand solution. Agitate sufficiently to make a uniform suspension.
6. Continue gentle agitation of the gel slurry overnight at 4°C.

7. After coupling, any remaining active esters are blocked by adding 0.1 mL 1*M* monoethanolamine, pH 8.0. Allow 1 h for completion of the blocking reaction at 4°C.
8. Transfer the 9E10-agarose slurry to a column and wash extensively with at least 10 bed volumes of PBS.
9. Proteins or other solutes not bound, or weakly bound by nonspecific interactions, must be washed off prior to elution. Wash with 2 mL glycine-NaCl, pH 2.2. Immediately wash the column with 10 bed volumes PBS until the pH of the eluate is around 7.0. The column is now equilibrated.

3.8.3. Purification of scFv Tagged with c-myc Peptide on 9E10-Affi-Gel 10 Column

1. Apply the plant extract to the column. Allow the antigen to bind for 10 min at 4°C.
2. Repeat binding by running the effluate eight times through the column. The following steps can be carried out at RT.
3. Wash the column three times with 10 bed volumes PBS.
4. The scFv is eluted with 3 mL glycine-NaCl, pH 2.2. It is important to remove the eluted antigen from the eluant as quickly as possible, to minimize damage or denaturation of the protein. Therefore, eight 400 µL fractions are collected in 100 µL 1*M* Tris-HCl, pH 8.0. Mix immediately after collection. Normally, the eluted antigen can be detected in the third, fourth, and fifth fraction.
5. Regenerate the column by washing with 1 mL eluant to remove retained proteins, and neutralize immediately with 0.1*M* Tris-HCl in PBS, pH 8.0.
6. Equilibrate the column with 2 × 10 bed volumes PBS.
7. Columns can be stored at 4°C in PBS containing 0.2% (w/v) sodium azide.

4. Notes

1. The assembled scFv DNA fragment is approx 0.8 kb in size.
2. The cells can be stored at –80°C for several months before the electroporation efficiency deteriorates.
3. These parameters work well with the Bio-Rad equipment, but it may be desirable to optimize the settings when using another system.
4. Cuvets can be reused if thoroughly rinsed with distilled water and disinfected with 70% ethanol.
5. Plants can also be grown in the greenhouse, but the leaves are then often more difficult to surface sterilize, especially if the leaves are damaged by insect pests.
6. Control explants can be put onto medium with kanamycin and cefotaxime in order to check the efficiency of the selection, and also onto medium without kanamycin and cefotaxime in order to check the efficiency of shoot production, and to provide suitable wild-type control material for further analysis (Southern, Western, and so on).
7. Transgenic plants should be grown in accordance with the local legal requirements for genetically manipulated organisms.

8. If the protein gel is used for Western blotting, the use of colored marker proteins (Bio-Rad) is recommended. The protein transfer has been successful if the marker proteins can be seen on the nitrocellulose.
9. This is only necessary if the protein extracts are not made with SDS-gel loading buffer.
10. Extraction with Triton X-100 is only necessary if the scFv has to be extracted from the ER.

Acknowledgments

This work was supported by the German Ministry of Education, Science, and Technology and the Deutsche Forschungsgemeinschaft.

References

1. Conrad, U. and Fiedler, U. (1994) Expression of engineered antibodies in plant cells. *Plant Mol. Biol.* **26,** 1023–1030.
2. Artsaenko, O., Peisker, M., zur Nieden, U., Fiedler, U., Weiler, E. W., Müntz, K., and Conrad, U. (1995) Expression of a single chain Fv antibody against abscisic acid creates a wilty phenotype in transgenic tobacco. *Plant J.*, **8,** 745–750.
3. Voss, A., Niersbach, M., Hain, R., Hirsch, H. J., Liao, Y., Kreuzaler, F., and Fischer, R. (1994) Reduced virus infectivity in *N. tabacum* secreting a TMV-specific full size antibody. *Mol. Br.* **1,** 15–26.
4. Tavladoraki, P., Benvenuto, E., Trinca, S., Martinis, D., Cattaneo, A., and Galeffi, P. (1993) Transgenic plants expressing a functional single-chain Fv antibody are specifically protected from virus attack. *Nature* **36,** 469–472.
5. Fiedler, U. and Conrad, U. (1995) High-level production and long-term storage of engineered antibodies in transgenic tobacco seeds. *Bio/Technology* **13,** 1090–1093.
6. Bird, R. E., Hardman, K. D., Jacobson, J. W., Johnson, S., Kaufman, B. M., Lee, S.-M., Lee, T., Pope, S. H., Riordan, G. S., and Whitlow, M. (1988) Single-chain antigen binding proteins. *Science* **242,** 423–426.
7. Huston, J. S., Levinson, D., Mudgett-Hunter, M., Tai, M., Novotny, J., Margolies, M. N., Ridge, R. J., Bruccoleri, R. E., Haber, E., Crea, R., and Oppermann, H. (1988) Protein engineering of antibody binding sites: recovery of specific activity in an anti-digoxin single-chain Fv analogue produced in *Escherichia coli. Proc. Natl. Acad. Sci. USA* **85,** 5879–5883.
8. Francisco, J. A., Campbell, R., Iverson, B. L., and Georgiou, G. (1993) Production and fluorescence-activated cell sorting of *Escherichia coli* expressing a functional antibody fragment on the external surface. *Proc. Natl. Acad. Sci. USA* **90,** 10,444–10,448.
9. Hoogenboom, H. R., Griffiths, A. D., Johnson, K. S., Chiswell, D. J., Hudson, P., and Winter, G. (1991) Multi-subunit proteins on the surface of filamentous phage: methodologies for displaying antibody (Fab) heavy and light chains. *Nucleic Acids Res.* **19,** 4133–4137.
10. Chester, K. A. and Hawkins, R. E. (1995) Clinical issues in antibody design. *TIBTECH* **13,** 294–300.

11. Biocca, S. and Cataneo, A. (1995) Intracellular immunization: antibody targeting to subcellular compartments. *Trends Cell Biol.* **5,** 248–252.

12. Orlandi, R., Güssow, D. H., Jones, P. T., and Winter, G. (1989) Cloning immunoglobulin variable domains for expression by the polymerase chain reaction. *Proc. Natl. Acad. Sci. USA* **86,** 3833–3837.

13. Clackson, T., Hoogenboom, H. R., Griffiths, A. D., and Winter, G. (1991) Making antibody fragments using phage display libraries. *Nature* **352,** 624–628.

14. Ge, L., Knappik, P., Pack, P., Freund, C., and Plückthun, A. (1994) "Antibody engineering: a practical approach." Expressing antibodies in *Escherichia coli.* Protein Engineering Using Recombinant Bacteriophages: Application To Coliclonal Antibodies. Atelier De Formation **61,** Inserm, Le Vesinet.

15. Artsaenko, O., Weiler, E. W., Müntz, K., and Conrad, U. (1994) Construction and functional characterisation of a single chain Fv antibody binding to the plant hormone abscisic acid. *J. Plant Physiol.* **144,** 427–429.

16. Munroe, S. and Pelham, H. (1986) An Hsp 70-like protein in the E.R.: identity with the 78 kD glucose-regulated protein and immunoglobulin heavy chain binding protein. *Cell* **46,** 291–300.

17. Hayashi, N., Kipriyanov, S., Fuchs, P., Welschof, M., Dörsam, H., and Little, M. (1995) A single expression system for the display, purification and conjugation of single-chain antibodies. *Gene* **160,** 129–130.

18. McCafferty, J., Griffiths, A. D., Winter, G., and Chiswell, D. J. (1990) Phage antibodies: filamentous phage displaying antibody variable domains. *Nature* **348,** 553–554.

19. Waterhouse, P., Griffiths, A. D., Johnson, K. S., and Winter, G. (1993) Combinatorial infection and In Vivo Recombination: a strategy for making large phage antibody repertoires. *Nucleic Acids Res.* **21,** 2265–2266.

20. Töpfer, R., Maas, C., Höricke-Grandpierre, C., Schell, J., and Steinbiss, H.-H. (1993) Expression vectors for high-level expression in dicotyledonous and monocotyledonous plants. *Meth. Enzymol.* **217,** 66–78.

21. Krebbers, E., Bosch, D., and Vandekerckhove, J. (1992) Prospects and progress in the production of foreign proteins and peptides in plants, in *Plant Protein Engineering* (Shewry, P. R. and Gutteridge, S., eds.), University Press, Cambridge, pp. 315–325.

22. Vandekerckhove, J., Van Damme, J., Van Lijesebettens, M., Botterman, J., De Block, M., Vandewiele, M., De Clercq, A., Leemans, J., Van Montagu, M., and Krebbers, E. (1989) Enkephalins produced in transgenic plants using modified 2S seed storage proteins. *Bio/Technology* **7,** 929–932.

23. Pen, J. and Sijmons, P. C. (1993) Protein production in transgenic crops: analysis of plant molecular farming, in *Transgenic Plants—Fundamentals and Applications* (Hiatt, A., ed.), Dekker, New York, pp. 238–251.

24. Bäumlein, H., Boerjan, W., Nagy, I., Panitz, R., Inze, D., and Wobus, U. (1991) Upstream sequences regulating legumin gene expression in heterologous transgenic plants. *Mol. Gen. Genet.* **225,** 121–128.

25. Owen, M. R. L., Gandecha, A., Cockburn, W., and Whitelam, G. (1992) Synthesis of a functional anti-phytochrome single-chain Fv protein in transgenic tobacco. *Bio/Technology* **10,** 790–794.

26. Bednarek, S. Y., and Raikhel, N. V. (1992) Intracellular trafficking of secretory proteins. *Plant Mol. Biol.* **20,** 133–150.

27. Firek, S., Draper, J., Owen, M. R. L., Gandecha, A., Cockburn, B., and Whitelam G. (1993) Secretion of a functional single-chain Fv protein in transgenic tobacco plants and cell suspension cultures. *Plant Mol. Biol.* **23,** 861–870.

28. Wandelt, C. I., Khan, M. R. I., Craig, S., Schroeder, H. E., Spencer, D., and Higgins, T. J. V. (1992) Vicilin with carboxy-terminal KDEL is retained in the endoplasmic reticulum and accumulates to high levels in the leaves of transgenic plants. *Plant J.* **2,** 181–192.

29. Zambryski, P., Joos, H., Gentello, J., Leemans, J., Van Montagu, M., and Schell, J. (1983) Ti-plasmid vector for introduction of DNA into plant cells without altering their normal regeneration capacity. *EMBO J.* **2,** 2143–2150.

30. An, G., Watson, B. D., Stachel, S., Gordon, M. P., and Nester, E. W. (1985) New cloning vehicles for transformation of higher plants. *EMBO J.* **4,** 277–284.

31. Bevan, M. W. (1984) Binary *Agrobacterium* vectors for plant transformation. *Nucleic Acids Res.* **12,** 8711–8721.

32. Frisch, D. A., Harris-Haller, L. W., Yokubaitis, N. T., Thomas, T. L., Hardin, D. H., and Hall, T. C. (1995) Complete sequence of binary vector Bin 19. *Plant Mol. Biol.* **27,** 405–409.

33. Saito, K., Kaneko, H., Yamazaki, M., Yoshida, M., and Murakoshi, I. (1990) Stable transfer and expression of chimeric genes in licorice (*Glycyrrhiza uralensis*) using an Ri plasmid binary vector. *Plant Cell Rep.* **8,** 718–721.

34. Parry, A. D., Blonstein, A. D., Babiano, M. J., King, P. J., and Horgan, R. (1991) Abscisic-acid-metabolism in a wilty mutant of *Nicotiana plumaginifolia. Planta* **183,** 237–243.

35. Koornneef, M., Jorna, M. L., Brinkhorst-van der Swan, D. L. C., and Karssen, C. M. (1982) The isolation of abscisic acid (ABA) deficient mutants by selection of induced revertants in nongerminating gibberellin sensitive lines of *Arabidopsis thaliana (L.) Heynh. Theor. App. Genet.* **61,** 385–393.

36. Tal, M. and Nevo, Y. (1973) Abnormal stomatal behaviour and root resistance, and hormonal inbalance in three wilty mutants of tomato. *Biochem. Genet.* **8,** 291–300.

37. Deblaere, R., Bytebier, B., de Greve, H., Deboeck, F., Schell, J., van Montagu, M., and Leemans, J. (1985) Efficient octopine Ti plasmid-derived vectors for *Agrobacterium*-mediated gene transfer to plants. *Nucleic Acids Res.* **13,** 4777–4788.

38. Hood, E. E., Helmer, G. L., Fraley, R. T., and Chilton, M. D. (1986) The hypervirulence of *Agrobacterium tumefaciens* A281 is encoded in a region of pTiBo542 outside of T-DNA. *J. Bacteriol.* **168,** 1291–1301.

39. Horsch, R. B., Fry, J. E., Hoffmann, N. L., Eichholtz, D., Rogers, S. G., and Fraley, R. T. (1985) A simple and general method for transferring genes into plants. *Science* **227,** 1229–1231.

10

Characterization and Applications of Plant-Derived Recombinant Antibodies

Rainer Fischer, Jürgen Drossard, Yu-Cai Liao, and Stefan Schillberg

1. Introduction

Expression of foreign proteins in plants has become a standard technique in plant molecular biology. Various plant species have been used to produce mammalian proteins, such as human interferon *(1)* and serum albumin *(2)*, as well as murine antibodies. Not only full-size antibodies *(3–6)* but also Fab fragments *(7)* and single-chain fragments (scFvs) *(8,9)* have been expressed successfully in tobacco or *Arabidopsis*, reaching expression levels as high as 1.3% of the total soluble protein *(3)*. ScFvs have also been expressed in plant-suspension cultures at levels of 0.5% total soluble protein *(10)*. The feasibility of expressing and targeting recombinant antibodies (rAbs) *(11,12)* has been achieved in different compartments of plants, including the cytoplasm (scFvs), endoplasmic reticulum, chloroplasts, and the intercellular space (full-size, scFvs, and single-domain antibodies) for various applications *(3–10)*. These results indicate the flexibility of the plant system for expression of rAbs, or fragments thereof, in various plant cell compartments.

Because of this flexibility, numerous applications of plant-derived rAbs are currently under investigation. Increased virus resistance in transgenic plants caused by binding of rAbs to virions has been demonstrated *(5,9)*. Further developments in this direction include the design of rAbs interfering with proteins that are involved in virus replication or movement. Furthermore, these rAbs are a novel tool to elucidate structure–function relationships during viral pathogenesis in vivo. The concept of antibody-based immunomodulation is

From: *Methods in Biotechnology, Vol. 3:*
Recombinant Proteins from Plants: Production and Isolation of Clinically Useful Compounds
Edited by: C. Cunningham and A. J. R. Porter © Humana Press Inc., Totowa, NJ

currently being employed in several laboratories to develop new strategies for fighting other plant pathogens, including bacteria, fungi, and nematodes.

Transgenic plants expressing full-size rAbs (or fragments) also offer the possibility for removal of environmental pollutants *(3,13)*. Full-size antibodies are too large to freely permeate plant cell walls, since the diameter of pores imposes a restriction on the size of molecules, which is 20–30 kDa for a globular protein. Consequently, expression and targeting of pollutant-specific, full-size antibodies to the apoplasm of plant cells is equivalent to engineering a binding and retention capacity within a semipermeable membrane, thus creating an in vivo biofilter system. Because of the apparent ease of application and safe method of decontamination, the use of transgenic plant material may present a breakthrough in treatment of contaminated sources.

Additional applications include the use of plants as bioreactors for mass production of rAbs with therapeutic *(12,14)* and diagnostic relevance. However, all described applications require detailed characterization of the transgenic plants and/or the recombinant proteins prior to obtaining approval for field trials and marketing a defined product. Validation procedures and comparative studies with the native product (e.g., from mice) are required during all steps of manufacturing to ensure purity, specificity, stability, safety, and efficacy of the recombinant protein. Furthermore, methods for defining glycosylation patterns *(15)* are important to ensure product consistency among different preparations over an extended period of plant cell cultivation and downstream processing. This analysis, in particular, is important for therapeutic proteins, because changes during cultivation and the environmental status of the cells result in different glycoforms, which in turn can have a dramatic impact on their therapeutic efficacy, plasma clearance, antigenicity, solubility, and resistance to proteases.

In this chapter, procedures are described that enable detection, localization, and characterization of plant-derived antibodies and comparison with their mammalian counterparts. These procedures have been used in our lab to analyze transgenic tobacco plants producing TMV-specific rAbs (full-size, F[ab']$_2$, Fab, scFvs), which are directed to either the apoplasm or the cytoplasm (scFvs). Total soluble proteins are isolated from leaves or suspension-culture cells for determination of production levels or analysis of protein folding and antibody specificity by ELISA. This assay is also employed to identify elite lines with stable expression of transgenes over several generations. In high-producing plants, export of the antibody is evaluated by ELISA-based analysis of the intercellular washing fluid (IWF) and protoplast cultures. Isolated antibodies are subjected to SDS-PAGE and immunoblotting to check purity, integrity, and stability. Immunoblotting, lectin overlay, and isoelectric focusing (IEF) techniques are used to gain additional information regarding glycosylation, microheterogenicity, and product conformity over an extended period of cultivation. Finally, the biological function of plant-derived antibodies in vivo is tested by assaying protection of transgenic plants against viral infections.

2. Materials
2.1. Plant Material

Nicotiana tabacum L. cv. Xanthi-nc plants, stably transformed with a plant expression vector containing the cDNAs of the TMV-specific rAb genes, are grown in phyto-chambers, as described (*see* **Subheading 3.6.**).

2.2. Virus Strain

Tobacco mosaic virus (TMV) strain *vulgare* is propagated on *N. tabacum* cv. Samsun and purified by isopycnic centrifugation in CsCl, as described *(5)*.

2.3. Antibodies, Lectins, and Enzymes

Unless otherwise indicated, antibodies were purchased from Jackson Immunoresearch (Via Dianova, Hamburg, Germany).

1. Goat antimouse serum: IgG-, H+L-, and Fc-specific.
2. Alkaline phosphatase-conjugated goat antimouse antibodies: IgG-, H+L-specific.
3. Rabbit-derived anti-idiotypic antibodies directed against TMV mAbs.
4. Alkaline phosphatase-conjugated goat antirabbit antibodies: IgG-, H+L-specific.
5. Chicken anti-TMV serum (Linans, Bettingen, Germany).
6. Biotin-conjugated *Ricinus communis* agglutinin (RCAI) (E-Y Labs, San Mateo, CA).
7. Streptavidin-alkaline phosphatase conjugate.

2.4. Special Laboratory Equipment and Supplies

1. Laminar flow hood (Flow Laboratories, Milano, Italy).
2. pH meter.
3. Osmometer.
4. Benchtop centrifuge (Heraeus, Düsseldorf, Germany).
5. Microtiter plates (M129B, Dynatech, Chantilly, VA).
6. Titertek Plus MS 2 Reader (ICN, Eschwege, Germany).
7. Orbital shaker (New Brunswick, Edison, NJ).
8. Growth chamber.
9. Gel electrophoresis system (Bio-Rad, Hercules, CA).
10. Electrophoretic transfer system (Bio-Rad).
11 Isoelectric focusing cell (Bio-Rad).
12. Hybond C nitrocellulose membrane (0.45 µm) (Amersham, Braunschweig, Germany).
13. Whatman 3MM filter paper (Whatman).

2.5. Buffers and Other Solutions

For preparation of buffers and solutions use only bidistilled water.

1. Protein-extraction buffer: 200 mM Tris-HCl, pH 7.5, 5 mM EDTA, 0.02% (w/v) sodium azide, and 0.1% (v/v) Tween-20. Filter-sterilize and store at 4°C.
2. Coating buffer: 15 mM Na$_2$CO$_3$, 35 mM NaHCO$_3$, and 0.02% (w/v) sodium azide. Adjust to pH 9.6 with acetic acid, filter-sterilize, and store at 4°C.
3. 10X PBS: 1.37M NaCl, 27 mM KCl, 81 mM Na$_2$HPO$_4$ and 15 mM KH$_2$PO$_4$. Adjust to pH 7.2 with 1M HCl, filter-sterilize, and store at room temperature.
4. PBS-T: 0.05% (v/v) Tween-20 in 1X PBS.
5. PS-solution: Dissolve 0.85% (w/v) NaCl in water and adjust to pH 7.2 with 1M HCl. Filter-sterilize and store at 4°C.
6. Sample buffer: 2% (w/v) polyvinylpyrrolidone, 0.02% (w/v) sodium azide, and 0.05% (v/v) Tween-20 in 1X PBS. Adjust to pH 6.0 with 1M HCl, filter-sterilize, and store at 4°C.
7. EMA buffer: 2% (w/v) polyvinylpyrrolidone (25–30 kDa), 0.2% (w/v) BSA, 0.02% (w/v) sodium azide, and 0.05% (v/v) Tween-20 in 1X PBS. Adjust to pH 7.4 with 1M HCl, filter-sterilize, and store at 4°C.
8. Substrate buffer: 0.1M diethanolamine, 1 mM MgCl$_2$. Adjust to pH 9.8 with HCl and filter-sterilize. Store at 4°C.
9. IWF buffer: 50 mM HEPES, 100 mM NaCl. Adjust to pH 7.2 with NaOH and filter-sterilize. Store at room temperature.
10. K3 medium: Add 10 mL each of the following stock solutions, 1 mL microelements, 1 mL vitamin solution I, 5 mL Fe in EDTA, 100 mg inositol, 250 mg xylose, and 136.92 g sucrose to 850 mL water, mix, adjust to pH 5.6, and adjust volume to 1000 mL. Adjust to 600 mosm with sucrose, filter-sterilize, and store at room temperature.
    a. Stock solutions for K3 medium: 125 mM NaH$_2$PO$_4$, 1.26M CaCl$_2$, 2.47M KNO$_3$, 312 mM NH$_4$NO$_3$, 102 mM (NH$_4$)$_2$SO$_4$, and 208 mM MgSO$_4$. Dissolve each chemical in 200 mL water, filter-sterilize, and store at 4°C.
    b. Microelements: 100 mM H$_3$BO$_3$, 100 mM MnSO$_4$, 37 mM ZnSO$_4$, 5 mM KI, 1 mM NaMoO$_4$, 0.1 mM CuSO$_4$, and 0.1 mM CoCl$_2$. Dissolve the components in water, filter-sterilize, and store at 4°C.
    c. Vitamin solution I: Dissolve 400 mg glycine, 400 mg nicotinic acid, 400 mg pyridoxine, and 20 mg thiamine-HCl in 200 mL water, filter-sterilize, and store at 4°C.
    d. Fe in EDTA: 20 mM FeSO$_4$ and 20 mM Na$_2$EDTA. Dissolve in water and boil for 5 min. Store at 4°C in a dark bottle.
11. W5 solution: 154 mM NaCl, 125 mM CaCl$_2$, 5 mM KCl, and 5 mM glucose. Dissolve in water and adjust to pH 5.8. Filter-sterilize and store at room temperature.
12. Separating-gel solution (*see* **Note 1**): 40% (v/v) acrylamide stock, 37.5% (v/v) Tris stock, pH 8.8, 21.15% (v/v) bidistilled H$_2$O; mix and degas; add 1% (v/v) SDS stock, 0.05% (v/v) TEMED, and 0.3% (v/v) APS stock; mix and pour immediately.

a. Acrylamide stock 30% (total monomer concentration [T] = 30 and crosslinker concentration [C] = 2.67): 29.2% (w/v) acrylamide, 0.8% (w/v) *N,N'*-methylene-bisacrylamide (*see* **Note 2**).

b. Tris stock (separating gel): 1*M* Tris-HCl, pH 8.8.

c. SDS stock: 10% (w/v) sodium dodecylsulphate.

d. APS stock: 20% (w/v) ammonium persulphate.

13. Stacking-gel solution: 13% (v/v) acrylamide stock, 12.5% (v/v) Tris stock, pH 6.8, 72.9% (v/v) bidistilled H_2O; mix and degas; add 1% (v/v) SDS stock, 0.1% (v/v) TEMED, 0.5% (v/v) APS stock; mix and pour immediately.

a. Tris stock (stacking-gel): 1*M* Tris-HCl, pH 6.8.

14. SDS-PAGE sample buffer (reducing): 62.5 m*M* Tris-HCl, pH 6.8, 10% (v/v) glycerol, 2% (w/v) SDS, 5% (v/v) β-mercaptoethanol, 0.05% (w/v) bromphenol blue.

15. 5X SDS-PAGE running buffer, pH 8.3: 125 m*M* Tris, 960 m*M* glycine, 0.5% (w/v) SDS.

16. Poinceau S solution for reversible protein staining (Sigma, St. Louis, MO).

17. Blocking solution: 0.45% (v/v) fish gelatin (Sigma).

18. AP buffer: 100 m*M* Tris-HCl, pH 9.6, 100 m*M* NaCl, 5 m*M* $MgCl_2$.

19. NBT/BCIP stable mix (Gibco BRL).

3. Methods

3.1. Extraction of Total Soluble Plant Proteins

To analyze antibody expression in transgenic tobacco plants, total soluble proteins must be extracted first.

1. Homogenize 0.5 g tobacco leaf tissue or plant suspension cells in a cooled mortar after addition of 1 mL protein extraction buffer.

2. Centrifuge the homogenate at 14,000*g* for 10 min at 4°C.

3. Transfer the clear supernatant to a new sample tube and store the plant extract at 4°C for further analysis (*see* **Notes 3** and **4**).

3.2. ELISA-Based Analysis of Plant-Derived Antibodies

Expression of full-size, TMV-specific antibodies is analyzed with two different ELISA procedures *(16,17)*. The first type of ELISA (ELISA-I) is used for determination of the total amount of antibodies in the plant extract, including nonassembled heavy and light chains *(5)*. The amount of TMV-specific—and therefore assembled and correctly folded—rAbs is determined with the second type of ELISA (ELISA-II). Analysis of scFv expression is carried out with ELISA-III.

3.2.1. ELISA-I

1. Coat microtiter plates with goat antimouse serum (2.5 µg/mL coating buffer) overnight at 4°C (*see* **Note 5**).

2. Thoroughly wash the plates three times by rinsing them with PBS-T and incubating for 5 min between washing steps.
3. Block with 1% (w/v) BSA in PS-solution for 1 h at room temperature.
4. Wash as before.
5. Add the crude plant extracts (*see* **Subheading 3.1.**) in serial dilutions to microtiter wells and incubate for 2 h at 37°C (*see* **Note 6**).
6. Wash as before.
7. Add alkaline phosphatase-conjugated goat antimouse antibodies (0.15 µg/mL EMA buffer) and incubate for 1 h at 37°C.
8. Wash as before.
9. Detect bound antibodies by adding the substrate *p*-nitrophenyl-phosphate (1 mg/mL substrate buffer). Incubate for 30–60 min at 37°C and measure substrate hydrolysis at 405 nm.

3.2.2. ELISA-II

1. Coat microtiter plates with polyclonal chicken anti-TMV serum (0.5 µg/mL coating buffer) overnight at 4°C (*see* **Note 7**).
2. Wash and block as described (*see* **Subheading 3.2.1.**, **steps 2** and **3**).
3. After washing add TMV (1 µg/mL sample buffer).
4. Continue as described (*see* **Subheading 3.2.1.**, **steps 4–9**).

3.2.3. ELISA-III

1. Coat microtiter plates, wash, block, and add TMV as described (*see* **Subheading 3.2.2.**, **steps 1–3**).
2. After washing, add crude plant extracts (*see* **Subheading 3.1.**) in serial dilutions to microtiter wells and incubate for 2 h at 37°C.
3. Wash as before.
4. Add rabbit-derived polyclonal anti-idiotypic antibodies (1 µg/mL EMA buffer) and incubate for 1 h at 37°C.
5. Wash as before.
6. Add alkaline phosphatase-conjugated goat antirabbit antibodies (0.15 µg/mL EMA buffer) and incubate for 1 h at 37°C.
7. Continue as described (*see* **Subheading 3.2.1.**, **steps 8** and **9**).

3.3. Preparation of Intercellular Washing Fluid

Recombinant antibodies containing a signal peptide are secreted to the apoplasm, which can be analyzed by recovery of the intercellular washing fluid (IWF). IWF of tobacco leaves is prepared essentially as described by Deverall and Deakin *(18)*.

1. Cut leaves into strips of 2–3 cm and rinse the segments in water for 15 min (*see* **Note 8**).

2. Vacuum infiltrate for three 5-min periods, interrupted by 30-s intervals at room temperature with IWF buffer (*see* **Notes 9** and **10**).
3. Dry leaf material between paper towels and roll into the barrel of a 10-mL plastic syringe (*see* **Note 11**). Place syringe in a 50-mL Falcon tube (Greiner, Germany) and centrifuge at 300*g* for 30 min at room temperature. IWFs are collected at the bottom of the Falcon tube.
4. Analyze IWF immediately by ELISA (or immunoblotting) or store at –80°C (*see* **Note 12**).

3.4. Analysis of Antibody Secretion in Protoplast Cultures

An alternative approach to test the secretion of antibodies is the analysis of protoplast cultures.

1. Prepare protoplasts from tobacco leaves as described in Chapter 13. Resuspend the protoplasts in K3 medium and adjust the suspension to 1×10^6 cells/mL.
2. Transfer 5.5 mL protoplast solution into a sterile Petri dish and incubate for 5 d at 24°C, while gently shaking (100 rpm) (*see* **Note 13**).
3. Add 1 vol of W5 solution to the suspension, transfer to a 10-mL centrifugation tube, pellet the cells at 100*g* for 10 min at room temperature, and keep the supernatant separate from the pellet (*see* **Note 14**).
4. Resuspend the cell pellet in 500 μL W5 solution. Lyse the protoplasts by freezing and thawing and pipeting several times up and down.
5. Centrifuge at 14,000*g* for 15 min at room temperature to pellet the cell debris.

Antibodies obtained either from protoplast pellet or culture supernatant are tested for antigen-binding in ELISA assays (*see* **Note 15**). Comparative analysis of antibody amounts in the pellet and the culture supernatant allows quantification of secreted rAbs.

3.5. Electrophoresis and Blotting

Standard methods for gel electrophoresis *(19)* and blotting have been described. Procedures for gel apparatus assembly and exact running conditions (voltage, running/transfer time, and so on) depend on the equipment being used and the protein(s) under investigation. Here we describe only modifications and improvements we have found to be useful for analysis of rAbs.

3.5.1. SDS-PAGE

Most proteins can be separated satisfactorily by discontinuous SDS-PAGE in a Tris-glycine buffer system under reducing conditions *(19)*.

1. Prepare the electrophoresis system and gels according to manufacturer's instructions (*see* **Note 16**). Dilute the running buffer (*see* **Subheading 2.5.**) 1:5-fold prior to use.

2. Adjust the protein concentration of the sample as required for subsequent application(s), add an equal volume of sample buffer, boil for 5 min (*see* **Note 17**), and put on ice.
3. Load samples and protein marker onto the gel and perform electrophoresis.
4. Stop electrophoresis when bromphenol blue dye reaches bottom of gel (typically, a run takes 45–60 min for two Mini-Protean II gels).
5. Disassemble gel sandwich and commence with protein staining (*see* **Note 18**) or blotting.

3.5.2. Immunoblot

The transfer of electrophoretically separated proteins onto porous membranes, mostly nitrocellulose, makes them accessible for sensitive and specific detection reactions, in particular, those utilizing specific antibodies or lectins directed against the protein of interest (*see* **Notes 19–21**).

1. Upon completion of gel electrophoresis, transfer the proteins to nitrocellulose or PVDF membrane, according to manufacturer's instructions (*see* **Note 22**).
2. Incubate membrane in Poinceau S solution for 5 min to check transfer efficiency. Prior to subsequent specific detection of transferred proteins, wash membrane several times with distilled water to remove stain.
3. Incubate membrane overnight at room temperature in 100 mL blocking solution, while gently shaking.
4. Wash membrane 3×10 min with PBS-T.
5. Dilute AP-labeled antibody in PBS-T according to manufacturer's instructions. Incubate membrane at room temperature for 2 h in AP-labeled antibody solution, while gently shaking.
6. Wash membrane 3×10 min with PBS-T and twice for 5 min in AP buffer.
7. Incubate in 0.1 mL of NBT/BCIP stable mix/cm^2 membrane at room temperature in the dark, without shaking. Stop the staining reaction by transferring the membrane into distilled water as soon as clear bands become visible, to avoid background staining.
8. Dry membrane between Whatman filter papers and store at room temperature in the dark. Photograph or scan blot for documentation as soon as possible.

3.5.3. Isoelectric Focusing (IEF)

The principle of isoelectric focusing is to set up a stable pH gradient in a gel and, under the influence of an electric field, allow proteins to migrate through this gradient to the point where the pH matches their IEP (*see* **Note 23**). IEF protocols, even more than SDS-PAGE, depend on the electrophoretic system and the type of gel that is used; therefore, only some general hints can be given here.

1. Initial experiments require a broad-range carrier ampholyte (pH range 3–10).
2. Prepare a polyacrylamide IEF gel with 5–6% T and 3% C.

3. All samples must be carefully desalted by dialysing against bidistilled H_2O.
4. Efficient cooling is essential for high resolution.
5. Good visualization is achieved by loading 1 μg of antibody per well and subsequent silver staining of the proteins.

3.5.4. Lectin Blot

The use of lectins *(15)* provides a fast and easy method to gain information about glycosylation patterns (*see* **Note 24**). Details of oligosaccharide structure are an important determinant of the physical and biological properties of a glycoprotein and must be studied in detail, especially if such proteins are used for therapy or diagnosis. The following experiment is carried out using biotinylated *R. communis* agglutinin (RCAI), a plant lectin with binding specificity for terminal galactose and N-acetylgalactosamine. A wide range of lectins with different specificities are commercially available and can be used in a similar fashion (*see* **Note 25**).

1. Perform SDS-PAGE using 1 μg of murine and plant-derived antibodies, as described (*see* **Subheading 3.5.1.**).
2. Transfer separated proteins onto nitrocellulose membrane, as described (*see* **Subheading 3.5.2.**).
3. Rinse membrane for 3 × 10 min with 1X PBS.
4. Block remaining binding sites with 2% (w/v) BSA in PBS-T for 1 h at room temperature.
5. Incubate with RCA1-biotin conjugate (5 μg/mL in 1X PBS) for 1 h at room temperature (*see* **Notes 26 and 27**).
6. Wash for 3 × 10 min with PBS-T.
7. Incubate with 1:3000-fold diluted alkaline phosphatase-labeled streptavidin for 1 h at room temperature.
8. Continue as described (*see* **Subheading 3.5.2., steps 6–8**).

3.6. Assays for Protection of Transgenic Plants Against Viral Infection

The functionality of plant-derived antibodies in vivo is analyzed by inoculating transgenic tobacco plants with TMV and determining the reduction of infection events.

1. Grow plants in a growth chamber at 23°C and 10,000 lx for 16 h, and at 20°C in the dark for 8 h (*see* **Note 28**).
2. Dust tobacco leaves with carborundum and inoculate with 200 μL TMV suspension (1 μg purified TMV/mL 1X PBS) per leaf using a glass spatula (*see* **Notes 29 and 30**).
3. Rinse leaves with tap water and incubate inoculated plants as described (*see* **Subheading 3.6., step 1**).

4. Count local lesion numbers 3–5 d after inoculation and calculate the percentage inhibition of infectivity (I) using the following equation:

$$I\ (\%) = [\text{lesion number (transgenic plant)}]/[\text{lesion number (nontransgenic plant)}]$$

4. Notes

1. All chemicals used for electrophoresis must be of the highest quality possible to ensure good results.
2. **Caution: These substances are neurotoxic!** Wear a mask and gloves while handling the chemicals and unpolymerized solutions in a fume hood.
3. Extracts of total soluble proteins can be stored for at least 3 mo at 4°C without significant decrease of antibody activity. For long-term storage, place extracts at –80°C. Once thawed, extracts should be stored at 4°C.
4. The concentration of soluble proteins in plant extracts can be determined with the Bio-Rad Protein Assay, using serial dilutions of BSA as a standard. Adaptation of the protein assay to microtiter plates *(21)* simplifies and speeds up the estimation of the protein concentration.
5. In each step, a volume of 100 µL is added to each well, except the blocking step, when 150 µL is used.
6. Extracts from nontransformed plants are used as a negative control and affinity-purified murine TMV-specific antibodies are used as a positive control. Always prepare dilutions using protein-extraction buffer.
7. Binding of TMV to the microtiter plate is performed indirectly through a polyclonal serum, because TMV dissociates into monomers in the alkaline-coating buffer *(22)*.
8. Leaf segments must be thoroughly rinsed with water to remove cytoplasmic contaminants from the cut area. Leaves smaller than 20 cm^2 are left intact and therefore washing is not mandatory.
9. Leaf material should be trapped below the IWF buffer with a weighted mesh during vacuum infiltration, to avoid air influx.
10. Vacuum infiltration can also be carried out at lower temperature to avoid loss of protein activity, and to reduce protease digestion.
11. After infiltration, the leaf tissue is easily damaged and should be handled with care.
12. IWF containing full-size antibodies can be concentrated using Centricon-100 units (Amicon). Fab and scFv fragments should be concentrated using Centricon-50 or Centricon-10 units, respectively (*see* Chapter 13).
13. The amount of secreted proteins that can be detected in the medium mainly depends upon the stability of the protein and, in the case of larger proteins, on the rapidity of cell wall resynthesis, because proteins larger than 20–30 kDa are retarded by cell walls. Furthermore, prolonged incubation can lead to an increase of cytosolic contaminants caused by cell death. Therefore, appropriate incubation times should be determined for each secreted protein and source of protoplasts from the corresponding transgenic plant variety.

14. The occurrence of cytosolic contaminants in the culture supernatant caused by damage of protoplasts is easily monitored by assaying for glucose-6-phosphate dehydrogenase activity.

15. Low amounts of antibodies in the supernatant require an additional concentration step using Centricon units (*see* **Note 12**).

16. For the separation of full-size antibodies, prepare gels with a total monomer concentration (T) of 12% (separating gel) and 4% (stacking gel). Alternatively, precast gradient gels (T = 4–10%) can be used.

17. Samples containing imidazole (e.g., for purification of His6-tagged scFvs) or urea (for isolation of total protein) should not be boiled, but treated at 50°C for 5 min, instead.

18. The method used for staining separated polypeptide chains depends on the desired assay sensitivity: Coomassie brilliant blue G-250 staining requires approx 200 ng of protein per band, while silver staining is 10–50 times more sensitive.

19. Plant-derived full-size rAbs and Fab fragments of murine origin are directly detected using alkaline phosphatase-labeled antimouse antibodies with specificity for constant domains of murine heavy or light chains (anti-H+L- or Fc-specific for murine IgG_1, IgG_{2a}, and IgG_{2b}).

20. For detection of scFv-fragments (or more detailed characterization of rAbs, described in **Note 19**), rabbit-derived polyclonal anti-idiotypic antibodies (raised by immunization with murine Fab fragments), V_L- and V_H-domain specific antibodies binding to framework four *(23)* or tag-specific antibodies (e.g., anti-c-myc or His6) *(12)* have to be used. These subsequently can be detected with alkaline phosphatase-labeled antirabbit antibodies.

21. Antibodies binding to plant-specific sugar epitopes *(24)* can be used for comparative analysis of *N*-glycosylation in plant-derived full-size rAbs and murine mAbs, in addition to lectin blots.

22. Always use forceps and wear gloves while handling the membrane, to avoid protein contamination.

23. Determination of the IEP of purified antibodies is important for several reasons, including: It is an important parameter for subsequent modification procedures (e.g., biotinylation); and antibodies, even of monoclonal origin, show a considerable amount of microheterogeneity, resulting in several distinct bands in the IEF gel over a range of 0.3–0.5 pH units. This pattern is reproducible for each monoclonal antibody and thereby provides an additional means for comparing, e.g., a murine mAb and its plant-derived counterpart. Furthermore, this analysis is an approved method during quality control for ensuring homogeneity of native and recombinant proteins over an extended period of cell cultivation.

24. For example, disease-associated changes in carbohydrate structures on human IgG *(25)* have been analyzed and glycosylation of plant-derived antibodies *(13)* has been investigated by lectin-binding assays. The exact elucidation of carbohydrate structures on proteins requires nuclear magnetic resonance (NMR), glycosequencer, or high-performance anion-exchange chromatography-pulsed

amperometric detection (HPA-EC-PAD) equipment. However, lectin blots provide sufficient data for comparative studies.

25. Some lectins require divalent cations for their activity and therefore need buffers modified according to manufacturer's instructions.

26. **Caution: RCAI is a very toxic biochemical. Always wear a mask and gloves while handling this lectin.** Do not allow to dry. Use 10% bleach for all subsequent clean-up. Soak all glassware in bleach after use.

27. When testing more than one lectin at a time, we recommend use of preparative gels prior to blotting. Upon completion of protein transfer, cut the membrane into strips and subsequently incubate these with different biotinylated lectins.

28. Infectivity assays are carried out with transgenic plant lines derived from *N. tabacum* cv. Xanthi-nc. These plants contain the N gene and respond hypersensitively upon virus infection. The initially infected cells and those adjacent become necrotic, leading to formation of clearly visible local lesions. Note: The N gene is not expressed above 28°C, and therefore the virus spreads systemically at higher temperatures.

29. The optimal TMV concentration should be determined for each new TMV preparation. Approximately 100–200 local lesions should develop per leaf of a nontransgenic plant, to guarantee a valid evaluation. Inoculate at least 10 plants (transgenic and nontransgenic) at the five-leaf stage to obtain statistically relevant data.

30. Some researchers use Sørensen phosphate buffer (SPB) instead of phosphate buffered saline (PBS) for inoculation of tobacco plants with TMV. However, we have not observed any difference while using PBS instead of SPB.

References

1. De Zoten, G. A., Penswick, J. R., Horisberger, M. A., Ahl, P., Schultze, M., and Hohn, T. (1989) The expression, localization, and effect of a human interferon in plants. *Virology* **172,** 213–222.

2. Sijmons, P. C., Dekker, B. M. M., Schrammeijer, B., Verwoerd, T. C., Van den Elzen, P. J. M., and Hoekema, A. (1990) Production of correctly processed human serum albumin in transgenic plants. *Bio/Technology* **8,** 217–221.

3. Hiatt, A., Cafferkey, R., and Bowdish, K. (1989) Production of antibodies in transgenic plants. *Nature* **342,** 469–470.

4. Stieger, M., Neuhaus, G., Momma, T., Schell, J., and Kreuzaler, F. (1990) Self assembly of immunoglobulins in the cytoplasm of the alga *Acetabularia mediterranea. Plant Sci.* **73,** 181–190.

5. Voss, A., Niersbach, M., Hain, R., Hirsch, H. J., Liao, Y. C., Kreuzaler, F., and Fischer, R. (1995) Reduced virus infectivity in *N. tabacum* secreting a TMV-specific full-size antibody. *Mol. Breeding* **1,** 39–50.

6. Düring, K., Hippe, S., Kreuzaler, F., and Schell, J. (1990) Synthesis and self-assembly of a functional monoclonal antibody in transgenic *Nicotiana tabacum. Plant Mol. Biol.* **15,** 281–293.

7. De Neve, M., De Loose, M., Jacobs, A., Van Houdt, H., Kaluza, B., Weidle, U., Van Montagu, M., and Depicker, A. (1993) Assembly of an antibody and its derived antibody fragment in *Nicotiana* and *Arabidopsis*. *Transgenic Res.* **2,** 227–237.
8. Owen, M., Gandecha, A., Cockburn, B., and Whitelam, G. (1992) Synthesis of a functional anti-phytochrome single-chain Fv protein in transgenic tobacco. *Bio/Technology* **10,** 790–794.
9. Tavladoraki, P., Benvenuto, E., Trinca, S., De Martinis, D., and Galeffi, P. (1993) Transgenic plants expressing a functional single-chain Fv antibody are protected from virus attack. *Nature* **366,** 469–472.
10. Firek, S., Draper J., Owen, M. R. L., Gandecha, A., Cockburn, B., and Whitelam, G. C. (1993) Secretion of a functional single-chain Fv protein in transgenic tobacco plants and cell suspension cultures. *Plant Mol. Biol.* **23,** 861–870.
11. Plückthun, A. (1991) Antibody engineering. *Curr. Opinion Biotechnol.* **2,** 238–246.
12. Winter, G. and Milstein, C. (1991) Man-made antibodies. *Nature* **349,** 293–299.
13. Hiatt, A. and Ma, J. K.-C. (1993) Characterization and applications of antibodies produced in plants. *Intern. Rev. Immunol.* **10,** 139–152.
14. Ma, J. K.-C., Hiatt, A., Hein, M., Vine, N. D., Wang, F., Stabila, P., Van Dolleweerd, D., Mostov, K., and Lehner, T. (1995) Generation and assembly of secretory antibodies in plants. *Science* **268,** 716–719.
15. Parekh, R. B. and Patel, T. P. (1992) Comparing the glycosylation patterns of recombinant glycoproteins. *TIBTECH* **10,** 276–280.
16. Clark, M. F. and Adams, A. N. (1977) Characteristics of the microplate method of enzyme-linked immunosorbent assay for the detection of plant viruses. *J. Gen. Virol.* **34,** 475–483.
17. Dore, I., Weiss, E., Altschuh, D., and Van Regenmortel, M. H. V. (1988) Visualisation by electron microscopy of the location of Tobacco Mosaic Virus epitopes reacting with monoclonal antibodies in enzyme immunoassay. *Virology* **162,** 279–289.
18. Deverall, B. J. and Deakin, A.-L. (1985) Assessment of *Lr20* gene-specificity of symptom elicitation by intercellular fluids from leaf rust-infected wheat leaves. *Physiol. Plant Pathol.* **27,** 99–107.
19. Laemmli, U. K. (1970) Cleavage of structural proteins during the assembly of the head of bacteriophage T4. *Nature* **227,** 680–683.
20. Lis, H. and Sharon, N. (1986) Lectins as molecules and as tools. *Ann. Rev. Biochem.* **55,** 35–67.
21. Redinbaugh, M. G. and Campell, W. H. (1985) Adaptation of the dye-binding protein assay to microtiter plates. *Anal. Biochem.* **147,** 144–147.
22. Van Regenmortel, M. H. V. (1986) Tobacco mosaic virus: antigenic structure, in *Plant Viruses*, vol. 2 (Van Regenmortel, M. H. V.), Plenum, New York, pp. 79–104.
23. Berry, M. J. and Davis, P. J. (1994) Assay and purification of Fv fragments in fermenter cultures: design and evaluation of generic binding. *J. Immunol. Methods* **167,** 173–182.

24. Faye, L., Gomord, V., Fichette-Lainé, A.-C., and Crispeels, M. J. (1993) Affinity purification of antibodies specific for Asn-linked glycans containing α1→3 fucose or β1→2 xylose. *Anal. Biochem.* **209,** 104–108.
25. Sumar, N., Bodman, K. B., Rademacher, T. W., Dwek, R. A., Williams, P., Parekh, R. B., Edge, J., Rook, G. A. W., Isenberg, D. A., Hay, F. C., and Roitt, I. M. (1990) Analysis of glycosylation changes in IgG using lectins. *J. Immunol. Methods* **131,** 127–136.

11

Production of Recombinant Antibodies in Plant Suspension Cultures

Jürgen Drossard, Yu-Cai Liao, and Rainer Fischer

1. Introduction

Monoclonal antibodies (MAbs) *(1)*, because of their binding specificity and stability both in vivo and in vitro, are extremely useful tools in medicine, biology, and organic chemistry. The combination of hybridoma technology and recombinant DNA techniques have given access, not only to full-size molecules, but also to various recombinant antibody fragments (RAbs) and fusion proteins *(2)*, thus broadening the range of possible applications. Recent improvements in heterologous gene expression systems *(3)* and the development of phage display technologies *(2)* have made it possible to design and express RAbs against almost any given antigen, and to fine-tune these molecules with respect to improved performance. Furthermore, incorporation of affinity tags has simplified the purification of heterologously expressed recombinant proteins *(4,5)*.

Although the production of MAbs and RAbs has become a standard technique, the ideal expression system for all purposes has yet to be developed. Currently, bacteria *(6)* and mammalian cell cultures *(7)* are the best-established systems; yeast and baculovirus-infected insect cells play a minor role *(8)*. However, all these systems have limitations. Cultivation of mammalian cells is delicate and requires expensive equipment and media. In downstream processing, care must be taken to remove oncogenic sequences or viral contaminations, if in vivo use is intended. Additionally, the use of animals as a source of MAbs is becoming more and more limited because of legal and ethical restrictions. Bacteria do not allow the production of glycosylated full-size antibodies and co-purified endotoxins are difficult to eliminate. Furthermore, recombinant

From: *Methods in Biotechnology, Vol. 3:*
Recombinant Proteins from Plants: Production and Isolation of Clinically Useful Compounds
Edited by: C. Cunningham and A. J. R. Porter © Humana Press Inc., Totowa, NJ

proteins often form inclusion bodies, making labor- and cost-intensive in vitro refolding necessary.

Based on most recent requirements for bulk quantities of functional, active recombinant proteins, alternatives to expression in microbes and animal cells are required. This applies in particular to RAbs needed for bioprocessing, therapy, and diagnosis, which have become major biotechnical markets in recent years *(9)*.

Transgenic plants *(10)* offer a useful alternative approach for the mass production of recombinant proteins, including antibodies *(11–13)*, because of their enormous biomass build-up *(14)*. Plants are easy to grow and, in contrast to bacteria or animal cells, their cultivation does not require special equipment or chemicals and largely avoids ethical problems. Furthermore, plants can be cultivated relatively cheaply as calli or cell suspensions under good laboratory practice and good manufacturing practice conditions *(9)*. Heterologous proteins can accumulate to high levels in plant cells and plant-derived antibodies have been shown to be virtually indistinguishable from those produced by hybridoma cells *(7,12)*. Protein synthesis, secretion, and folding, as well as posttranslational modifications such as signal peptide cleavage, disulfide-bond formation, and initial glycosylation, are very similar between plant and animal cells. However, co-purification of host-derived antibodies, as well as blood-borne pathogens, is excluded during downstream processing of recombinant proteins from plants. This in turn should result in a more homogenous and safe recombinant product.

Although plants have been used for the production of several mammalian proteins, to date no reports regarding large-scale production and downstream processing of plant-derived recombinant proteins are available. We describe here a protocol for the establishment of suspension cultures from transgenic *Nicotiana tabacum* cv. Petit Havana SR1 plants for production and purification of an anti-tobacco mosaic virus (TMV) antibody. Expression of cloned full-size cDNAs of heavy and light chains resulted in assembling of functional antibodies that were exported to the intercellular space *(12)*. TMV-specificity and affinity of the plant-produced antibody were not altered in comparison to the original murine MAb. Suspension cell lines were established from elite plants that produced up to 15 mg of RAb per kilogram wet weight under nonoptimized culture conditions. These suspension cell lines were used to analyze physicochemical parameters of plant cell cultures on the shake-flask level and to study the effects of varying media supplements (phytohormones, vitamins, and essential/nonessential amino acids). Furthermore, strategies for rapid and efficient purification were developed that preserved antibody activity. Scale-up was performed to a level of 5 kg of plant material per batch.

The purification scheme presented for full-size antibodies consists of two chromatographic steps that yield a product with a high degree of purity (>95%) because of the selective first step based on Protein A-affinity chromatography. The only contamination detectable by gel electrophoresis and subsequent protein staining or immunoblotting has been identified as a fragment of the antibody heavy chain itself. Critical (e.g., therapeutic) applications for plant-derived RAbs would therefore require at least one additional purification step, based on ion exchange, hydrophobic interactions, or hydroxyapatite. The purification of RAb fragments requires different approaches, because these molecules usually do not bind Protein A (or Protein G). Such molecules can be purified by immobilized metal-affinity chromatography (IMAC), using a C-terminal His_6 affinity tag attached to the recombinant protein *(4,5)*.

Tobacco suspension cultures were established as a model system to demonstrate their potential for production and downstream processing of full-size RAbs. Based on this experience, modifications and improvements are currently under investigation, including: expression of other murine isotypes ($IgG_{2a}/_\kappa$ and $IgG_1/_\kappa$); expression of RAb fragments (F[ab']$_2$, Fab, and scFvs [*15*]); expression of RAbs with diagnostic and/or therapeutic interest; increase expression at the genetic level by modifying regulatory elements in the plant expression vector; expression in different plant species with faster biomass production and reduced levels of secondary metabolites; and pilot-scale production by using appropriate fermentation technologies (e.g., stirred tank reactor or air-lift-fermenter) *(8,16)*.

2. Materials

2.1. Plant Material

Nicotiana tabacum cv. Petit Havana SR1 plants are grown as sterile axenic shoot cultures as described in **Subheading 3.1.** This cultivar is easily propagated also as callus or in suspension (*see* **Note 1**). Leaf disks from plants grown in soil can be used for transformation and initiation of callus cultures. However, sterile grown plants are preferred because of reduced risk of contamination and faster growth.

2.2. Bacteria

Escherichia coli and *Agrobacterium tumefaciens* strains are transformed with the respective binary plant expression vectors and cultivated as described elsewhere *(12,17,18)*.

2.3. Special Laboratory Equipment and Supplies

1. Plant-culture room with illumination system.
2. Laminar flow hood (Flow Laboratories, Milano, Italy).

3. Steel punch $\varnothing$ 5 mm.
4. 30–37°C incubator (Heraeus, Düsseldorf, Germany).
5. Whatman 3MM filter paper (Whatman, Clifton, NJ).
6. Miracloth (Calbiochem, La Jolla, CA).
7. Rotary shaker (New Brunswick, Edison, NJ).
8. ProVario3 cross-flow filtration system (Filtron, Karlstein, Germany).
9. Econo low-pressure chromatography system (Bio-Rad, Munich, Germany).

2.4. Tissue Culture Chemicals and Media Composition

All tissue culture basal media and additives are obtained from Sigma, as "plant cell culture tested" quality, unless otherwise stated.

2.4.1. Chemicals

1. Murashige and Skoog (MS) basal salts with minimal organics (MSMO) *(19)*.
2. Gamborg's B-5 basal medium with minimal organics (B-5) *(20)*.
3. Sucrose.
4. 2,4-Dichlorophenoxyacetic acid (2,4-D).
5. Dimethyl sulfoxide (DMSO).
6. α-Naphthaleneacetic acid (NAA).
7. Kinetin.
8. Agar.
9. Kanamycin sulfate.
10. Claforan (Hoechst, Frankfurt, Germany).
11. MEM amino acid solution.
12. MEM nonessential amino acid solution.

2.4.2. Stock Solutions and Composition of Media

All media are prepared immediately before use with ultrapure water (>18 MΩ/cm) and autoclaved or filter-sterilized (0.2 µm); heat-sensitive components (antibiotics) are added after cooling to below 50°C. All quantities given below are for 1 L of medium. We use 20 mL of MS medium for preparing Petri dishes or 100 mL for culture pots. Stock solutions (*see* **Note 2**) are filter-sterilized and stored in aliquots at 4°C or –20°C, according to manufacturers' instructions.

1. Kanamycin-sulfate stock: 100 mg/mL in H_2O.
2. Claforan stock: 200 mg/mL in H_2O.
3. 2,4-D stock: 2 mg/mL in DMSO.
4. Kinetin stock: 1 mg/mL in H_2O.
5. NAA stock: 1 mg/mL in H_2O.
6. $CaCl_2$ stock: 1*M* in H_2O.
7. MS: 4.62 g MSMO, 20 g sucrose, 8 g agar; before adding agar, adjust pH to 6.0 with 1*N* KOH (*see* **Note 3**).
8. MSH: MS plus 1 mL NAA stock, 0.2 mL kinetin stock.

9. MSHC: MSH plus 2.5 mL Claforan stock.
10. MSHK: MSH plus 0.75 mL kanamycin-sulfate stock.
11. MSHKC: MSHK plus 2.5 mL Claforan stock.
12. MSHKCr: MSHK plus 0.25 mL Claforan stock.
13. Suspension culture medium: 3.2 g B-5, 20 g sucrose, 1 mL 2,4-D stock, 0.1 mL kinetin stock, 0.75 mL kanamycin-sulfate stock; adjust pH to 5.5 with 1N KOH (*see* **Note 4**).

2.5. Buffers and Other Solutions

All buffers are made up from components at the concentrations specified below. Chemicals are analytical grade or molecular biology grade, unless otherwise mentioned. Aqueous solutions are prepared with deionized water; chromatography buffers are filter-sterilized (0.2 μm) and degassed before use.

1. Extraction buffer: 200 mM Tris-HCl, pH 8.0, 100 mM NaCl, 10 mM EDTA, 10 mM DTT, 0.15% (w/v) pectinase (technical grade, ICN); prepare fresh buffer before use (*see* **Note 5**).
2. PBS: 137 mM NaCl, 2.7 mM KCl, 8.1 mM Na_2HPO_4, 1.5 mM KH_2PO_4.
3. Binding buffer: PBS plus 500 mM NaCl.
4. Elution buffer 1: 100 mM citrate, pH 5.0.
5. Elution buffer 2: 100 mM citrate, pH 3.0.

2.6. Chromatography Media

1. Protein A Hyper D (BioSepra, Frankfurt, Germany).
2. Sephacryl S300 HR (Pharmacia, Freiburg, Germany).

3. Methods

Manipulation of plants, leaf disks, calli, and suspension cultures must be performed in a laminar flow hood, taking standard laboratory precautions against microbial contamination using sterilized tools and equipment. Glassware must be cleaned with bidestilled water only (no detergents), covered with aluminium foil (*see* **Note 6**), and sterilized for 4 h at 180°C.

3.1. Cultivation Conditions

Unless otherwise stated, sterile axenic shoots, leaf disks, and calli are cultivated at 24°C, 3000 lx, with a 16 h daily photoperiod. Suspension cultures are maintained at lower light intensities (500 lx).

3.2. Leaf Disk Transformation

This protocol provides a reliable and easy way for transferring T-DNA carrying the gene of interest into plants. Simultaneously, by cultivating leaf disks of untransformed plants on MSH-medium, nontransgenic callus cultures are obtained as negative controls. Additional methods for transformation of plant cells or tissue are described in Chapter 13.

3.2.1. Preparation of Sterile Axenic Shoot Cultures

1. Wash tobacco seeds briefly with sterile H_2O in a Petri dish.
2. Surface-sterilize seeds in 70% (v/v) ethanol for 5 min.
3. Wash seeds once with sterile H_2O.
4. Transfer seeds to Petri dishes containing MS medium (10 seeds per dish).
5. Seal Petri dishes with parafilm (*see* **Note 7**).
6. Incubate dishes, as described (*see* **Subheading 3.1.**); seeds will germinate after approx 2 wk.
7. As soon as the seedlings touch the lids of the Petri dishes, transfer them to culture pots containing MS medium (three seedlings per pot) and continue cultivation until leaves are used for leaf disk transformation or the initiation of wild-type callus cultures used as negative controls.

3.2.2. Preparation of Leaf Disks

1. After 3–4 wk of axenic shoot growth in the culture pots, remove fully developed, healthy looking leaves with a scalpel and forceps, place them (one at a time) in a glass Petri dish containing an autoclaved Whatman 3MM filter paper, and cut out leaf disks using sterile punches.
2. Remove the leaf disks with forceps and place them with the lower epidermis toward solidified media as follows:
   a. Approx 250 disks on 10 MSH dishes for infection with *A. tumefaciens* (*see* **Subheading 3.2.3.**).
   b. 20 Disks on two MSHC dishes as regeneration control.
   c. 20 Disks on two MSHKCr dishes as selection control. **Dishes b and c are not infected with *Agrobacteria*!**
3. Incubate dishes *b* and *c* as described. Disks on dishes b should regenerate within 2–3 wk, otherwise the concentration of phytohormones is not correct; disks on dishes c should turn yellow within the same time, otherwise the kanamycin concentration is not correct or the kanamycin is inactive.

3.2.3. Infection with A. tumefaciens

1. Add 5 mL of a recombinant *Agrobacteria* overnight culture into each of dishes *a*. (*see* **Subheading 3.2.2.**). Ensure good contact between bacteria and leaf disks. Incubate for 2 min at room temperature.
2. Carefully remove the *Agrobacteria* suspension using a Pasteur pipet.
3. Seal dishes with parafilm and incubate for 2 d.
4. Incubate for 1 d at 30°C in an incubator with a 25W bulb, 16 h photoperiod.

3.2.4. Selection of Transgenic Tissue

1. Place all infected leaf disks in a tea sieve and wash twice with bidestilled water for 5 min, while shaking them gently.
2. Dry disks briefly between sterile Whatman 3MM filter paper.

3. Place disks on MSHC dishes (with the lower epidermis toward the agar), seal, and continue incubation at 30°C for another 3 d.
4. Transfer disks onto MSHKCr dishes (10 disks per dish) for selection. Cultivate at 24°C, 3000 lx with 16 h illumination from now on. After approx 2 wk, callus tissue will begin to form, but shoots will require at least another 2 wk for development (*see* **Note 8**).

3.3. Propagation of Transgenic Calli

1. Remove approx 0.3 cm³ pieces of callus tissue using a scalpel and forceps and transfer them onto fresh MSHKCr dishes. Try to transfer friable, dedifferentiated tissue only.
2. Subcultivate pieces of about 1 cm³ every 4–5 wk. Omit Claforan after the first round of subcultivation.

3.4. Regeneration of Transgenic Plants

After leaf disk transformation, whole plants can be regenerated for analysis and breeding (*see* **Notes 9** and **10**).

1. When internodes have formed on shoots (*see* **Subheading 3.2.4.**), dissect the shoots from the calli and transfer them to fresh MSHKCr dishes. Cut off shoots right against the callus, but without co-transferring callus tissue. Cultivation is continued at 24°C as described in **Subheading 3.1.** Transgenic shoots will survive, but wild-type shoots will bleach.
2. As soon as the shoots touch the lids of the dishes, transfer them to MSKCr pots (three plants per pot) for root induction. Cultivation conditions are unchanged. Roots should form within 2 wk, otherwise accidentally co-transferred callus tissue may have to be removed.
3. In order to obtain seeds, cultivate the transgenic plants for another 3–4 wk, before transferring into soil (*see* **Note 11**).

3.5. Establishment of Suspension Cultures

Suspension cultures are established using friable, healthy transgenic callus tissue for inoculating a small volume of suspension-culture medium.

1. Under sterile conditions, transfer at least 3–4 g of callus tissue into a 250 mL Erlenmeyer flask containing 50 mL of suspension-culture medium (*see* **Note 12**).
2. Incubate for 2 wk on a rotary shaker (120 rpm), as described (*see* **Subheading 3.1.**).
3. Using a serological pipet, transfer 10 mL of the suspension into a new 250 mL flask; dilute 1:5 with suspension-culture medium and continue plant cell cultivation.
4. Repeat this process weekly; after approximately three rounds of subcultivation, a stable suspension culture consisting of single cells and small aggregates will have been established.

3.6. Batch Production of Cell Suspension in Shake Flasks

The procedure described below yields 10 L of plant cell suspension, resulting in 3–4 kg wet cell weight after filtration. Amino acid supplementation significantly increases production of full-size RAbs (from ~15 mg/kg to more than 50 mg/kg RAb in our cell line, P9 [12]). Addition of calcium chloride is necessary to facilitate amino acid uptake by the suspension cells. Cell suspensions needed for subcultivation are not supplemented with $CaCl_2$ or amino acids (5–6 flasks are prepared in this way with every batch).

1. Prepare suspension culture medium as described **in Subheading 2.4.2.**
2. Add 1 mL of $CaCl_2$ stock/L of medium (*see* **Note 13**).
3. Prepare 20 1-L Erlenmeyer flasks by diluting 100 mL of densely grown cell suspension with 400 mL of fresh suspension-culture medium.
4. Incubate for approx 10 d, until a settled cell volume (SCV) of 60–70% is reached. Determine the SCV by using a 100-mL graduated cylinder and allowing cells to settle for 30 min.
5. Add 1 mL of MEM amino acid solution and 0.5 mL of MEM nonessential amino acid solution per flask (*see* **Note 14**).
6. Shake for another 8–12 h (*see* **Note 15**).
7. Filter cell suspension through two layers of Miracloth using a vacuum pump; discard filtrate (*see* **Note 16**) and use the plant cells for antibody purification.

3.7. Purification of Recombinant Full-Size Antibodies

Isolation of recombinant proteins from transgenic plant cells faces unique problems, because of release of proteases and oxidizing agents (phenols, tannins) from subcellular compartments after cell disruption. Furthermore, all cell debris must be removed from the crude extract and a large volume has to be processed during the first downstream processing step. To circumvent these problems we have developed the protocol described below by using an enzymatic method for cell lysis, cross-flow filtration for clarification, and fast-flow affinity chromatography for rapid processing (21). Combination of these techniques speeds up processing significantly and minimizes product loss during initial purification (*see* **Note 17**).

1. Add 1 vol of extraction buffer (v/w) to the vacuum-filtrated cells and mix.
2. Pour the suspension into an appropriate vessel (e.g., 5-L Erlenmeyer flask for 2.5 L of suspension).
3. Shake at 120 rpm for 4 h at 37°C.
4. Filter through two layers of Miracloth as above; keep filtrate on ice from now on.
5. Filter through 0.2 μm membrane in cross-flow filtration system (*see* **Note 18**).
6. Meanwhile, equilibrate a prepacked Protein A Hyper D column (20 mL matrix) with 5 column volumes of PBS.

7. Pass filtrate (*see* **Subheading 3.7., step 5**) through column at ~300 cm/h using a low-pressure chromatography system.
8. Wash column with 5 column volumes of binding buffer.
9. Wash with elution buffer 1 until the baseline is reached (*see* **Note 19**).
10. Elute bound RAbs with 3 column volumes of elution buffer 2, pool antibody-containing fractions and immediately adjust to pH 7.2 with 1*M* Tris. Re-equilibrate affinity matrix with 5 column volumes of PBS immediately.
11. Pass antibody-containing fractions through Sephacryl S300 HR gel filtration column (3.4 × 80 cm) for further purification, removal of aggregates (*see* **Note 20**), and transfer of the purified product into a suitable buffer for final processing and storage.
12. Pool antibody-containing fractions, filter-sterilize, and use for characterization (*see* **Chapter 10**), or deposit aliquots at –20°C for long-term storage (*see* **Note 21**).

4. Notes

1. One of our suspension cell lines (P9) was established 2 yr ago and still shows production of recombinant full-size antibodies with almost no loss of productivity.
2. Some chemicals (e.g., kanamycin-sulfate) may require warming to 37°C and/or stirring before they are dissolved. Before use, briefly vortex all stock solutions.
3. The use of (expensive) glucose as a carbon source, although recommended by some authors, showed no significant antibody production increase in our hands.
4. The relatively high concentration of 2,4-D in suspension-culture medium is suitable for cultivation of *N. tabacum* cv. Petit Havana SR1 only. Other tobacco cultivars may require different concentrations of phytohormones. In particular, 2,4-D may have to be reduced.
5. The extraction buffer has been optimized for *N. tabacum* cv. Petit Havana SR1. It prevents oxidation and reduces degradation and inactivation of RAbs in crude plant cell extracts. Other tobacco varieties or plant species may require different buffer compositions.
6. The use of expensive covers for the culture vessels is not necessary. Aluminium foil is easily co-sterilized and provides a satisfactory safety level against contamination.
7. Axenic shoot cultures and calli grow slightly better without parafilm (better oxygen transfer), but the risk of contamination is increased.
8. In some cases, shoot induction is better if NAA concentration is reduced or the hormone is completely omitted. This should be tested in parallel.
9. Propagation of transformed calli results in heterologous cell populations with varying levels of RAb production as a result of position effects and copy number. Regenerated plants, however, are regarded as offspring derived from a single cell. Transgenic plants can be self-fertilized to produce homozygous offspring in order to increase production levels. Furthermore, crossing independent high producers results in double transgenic plants with further improved yields of recom-

binant protein, unless silencing downregulates the transgene *(22,23)*. Callus cultures can then be initiated from young leaves of these plants by using leaf disks cultivated on MSHK medium; seeds can be used for almost infinite storage of the genetic material at ambient temperatures.

10. Establishment of suspension cultures from regenerated transgenic plants often requires more than 1 yr, but working with primary transformed calli takes only 6–8 wk. Furthermore, we sometimes observed higher expression levels in suspension cultures derived from these calli compared to transgenic plants.

11. Seeds of *N. tabacum* cv. Petit Havana SR1 can be obtained after a relatively short regeneration time (4 mo). Other varieties may need more than 7 mo.

12. Do not pass callus tissue through a sieve for establishment of suspension cultures. Cells grow better after initial inoculation if the suspension contains some aggregates. Make sure that at least 3–4 g of callus material are transferred.

13. $CaCl_2$ should be added at least 24 h before amino acid supplementation. However, we also obtained comparable results by adding $CaCl_2$ immediately after autoclaving the medium and cooling below 50°C.

14. Essential and nonessential amino acids are added at 10 times lower concentrations than recommended by the manufacturer. Supplementation with higher concentrations showed adverse effects on suspension cells, sometimes resulting in complete loss of antibody production.

15. Avoid longer incubation, because antibody production will be reduced to original level within less than 24 h.

16. SR1 cells do not secrete full-size antibodies into the medium under the culture conditions described above. RAbs (e.g., scFv) are partly secreted but a substantial amount is also retained within the cell. Other tobacco cultivars may be different in this respect, so the filtrate should be tested for antibody content whenever a new cultivar or another plant species is used.

17. Addition of protease inhibitors other than EDTA, or of anti-oxidants such as PVPP, during the first downstream processing step is not necessary and does not eliminate cleavage of the heavy-chain product (*see* **Subheading 1.**). Furthermore, we do not recommend the use of PVPP, because it clogs filter membranes during clarification.

18. The use of cross-flow filtration is highly recommended for clarification, because conventional membrane filters (even with large cross-sectional areas) are quickly clogged by plant cell debris, even after time-consuming centrifugation of the crude extract at high speed. With cross-flow filtration, however, no centrifugation at all is necessary.

19. The elution at pH 5.0 reveals one peak at 280 nm. These fractions, however, do not contain antibody (probably pectinase or some other protein binding nonspecifically to the affinity matrix).

20. The cleaved heavy-chain product (*see* **Note 17**) tends to be associated with the affinity purified full-size RAbs and cannot be removed by gel filtration chromatography.

21. Store aliquots of purified recombinant antibodies at 4°C for subsequent characterization (ELISA, gel electrophoresis, IEF, Western Blot). Avoid repeated freeze–thaw cycles, as these will cause protein degradation and loss of activity.

References

1. Köhler, G. and Milstein, C. (1975) Continuous cultures of fused cells secreting antibody of predefined specificity. *Nature* **256,** 495–497.
2. Winter, G. and Milstein, C. (1991) Man-made antibodies. *Nature* **349,** 293–299.
3. Plückthun, A. (1991) Antibody engineering. *Curr. Opinion Biotechnol.* **2,** 238–246.
4. Ford, C. F., Suominen, I., and Glatz, C. E. (1991) Fusion tails for the recovery and purification of recombinant proteins. *Prot. Express. Purific.* **2,** 95–107.
5. Nygren, P. A., Stahl, S., and Uhlen, M. (1994) Engineering proteins to facilitate bioprocessing. *Tibtech* **12,** 184–188.
6. Skerra, A. (1993) Bacterial expression of immunoglobulin fragments. *Curr. Opinion Immunol.* **5,** 256–262.
7. Bebbington, C. R., Renner, G., Thomson, S., King, D., Abrams, D., and Yarranton, G. T. (1992) High-level expression of a recombinant antibody from myeloma cells using a glutamine synthetase gene as an amplifiable selectable marker. *Bio/Technology* **10,** 169–175.
8. Taticek, R. A., Lee, W. C. T., and Shuler, M. L. (1994) Large scale insect and plant cell culture. *Curr. Opinion Biotechnol.* **5,** 165–174.
9. Manohar, V. and Hoffman, T., (1992) Monoclonal and engineered antibodies for human parenteral clinical use: regulatory considerations. *Tibtech* **10.**
10. Potrykus, I. (1991) Gene transfer to plants: assessment of published approaches and results. *Ann. Rev. Plant Phyiol. Plant Mol. Biol.* **42,** 205–225.
11. Hiatt, A., Cafferkey, R., and Bowdish, K. (1989) Production of antibodies in transgenic plants. *Nature* **342,** 469–470.
12. Voss, A., Niersbach, M., Hain, R., Hirsch, H. J., Liao, Y.C., Kreuzaler, F., and Fischer, R. (1995) Reduced virus infectivity in *N. tabacum* secreting a TMV-specific full-size antibody. *Mol. Breeding* **1,** 39–50.
13. Ma, J. K.-C., Hiatt, A., Hein, M., Vine, N. D., Wang, F., Stabila, P., Van Dolleweerd, D., Mostov, K., and Lehner T. (1995) Generation and assembly of secretory antibodies in plants. *Science* **268,** 716–719.
14. Scragg, A. H. (1992) Large scale plant cell culture: methods, applications and products. *Curr. Opinion Biotechnol.* **3,** 105–109.
15. Firek, S., Draper J., Owen, M. R. L., Gandecha, A., Cockburn, B., and Whitelam, G. C. (1993) Secretion of a functional single-chain Fv protein in transgenic tobacco plants and cell suspension cultures. *Plant Mol. Biol.* **23,** 861–870.
16. Endreβ, R. (1994) *Plant Cell Biotechnology.* Springer-Verlag, Berlin.
17. Horsch, R. B., Fry, J. E., Hoffman, N. L., Eicholtz, D., Rogers, S. G., and Fraley, R. T. (1985) A simple and general method for transfering genes into plants. *Science* **227,** 1229–1231.

18. Sambrook, J., Fritsch, E. F., and Maniatis, T. (1989) *Molecular Cloning: A Laboratory Manual*, 2nd ed. Cold Spring Harbor Laboratory, Cold Spring Harbor, NY.
19. Murashige, T. and Skoog, F. (1962) A revised medium for rapid growth and bioassays with tobacco tissue cultures. *Physiol. Plant.* **15**, 473–497.
20. Gamborg, O. L., Miller, R. A., and Ojima, K. (1968) Nutrient requirements of suspension cultures of soybean root cells. *Exp. Cell Res.* **50**, 151–158.
21. Fulton, S. P. (1994) Large-scale processing of macromolecules. *Curr. Opinion Biotechnol.* **5**, 201–205.
22. Hobbs, S. L. A., Warkentin, T. D., and DeLong, C. M. O. (1993) Transgene copy number can be positively or negatively associated with transgene expression. *Plant Mol. Biol.* **21**, 17–26.
23. Matzke, M. A. and Matzke, A. J. M. (1995) How and why do plants inactivate homologous (trans)genes? *Plant Physiol.* **107**, 679–685.

12

Production of Foreign Proteins in Tobacco Cell Suspension Culture

Véronique Gomord, Anne-Catherine Fitchette-Lainé, Lise-Anne Denmat, Dominique Michaud, and Loïc Faye

1. Introduction

The transfer of foreign genes to plant cells is most often performed using the biological vector *Agrobacterium tumefaciens (1)*. This *Agrobacterium*-mediated introduction of cloned genes into plants has been widely used for promoter characterization using reporter genes, analysis of gene function, and protein targeting. *Agrobacterium*-mediated gene transfer has also been a method of choice for introducing genes that increase crop resistance to insects or viruses, or genes that encode foreign proteins of industrial or bio-pharmaceutical interest.

The *Agrobacterium*-mediated technique used for plant transformation is based on the co-cultivation of plant cells with *A. tumefaciens* and the subsequent selection of transformed plant cells by plating on an appropriate selective medium *(1)*. Co-cultivation of *A. tumefaciens* with wounded plant parts, such as leaf disks, is probably the simplest gene transfer protocol. An efficient transformation of plant cells can also be obtained using *Agrobacterium*-mediated gene transfer into protoplasts prepared from plant tissues. However, severe limitations, notably associated with the preparation of protoplasts and the regeneration of the plant cell wall, restrict the use of this procedure (*see* **Note 1**).

As an alternative, we have developed a rapid and efficent system for tobacco cells (*Nicotiana tabacum*, cv. Bright Yellow 2) transformation. Using this technique, the expression of recombinant proteins can be analyzed within 3 d of co-culture. This timing is similar to a transient expression using a physical

From: *Methods in Biotechnology, Vol. 3:*
Recombinant Proteins from Plants: Production and Isolation of Clinically Useful Compounds
Edited by: C. Cunningham and A. J. R. Porter © Humana Press Inc., Totowa, NJ

gene transfer by polyethelene glycol (PEG) or electroporation, but it does not require protoplast preparation. In addition, stable transgenic calli and suspension-cultured cells are obtained, respectively, at 3–4 wk and at 6–7 wk after co-culture with *A. tumefaciens*. BY2 tobacco cells are widely used in plant cellular biology because of their exceptionally high growth rate and protein production level *(2)*. The procedure described here results in the rapid generation of stably transformed BY2 tobacco cells, making this cellular model a powerful tool for studying the targeting, secretion, maturation, and stability of heterologous recombinant proteins.

2. Materials

2.1. Plant Material

Tobacco suspension-cultured cells (*N. tabacum* cv BY2) are grown in MS medium (*see* **Subheading 2.5.**) in the dark, at 25°C, with orbital shaking. Calli are grown in the same medium, supplemented with 8% agar, and containing 100 µg/mL kanamycin and 250 µg/mL cefotaxim.

2.2. Bacterial Strains

A. tumefaciens strain LBA4404 is used for gene transfer. Prior to inoculation, bacteria are grown for 24 h at 28°C in YEB medium (*see* **Subheading 2.5.**) containing 300 µg/mL streptomycin and 100 µg/mL spectinomycin.

2.3. cDNA Material

The seed vacuolar lectin phytohemagglutinin (PHA), from the common bean (*Phaseolus vulgaris*), is used as a protein model to illustrate the procedure. The coding sequence of PHA-L subunit *(3)* was inserted in an expression cassette containing the CaMV 35S promoter as a 533 bp EcoRI-SphI fragment from pDH51 *(4)* and the 3' end from octopin synthase as a 215 bp PvuII-RsaI fragment *(5)*. This expression cassette was then ligated to the binary plant transformation vector, pDE1001 *(6)*, which carries the neomycin phosphotransferase gene as a selectable marker for plant cell transformation. Alternatively, the coding sequence of sporamin, a storage protein of sweet potato, was inserted in the same expression cassette and then introduced into pDE1001.

2.4. Special Laboratory Tools and Materials

1. A tissue-culture room, maintained at 25°C in the dark.
2. An orbital shaker.
3. General tissue culture and plant culture materials.
4. Liquid nitrogen.

2.5. Media and Buffers

All media and buffers are made up as aqueous preparations. The media for tissue culture are used after sterilization using standard procedures.

1. YEB medium: 1.3% (w/v) nutrient broth (Difco, Detroit, MI), 0.1% (w/v) yeast extract (Difco), 0.5% (w/v) sucrose, and 2 mM MgSO$_4$.
2. Agar-YEB medium: YEB medium supplemented with 1.4% (w/v) Bactoagar (Difco), 300 µg/mL streptomycin (Sigma, St. Louis, MO), and 100 µg/mL spectinomycin (Sigma).
3. Murashige and Skoog (MS) medium: 4.3 mg/mL of MS mineral salt mixture (ICN, Costa Mesa, CA), 3% (w/v) sucrose, 0.05 mg/mL myo-inositol, 1 mg/mL thiamine-HCl, 0.2 µg/mL 2,4 dichlorophenolacetic acid, and 2 g/L KH$_2$PO$_4$.
4. Agar-MS medium: MS medium supplemented with 0.8% (w/v) Bactoagar and 100 µg/mL kanamycin (Sigma) and 250 µg/mL cefotaxim (Sigma).
5. Denaturation buffer: 62.5 mM Tris-HCl, pH 6.8, containing 5% (v/v) β-mercaptoethanol, 10% (v/v) glycerol, and 2% (w/v) SDS.

2.6. Gel Electrophoresis and Blotting Systems

All electrophoretic gels and solutions are made up as aqueous preparations. The procedures for standard sodium dodecyl sulfate-polyacrylamide gel electrophoresis (SDS-PAGE) and immunoblotting analysis are described in **refs. 7** and **8**, respectively.

1. SDS-PAGE separation gel: 15% (w/v) acrylamide (acrylamide:*bis*-acrylamide 30:0.8), 375 mM Tris, pH 8.8, 0.1% (w/v) SDS.
2. SDS-PAGE stacking gel: 5% (w/v) acrylamide (acrylamide:*bis*-acrylamide 30:1.5), 175 mM Tris, pH 6.8, 0.1% (w/v) SDS.
3. SDS-PAGE sample buffer: 62.5 mM Tris-HCl, pH 6.8, containing 5% (v/v) β-mercaptoethanol, 10% (v/v) glycerol, and 2% (w/v) SDS.
4. SDS-PAGE running buffer: 25 mM Tris-HCl, 192 mM glycine, 0.1 (w/v) SDS.
5. Gel-staining solution: 0.25% (w/v) Coomassie brilliant blue R250 in 50% (v/v) methanol, 5% (v/v) acetic acid.
6. Gel-destaining solution: 25% (v/v) methanol, 9% (v/v) acetic acid.
7. Blotting-transfer buffer: 25 mM Tris, pH 8.3, 192 mM glycine, and 10% (v/v) methanol.
8. Membrane-staining solution: 0.1% (w/v) Ponceau Red in 3% (v/v) trichloroacetic acid. Destain with TBS buffer.
9. Membrane saturation solution: 3% (w/v) gelatin in TBS buffer.
10. TBS buffer: 20 mM Tris-HCl, pH 7.5, 500 mM NaCl.

3. Methods

3.1. Agrobacterium tumefaciens *Transformation*

1. Transfect the pDE1001-derived constructs into *A. tumefaciens* (LBA 4404) according to Höfgen and Willmitzer (**9**), and select the transformants on YEB medium.

2. Initiate liquid cultures of *A. tumefaciens* by inoculating single colonies picked from stock plates into 5 mL YEB medium supplemented with streptomycin and spectinomycin. Incubate at 28°C on a rotary shaker (200 rpm).
3. After 24 h, inoculate 30 µL of growing bacteria into 3 mL of fresh medium, incubate overnight as before, then use to transform tobacco cells.

3.2. Tobacco Cell Suspension Transformation

1. Grow tobacco suspension-cultured cells in 150 mL MS medium at 25°C, in the dark, with continuous shaking in a 500-mL Erlenmeyer flask. For transformation, 3-d-old cells are used at the logarithmic phase of growth.
2. Coincubate 1 mL of tobacco cells suspension with 50 µL of the overnight *A. tumefaciens* culture in small Petri dishes for 2–3 d in the dark at 25°C.
3. Wash the tobacco cells three times with MS medium by centrifugation for 3 min, at 50*g*.
4. Analyze the cells for expression or plate them onto agar-MS medium containing kanamycin to produce stable transgenic cells.

3.3. Detection of the Recombinant Proteins from Tobacco Cells

3.3.1. Immunodetection Three Days After Transformation

The expression of PHA in tobacco cells is monitored after 0, 1, 2, and 3 d of co-culture with *A. tumefaciens*, using standard SDS-PAGE and immunoblotting procedures *(7,8)*.

1. Resuspend the transformed cells (*see* **Subheading 3.2.**) in 100 µL of denaturation buffer and boil for 10 min to allow complete denaturation of the proteins.
2. The resulting extracts (20 µL) are electrophoresed through a 15% (w/v) SDS-polyacrylamide gel *(7)*, and the resolved proteins are transferred onto a nitrocellulose sheet *(8)*.
3. The recombinant protein is immunodetected by standard immunostaining *(10)*, using a rabbit antiserum against PHA. As illustrated in **Fig. 1**, two PHA forms are detected after 3 d of co-culture, demonstrating the usefulness of the procedure for the rapid analysis of recombinant protein expression in plant cells. Both forms also react with antibodies that were specific for plant complex N-glycans (not shown) *(10)*. See **Note 2** for the significance of this latter immunoreaction.

3.3.2. Detection from Transgenic Calli

1. After co-culture with *A. tumefaciens*, resuspend the cells in 200 µL of MS medium to obtain larger amounts of transformed tobacco cells, and plate onto MS-agar medium.
2. Select transformed microcalli on MS-agar medium containing 100 µg/mL kanamycin and 250 µg/mL cefotaxime. Kanamycin-resistant calli are identified after incubation at 25°C for 3 wk by active growth and white color, compared to untransformed cells that turn dark brown.

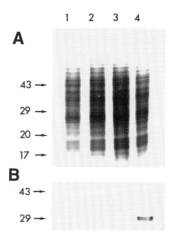

Fig. 1. Detection of PHA expressed in transgenic tobacco cells. Proteins obtained from mock-transformed tobacco cells (lane 1) or transformed cells after 1 d (lane 2), 2 d (lane 3) or 3 d (lane 4) of coculture with *A. tumefaciens* were separated by SDS-PAGE **(A)** and transferred to nitrocellulose **(B)**. On B, recombinant PHA was immunodetected on nitrocellulose with antibodies prepared against deglycosylated PHA and a second antibody coupled to alkaline phosphatase. Arrowheads on the left indicate position of molecular weight markers.

3. Analyze the putative transformed calli for PHA expression using immunoblotting, as described in **Subheading 2.6.** Select calli exhibiting high levels of lectin expression to establish suspension-cultured cells. Maintain the other calli on the selective MS-agar medium.

3.3.3. Detection from Transgenic Suspension Cell Cultures

1. For analysis of the recombinant protein expression level in transgenic suspension-cultured cells, propagate 2-wk-old, transformed calli in 150 mL MS liquid medium in a 500-mL Erlenmeyer flask. Subculture the cell lines weekly with a 2% inoculum of 7-d-old cells. The selection agent, kanamycin, is maintained for 3–4 subcultures (*see* **Note 3**). Analyze the expression of recombinant PHA by SDS-PAGE before and after purification.
2. Monitor the expression of recombinant PHA by immunoblotting, as described previously (*see* **Fig. 2** as an example, and **Note 4** for additional comments).

3.4. Simultaneous Expression of Two Recombinant Proteins

The transformation protocol described in **Subheading 3.2.** may be used to simultaneously express two recombinant proteins in the same cell. As an example, the coding sequence of sporamin, a vacuolar storage protein from sweet potato *(12)*, was inserted in the plant expression cassette previously used

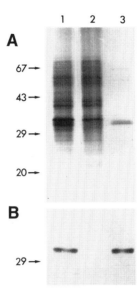

Fig. 2. PHA purification from transgenic tobacco cell cultures. Proteins extracted from suspension-cultured tobacco cells were analyzed using SDS-PAGE and Coomassie blue staining (**A**), or transferred to nitrocellulose and immunodetected with antibodies directed against deglycosylated PHA and a second antibody coupled to alkaline phosphatase (**B**). Lane 1: crude protein extract; lane 2: proteins unretained on thyroglobulin-Sepharose column; lane 3: affinity-purified recombinant PHA, retained on thyroglobulin-Sepharose column. Arrowheads on the left indicate the position of molecular weight markers.

Large amounts of PHA were purified from transformed tobacco suspension-cultured cells by affinity chromatography on a porcine thyroglobuline-Sepharose column *(3)*. As illustrated on lane 3, affinity-purified PHA from suspension-cultured cells could be resolved into two polypeptides of Mr 34,000 and Mr 35,500 by SDS-PAGE. These polypeptides correspond to two PHA forms that differ in their proteolytic maturation *(11)*. From densitometry of SDS-PAGE gels and from protein assays using crude extracts and affinity-purified fractions, we estimate that recombinant PHA represents 4% of the total proteins synthesized by suspension-cultured tobacco cells transformed according to the procedure described in **Subheading 3.2.**

for PHA (*see* **Subheading 3.1.**), and the resulting construct was introduced into the *A. tumefaciens* strain LBA 4404. *Agrobacterium* lines containing either the PHA construct or the sporamin construct were then simultaneously co-cultured with the cells to be transformed. The following steps describe the double transformation protocol and the analysis of the recombinant proteins simultaneously expressed.

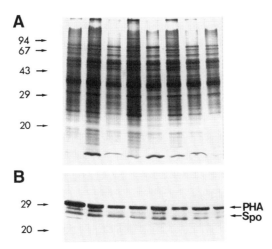

Fig. 3. Detection of phytohemagglutinin and sporamin in eight transgenic callus lines 3 wk after cotransformation. Protein extracted from transformed calli were analyzed using SDS-PAGE (**A**), or blotted and immunodetected with a mixture of anti-PHA and anti-Sporamin antibodies (**B**). We observed that 70–90% of transformed calli co-expressed both recombinant proteins after this co-transformation.

1. Incubate 1 mL of 3-d-old suspension-cultured cells for 2 d simultaneously with 50 µL of overnight *A. tumefaciens* culture encoding PHA, and with 50 µL of overnight *A. tumefaciens* culture encoding sporamin, as described in **Subheading 3.2.**
2. Wash the tobacco cells three times with MS medium by centrifugation for 3 min at 50*g*, and plate onto agar MS-medium containing 100 µg/mL kanamycin, as in **Subheading 3.2.**
3. Analyze kanamycin-resistant calli by standard immunoblotting, using appropriate specific primary antibodies (*see* **Fig. 3**).
4. To obtain further evidence that single transformed cells produce either recombinant proteins, a suspension culture induced from a single callus co-expressing high amounts of both PHA and sporamin (determined by immunoblotting, **step 3**) should then be analyzed by cell-printing immunoblotting. Plate a 1:50 dilution of the transgenic cell culture and press between two nitrocellulose sheets. One sheet is used for immunodetecting the first protein (e.g., PHA) and the other is used to detect the second protein (e.g., sporamin). For cells expressing both proteins, each signal obtained on the first blot should corresponds with a signal on the second blot (*see* **Fig. 4** as an example) demonstrating the expression of the two recombinant proteins in the same plant cell (*see* **Note 5**).

4. Notes

1. The plant cell wall has been described as an efficient barrier and trap for DNA molecules the size of a functional gene. We describe here a method to transform

A   **B**

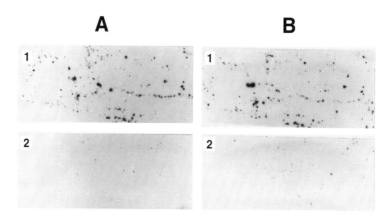

Fig. 4. Immunodetection of phytohemagglutinin and sporamin on cell prints. After dilution, a control (unstransformed) (blot 2) or a transgenic cell culture expressing PHA and sporamin (blot 1) were plated and pressed between two sheets of nitrocellulose. Blots 1 and 2 on **A** were treated for PHA immunodetection, and blots 1 and 2 on **B** for sporamin immunodetection using the ECL amplification method. Within the limits of the approach related to inequal amounts of the protein transferred on each blot, it clearly appears that each single cell produces both proteins.

actively growing tobacco cells by *A. tumefaciens* without the need to prepare protoplasts. This protocol, which leads to the stable expression of a recombinant protein, is highly efficient and allows the transgenic tobacco BY2 cells to be obtained within 3 d after the initial transformation step. Consequently, the expression of recombinant proteins in these stable transformants is detected almost as rapidly as using transient expression in transformed tobacco protoplasts, but the technical difficulties associated with protoplast preparation are avoided.

2. In the present study, both forms of the recombinant PHA detected **(Fig. 1)** are glycosylated, indicating that these proteins are expressed in the plant cells and not in the bacteria, which lack the necessary enzymatic capabilities.

3. Large amounts of transgenic BY2 cells can be produced within 7 wk after transformation in suspension cultures induced from a single callus. This large-scale production of transgenic tobacco BY2 cells has facilitated our recent studies on sporamin targeting *(13)*, without the usual burden of cellular fractionation required with plant organs.

4. The high level of recombinant protein in BY2 tobacco cells also facilitates the biochemical analysis of the protein maturation steps. For example, we have obtained sufficient amounts of recombinant PHA produced in transgenic suspension-cultured BY2 cells to perform a detailed structural analysis of its N-glycans *(11)*. The same protocol could efficiently be applied to multiple cultivars in order

to obtain transgenic suspension-cultured cells and plants expressing the same recombinant protein.

5. As illustrated here with the production of both PHA and sporamin in transgenic BY2 cells, this technique allows the generation of double transformants that express high amounts of two different recombinant proteins. The latter eventually accumulate in different compartments of the secretory pathway, thus providing new cellular models for a study of secretory protein sorting in plants.

Acknowledgments

This study was partially supported by a grant from Société d'Exploitation Industrielle des Tabac et Allumettes (SEITA), France. We are indebted to M. J. Chrispeels (University of California, San Diego) and to K. Nakamura (University of Nagoya, Japan) for the gift of PHA cDNA and sporamin cDNA, respectively. We also wish to express our thanks to J. P. Salier and B. Tague for comments on the manuscript.

References

1. Hooykaas, P. J. J. and Schilperoort, R. A. (1992) *Agrobacterium* and plant genetic engineering. *Plant Mol. Biol.* **19,** 15–38.
2. Kato, K., Matsumoto, T., Koiwai, A., Mizusaki, S., Nishida, K., Noguchi, M., and Tamaki, E. (1972), in *Fermentation Technology Today* (Terui, G., ed.), pp. 689–695.
3. Sturm, A., Voelker, T. A., Herman, E. M., and Chrispeels, M. J. (1988) Correct glycosylation, Golgi-processing and targeting in protein bodies of vacuolar protein phytohemagglutinin in transgenic tobacco. *Planta* **175,** 170–183.
4. Pietrzak, M., Shillito, R. D., Hohn, T., and Potrykus, Y. (1986) Expression in plants of two bacterial antibiotic resistance genes after protoplast transformation with a new plant expression vector. *Nucleic Acids Res.* **14,** 5857–5868.
5. Barker, R. F., Idler, K. B., Thompson, D. V., and Kemps, J. D. (1983) Nucleotide sequence of the T-DNA region of *Agrobacterium tumefaciens* octopine Ti plasmid pTil5955. *Plant Mol. Biol.* **2,** 335–350.
6. Denecke, J., Botterman, J., and Deblaere, R. (1990) Protein secretion in plant cells can occur via a default pathway. *Plant Cell* **2,** 51–59.
7. Laemmli, U. (1970) Cleavage of structural proteins during the assembly of the head of bacteriophage T4. *Nature* **227,** 680–685.
8. Towbin, H., Staehelin, T., and Gordon, J. (1979) Electrophoretic transfer of proteins from polyacrylamide gels to nitrocellulose sheets: procedure and some applications. *Proc. Natl. Acad. Sci., USA* **76,** 4350–4354.
9. Höfgen, R. and Willmitzer, L. (1988) Storage of competent cells for *Agrobacterium* transformation. *Nucleic Acids Res.* **16,** 9877.

10. Faye, L., Gomord, V., Lainé, A.-C., and Chrispeels, M-J. (1993) Affinity purification of antibodies specific for Asn-linked glycans containing a1→3 fucose or b1→2 xylose. *Anal. Biochem.* **209,** 104–108.
11. Rayon, C., Gomord, V., Faye, L., and Lerouge, P. (1996) N-glycosylation of phytohemagglutinin expressed in bean cotyledon or in transgenic tobacco cells. *Plant Physiol. Biochem.,* **34,** 273-281.
12. Matsuoka, K., Matsumoto, S., Hattori, T., Machida, Y., and Nakamura, K. (1990) Vacuolar targeting and posttranslational processing of the precursor to the sweet potato tuberous root storage protein in heterologous plant cells. *J. Biol. Chem.* **265,** 19,750–19,757.
13. Gomord, V., Denmat, L. A., Fitchette-Laine, A. C., Satiat-Jeunemaitre, B., Hawes, C., and Faye, L. (1997) The C-terminal HDEL sequence is sufficient for retention of secretory proteins in the endoplasmic reticulum but promotes vacuolar targeting of proteins that escape the endoplasmic reticulum. *Plant J.* **11,** 101–103.

13

Transient Gene Expression in Plant Protoplasts

Stefan Schillberg, Sabine Zimmermann, Dirk Prüfer, Detlef Schuman, and Rainer Fischer

1. Introduction

Transient expression of genes in plant protoplasts is a powerful tool in plant molecular biology that allows quick screening and analysis of engineered proteins prior to stable transformation. Although stably transformed plant material may be preferable in many instances, transformation and regeneration of plants is time-consuming, tedious, and requires plenty of space. Transient expression techniques enable measurement of gene expression, as well as analysis of protein stability and activity, very shortly after DNA uptake; therefore, a large number of samples can be analyzed in a short period of time. Moreover, most of the introduced plasmid DNA remains extra-chromosomal during the transient assay *(1)* and, consequently, gene activity is not biased by position effects, as observed in stably transformed plants. Because of its speed, convenience, and flexibility, transient gene expression has been widely used for analysis of promoter and regulatory elements involved in transcription and translation *(2–6)*, induction of gene expression by exogenous stimuli *(7,8)*, and for verifying functionality of cloned genes or cDNAs.

The delivery of foreign genes into protoplasts is usually carried out by electroporation *(9)* or treatment with polyethylene glycol (PEG) *(10,11)*. In addition, analysis of transient expression is also feasible in intact plant cells, using particle bombardment for introducing the gene(s) of interest *(12,13)*. PEG-mediated transformation offers the advantage that only common, inexpensive laboratory supplies and equipment are required. Electroporation yields highly reproducible results, and hence this method is often the preferred technique for direct gene transfer into plant protoplasts. These transformation meth-

From: *Methods in Biotechnology, Vol. 3:*
Recombinant Proteins from Plants: Production and Isolation of Clinically Useful Compounds
Edited by: C. Cunningham and A. J. R. Porter © Humana Press Inc., Totowa, NJ

ods also overcome host-range limitations of *Agrobacteria*-mediated gene transfer, since DNA uptake by protoplasts is either promoted by reversible permeabilization of the plasma membrane in high-intensity electric fields or by chemical treatment with PEG.

Our group is interested in investigating molecular events that lead to plant virus infection, replication, and movement, in order to understand the mechanisms of plant–pathogen interactions in more detail and to develop novel strategies for improving plant virus resistance. This approach requires the development of recombinant antibody fragments (RAbs) binding to different viral proteins, such as coat proteins, movement proteins, and replicases, to immunomodulate these targets in vivo. Fast and reliable methods to deliver genes coding for rAbs and to analyze their biological activity in the correct plant cell compartment are a major prerequisite for achieving these goals.

This chapter illustrates the application of transient protoplast expression for functional analysis of cloned multimeric full-size antibodies or fragments thereof *(14)*. A major requirement for transient expression assays is the preparation of viable protoplasts. The protocols presented here can be adapted to a wide range of plant species and tissue sources used for protoplast preparation. It should be noted, however, that independent preparations of protoplasts originating from the same plant can be of different quality. Therefore, it is always necessary to determine the efficiency of DNA delivery in the isolated protoplasts by monitoring transient expression of reporter enzymes such as β-glucuronidase (GUS). PEG-mediated transformation and electroporation are described as tools for DNA delivery into plant protoplasts. Evaluation of transiently expressed antibodies, which are either located in the cytoplasm or, in the case of cDNA constructs possessing a signal peptide, secreted into the regeneration medium or retarded in the endoplasmic reticulum, is feasible both for PEG-mediated transformation and electroporation. Recombinant antibody production can be analyzed in an enzyme-linked immunosorbent assay (ELISA) using either lysed protoplast extracts or concentrated protoplast supernatants.

2. Materials

2.1. Plant Material

Suspension cultures of *Nicotiana tabacum* cv. Petit Havana SR1 (*see* Chapter 11).

2.2. DNA Material

1. cDNAs of TMV-specific antibody genes and fragments thereof cloned in the plant expression vector pSS *(14)*.

2. β-glucuronidase gene (GUS) from *Escherichia coli* *(15)* cloned into the pRT vector *(16)*.

2.3. Special Laboratory Equipment and Supplies

1. Laminar flow hood.
2. pH Meter.
3. Osmometer.
4. Minisart syringe filter units, 0.2 μm pore size (Sartorius, Goettingen, Germany).
5. Orbital shaker.
6. Sterile filter unit containing a 250-mesh nylon screen.
7. Varifuge RF (Heraeus, Hanau, Germany).
8. Micro pipets, 200μl (Brand, Wertheim, Germany).
9. Hemocytometer.
10. Light microscope.
11. Water bath.
12. Gene Pulser apparatus and Capacitance Extender (Bio-Rad, Richmond, CA).
13. Gene Pulser Cuvettes, 0.4 cm electrode gap (Bio-Rad).
14. Benchtop centrifuge.
15. Spectrofluorometer (Jobin Yvon, Grasbrunn, France).
16. Centricon concentrator (Amicon, Beverly, MA).

2.4. Buffers and Other Solutions

For preparation of solutions and media, use only bidestilled water.

1. K3 medium: Add 10 mL each of the following stock solutions, 1 mL microelements, 1 mL vitamin solution I, 5 mL Fe in EDTA, 100 mg inosit, 250 mg xylose, and 136.92 g sucrose to 850 mL water, mix, adjust to pH 5.6, and adjust volume to 1000 mL. Finally, adjust to 600 mOsm with sucrose. Filter-sterilize and store at room temperature.
   a. Stock solutions for K3 medium: 125 mM NaH_2PO_4, 1.26M $CaCl_2$, 2.47M KNO_3, 312 mM NH_4NO_3, 102 mM $(NH_4)_2SO_4$, and 208 mM $MgSO_4$. Dissolve each chemical in 200 mL water, filter-sterilize, and store at 4°C.
   b. Microelements: 100 mM H_3BO_3, 100 mM $MnSO_4$, 37 mM $ZnSO_4$, 5 mM KI, 1 mM $NaMoO_4$, 0.1 mM $CuSO_4$, and 0.1 mM $CoCl_2$. Dissolve these components in water, filter-sterilize, and store at 4°C.
   c. Vitamin solution I: Dissolve 400 mg glycine, 400 mg nicotinic acid, 400 mg pyridoxine, 20 mg thiamine-HCl in 200 mL water, filter-sterilize and store at 4°C.
   d. Fe in EDTA: 20 mM $FeSO_4$ and 20 mM Na_2EDTA. Dissolve in water and boil for 5 min. Store at 4°C in a dark bottle.
2. Digestion enzyme solution in K3 medium: Dissolve 0.5% (w/v) macerozyme R-10 (Serva, Heidelberg, Germany) in 100 mL K3 medium and stir gently to avoid foaming. Add 1.5% (w/v) cellulase Onozuka R-10 (Serva) and stir gently

for 3 h. Adjust the pH to 5.7, stir for another hour, and check pH. Filter-sterilize. Prepare fresh solution before each use.

3. NAA: Dissolve 80 µg/mL NAA (naphtalen acetic acid) in water and filter-sterilize. Store at 4°C.

4. Claforan: Dissolve 250 mg/mL Claforan (Hoechst, Frankfurt a.M., Germany) in water and filter-sterilize. Store at –20°C.

5. W5: 154 mM NaCl, 125 mM CaCl$_2$, 5 mM KCl, and 5 mM glucose. Dissolve in water and adjust to pH 5.8. Filter-sterilize and store at room temperature.

6. Transformation buffer: Dissolve 450 mM mannitol, 15 mM MgCl$_2$, and 0.1% (w/v) MES in water and adjust to pH 5.6. Filter-sterilize and store at room temperature.

7. PEG solution: Dissolve 100 mM Ca(NO$_3$)$_2$, 400 mM mannitol, and 40% (w/v) PEG 4000 in water. Adjust to pH 7–9, filter-sterilize and store at room temperature.

8. GUS-extraction buffer: Dissolve 50 mM NaHPO$_4$, pH 7.0, 10 mM β-mercapto-ethanol, 10 mM Na$_2$EDTA, and 0.1% (v/v) Triton X-100 in water and filter-sterilize. Store at –20°C. β-mercaptoethanol must be freshly added prior to use.

9. Stop buffer: Dissolve 0.2M Na$_2$CO$_3$ in water and autoclave. Store at room temperature.

10. Substrate solution: Dissolve 10 mM 4-methyl umbelliferyl glucuronide (Sigma) in water and filter-sterilize. Store at –20°C.

11. Phosphate-buffered saline (PBS): 137 mM NaCl, 2.7 mM KCl, 8.1 mM Na$_2$HPO$_4$, and 1.5 mM KH$_2$PO$_4$. Dissolve in water, autoclave, and store at room temperature.

12. MS medium: 0.47% (w/v) MS salts (Serva), 2% (w/v) sucrose, and 0.8% (w/v) agar dissolved in 1000 mL water. Adjust the pH to 5.7–5.8. After auto-claving and cooling to 60°C, add 500 µL vitamin solution II and 100 µL vitamin solution III.

    a. Vitamin stock solution II: Dissolve 0.4% (w/v) glycine, 0.4% (w/v) nicotine acid, and 0.1% (w/v) pyridoxine in 100 mL water, filter-sterilize, and store at 4°C.

    b. Vitamin stock solution III: Dissolve 0.4% (w/v) thiamine-HCl in 10 mL water, filter-sterilize, and store at 4°C.

3. Methods

3.1. Preparation of Protoplasts

Protoplasts are isolated from cell suspension cultures 4–5 d after transfer-ring to fresh medium (*see* **Notes 1** and **2**). The establishment of tobacco sus-pension cultures is described in Chapter 11. Preparation of protoplasts should always be carried out under aseptic conditions in a laminar flow hood. Proto-plasts are fragile and should be handled with care at all stages. Use pipets with wide openings to minimize protoplast damage.

1. Centrifuge 40 mL of an exponentially growing suspension culture at 200g for 10 min at room temperature. Remove the supernatant and resuspend the cells in 100

mL of digestion enzyme solution in K3 medium (*see* **Note 3**). Transfer the solution to a 500-mL Erlenmeyer flask and incubate for 16 h at 26°C in the dark.

2. Upon completion of digestion, gently shake (100 rpm) the cell suspension for 20 min at room temperature to dislodge protoplasts from partially digested cell clumps.

3. Pass the dark green protoplast solution carefully through a sterile filter unit containing a 250-mesh nylon screen.

4. Transfer the filtrate containing the protoplasts into two 50-mL Falcon tubes and centrifuge at 250g for 5 min at room temperature to separate the floating protoplasts from cell debris.

5. Push a 200-μL micropipet through the protoplast layer, and carefully aspirate the cell debris and enzyme solution, except for the top 2 mL layer containing the protoplasts (*see* **Note 4**).

6. Combine both protoplast solutions in one 15-mL Falcon tube and fill the tube with W5 solution. Mix gently and pellet the protoplasts by spinning at 50g for 2 min at room temperature.

7. Discard the supernatant, except for the bottom 1 mL. Resuspend the protoplasts gently by swirling, add 14 mL transformation buffer, and spin at 50g for 2 min at room temperature.

8. Discard the supernatant and resuspend the protoplasts in 1 mL transformation buffer for subsequent PEG-mediated transformation (*see* **Subheading 3.2.**) or electroporation (*see* **Subheading 3.3.**) (*see* **Notes 5** and **6**).

3.2. PEG-Mediated Transformation

The following procedure was adapted from Negrutiu et al. *(11)*. All steps described below should be performed under aseptic conditions.

1. Determine the protoplast number using a hemocytometer. Adjust the protoplast suspension to 1×10^6 cells per mL transformation buffer. For each transformation reaction, carefully transfer 330 μL protoplast solution into a 15-mL Falcon tube.

2. Incubate the protoplasts in a water bath for 5 min at 42°C (*see* **Note 7**). After heat shock, place the cells on ice for 1 min and subsequently allow to stand at room temperature for 5 min.

3. Add 10–20 μg of the plasmid DNA to be transformed in a maximum volume of 15 μL (*see* **Note 8**). Mix carefully and let stand at room temperature for 10 min.

4. Add an equal volume of PEG solution (345 μL), mix gently, and incubate for 25–30 min at room temperature (*see* **Note 9**).

5. Dilute each transformation sample with 4 mL of K3 medium containing freshly added NAA and Claforan at final concentrations of 1 μg/mL and 500 μg/mL, respectively. For transient expression, store the sample tubes in a horizontal position to obtain an optimal distribution. Incubate for 48 h at 26°C in the dark (*see* **Note 10**).

6. Centrifuge the sample tubes at 50g for 2 min at room temperature and store supernatant and protoplasts separately at –80°C for further analysis.

3.3. Electroporation

Tobacco protoplasts are transformed by electroporation, essentially as described by Watanabe et al. *(17)* (*see* **Note 11**).

1. Prepare protoplasts, as described (*see* **Subheading 3.1.**), and adjust their number to 4×10^5 per mL transformation buffer. Use 500 µL of this protoplast suspension per electroporation.
2. Add 30–40 µg plasmid DNA to the protoplasts and immediately transfer the suspension to a prechilled cuvet (0.4 cm electrode gap). Mix the cuvet contents with a 1 mL plastic pipet to avoid sedimentation of protoplasts. Place the cuvet in the cuvet holder and subject the protoplasts to a 12-ms high-voltage pulse provided by discharge of a 125-µF capacitor set to 300 V (*see* **Notes 12–14**).
3. Incubate protoplasts for 30 min on ice (*see* **Note 15**).
4. Centrifuge the sample tubes at 50g for 2 min at room temperature, discard the supernatant, and resuspend the pellet in 4 mL of K3 medium containing NAA (1 µg/mL) and Claforan (500 µg/mL).
5. Incubate for 48 h at 26°C in the dark (*see* **Note 10**) and harvest protoplasts and supernatant as described (*see* **Subheading 3.2., step 6**).

3.4. β-Glucuronidase (GUS) Assay

The quality of protoplast preparations and the efficiency of DNA delivery should be examined by transient expression of the GUS gene. The GUS assay is reproducible, sensitive, and simple *(15,18)*, and therefore recommended as a control experiment.

1. Thaw and resuspend the frozen protoplasts, which had been transformed with a GUS-gene expression construct either by PEG-mediated transformation (*see* **Subheading 3.2., step 6**) or electroporation (*see* **Subheading 3.3., step 5**), in 50 µL GUS-extraction buffer.
2. Lyse protoplasts by pipeting up and down several times with a 10 µL pipet tip.
3. Centrifuge lysate at 1000g for 5 min at room temperature. Dispense 10-µL aliquots of the supernatant into two tubes for time-points t_0 and t_{30}.
4. Add 1 mL stop buffer to the sample t_0.
5. Place 5 µL substrate solution and 35 µL GUS-extraction buffer to both samples. Mix and incubate the tube t_{30} at 37°C for 30 min.
6. Stop the t_{30} reaction by adding 1 mL stop buffer.
7. Determine the GUS-catalyzed hydrolysis of the substrate 4-methyl umbelliferyl glucuronide (MUG) by measuring the concentration of the fluorescent end product methyl umbelliferone (MU), using a spectrofluorometer with an excitation setting at 365 nm and emission at 455 nm.

3.5. Isolation and Concentration of RAbs

RAbs transiently expressed either in the cytoplasm of tobacco protoplasts, or retarded in the endoplasmic reticulum, can be isolated as follows:

1. Thaw and resuspend the frozen protoplast pellet obtained by PEG-mediated transformation (*see* **Subheading 3.2., step 6**) or electroporation (*see* **Subheading 3.3., step 5**) in 50 µL PBS. Lyse the protoplasts by pipeting several times up and down with a 10 µL pipet tip.
2. Pellet the cell debris at 8000*g* for 5 min at room temperature and transfer the supernatant into a fresh tube.

RAbs that possess a signal peptide use the default pathway and are secreted into the regeneration medium. However, because of the large K3-medium volume, RAbs must be concentrated and transferred into a suitable buffer, such as PBS, prior to subsequent analysis.

3. Thaw the frozen supernatant obtained by PEG-mediated transformation (*see* **Subheading 3.2., step 6**) or electroporation (*see* **Subheading 3.3., step 5**). Spin at 10,000*g* for 10 min at room temperature to remove cell debris and insoluble components.
4. Transfer the supernatant to a Centricon concentrator (Amicon) that was prerinsed with bidistilled water to remove glycerine from the ultrafiltration membrane. Spin the concentrator unit at 4°C using a fixed angle rotor (*see* **Note 16**).
5. Following concentration, wash the rAb containing retentate twice, using 500 mL PBS.

The isolated and concentrated recombinant antibodies obtained from either protoplasts or supernatants can be serially diluted in PBS and used for testing antigen-binding in ELISA assays *(14)*.

3.6. Comments

The advantages of transient expression in plant protoplasts are evident for quick identification of recombinant antibodies. In particular, single-chain Fv fragments have proven to be a valuable new tool to inactivate plant pathogens by immunomodulation via recombinant antibodies binding to key targets that are involved in pathogenesis. Because of time constraints and labor required, antibodies should be expressed transiently in protoplasts first to characterize RNA processing and translation, protein folding, biological activity, and targeting of the engineered proteins prior to generation of stably transformed transgenic plants.

The presented protocols for transient expression of RAbs in tobacco protoplasts are applicable, with slight modifications, to many other cloned genes and plant species. The use of different protoplast sources is particularly advisable for analyzing other RAbs in compatible pathosystems, or when endogenous protoplast products interfere with the heterologous gene product. Protocols for transient expression in plant protoplasts are likewise described for *Arabidopsis (19)*, bean *(20)*, maize *(8)*, parsley *(21)*, pea *(22)*, potato *(23)*, and rice *(7)*. Furthermore, protocols for additional plant species are provided by Hauptmann et al. *(24)* and Rasmussen and Rasmussen *(25)*.

Although transient expression in protoplasts is a valuable tool for quick functional analysis of foreign gene(s), investigations at the level of whole plants or tissues are still required to study complex functions in vivo, because protoplasts do not represent the situation present in intact plants and the transformed DNA is usually not integrated into the plant genome.

4. Notes
4.1. Preparation of Protoplasts

1. For consistent results, maintain uniform growth conditions, because the physiological state of the plant cells is an important factor influencing protoplast yield and transformation efficiency. At the time of protoplast preparation, the cells should be actively growing, because stationary-phase cells possess low levels of transcriptional activity.

2. The presented procedure can also be used for isolation of protoplasts from sterile shoot cultures of tobacco or potato growing on MS medium. For those preparations, cut either tobacco leaves (approx 6 g) into small segments or 20 potato plantlets (4 wk old, without roots). Transfer the plant material into 100 mL of digestion-enzyme solution in K3 medium and proceed as described.

3. Best release of protoplasts is obtained with digestion enzymes dissolved in freshly prepared K3 medium. Because of batch to batch variations, these enyzmes have to be tested for suitability prior to use.

4. After centrifugation, protoplasts form a circular layer on the surface of the enzyme solution. Insert the micropipet as fast as possible through the circular layer to avoid aspiration of the floating protoplasts.

5. Prepared protoplasts can be stored, at the most, for 3 h at 4°C. Store the tube obliquely to avoid aggregation of protoplasts.

6. Viable protoplast preparations are essential for subsequent transformation experiments. Healthy protoplasts are spherical, display cytoplasmic streaming, and show cell wall resynthesis *(26)*. The protoplast viability can be determined either through exclusion of Evan's blue or staining with fluorescein diacetate (FDA) *(26)*.

4.2. PEG-Mediated Transformation

7. Incubation of the protoplasts at 42°C is not strictly necessary. However, for some constructs, higher expression values were obtained after heat shock *(27)*.

8. According to our experience, high expression levels are obtained when using supercoiled plasmid DNA isolated by the alkaline extraction procedure *(28)*. Further purification of DNA by PEG precipitation or CsCl centrifugation appears to be unnecessary.

9. Use only freshly prepared and filter-sterilized PEG solution (do not autoclave). Check the pH prior to use.
10. The suitable incubation time for transient expression should be determined for each DNA construct and source of protoplasts.

4.3. Electroporation

11. Protoplasts from other plant species might require alteration of parameters such as protoplast preparation, electroporation buffer, temperature, duration of pulse, voltage, number of protoplasts, and DNA concentration.
12. Prior to electroporation, optimal pulse parameters that kill less than 50% of the protoplasts have to be determined. Protoplast viability should be analyzed during the first 24 h after the pulse, as described (*see* **Note 6**).
13. Try to electroporate the protoplasts before they settle at the bottom of the cuvet.
14. Chilling protoplasts during electroporation is not really necessary, as long as the duration of the pulse does not exceed 20 ms. However, prechilling avoids cell damage through local heating during pulsing.
15. Resting on ice after pulsing prolongs the period when the pores are open, so that DNA may enter over an extended period of time.

4.4. Isolation and Concentration of Recombinant Antibodies

16. The size of the antibody implies the cut-off range of the Centricon concentrator. Full-size antibodies can be concentrated using a Centricon-100 unit at 1000*g*. For the concentration of antibody fragments, such as single-chain Fvs or Fab fragments, Centricon-10 with a maximum *g*-force of 5000 must be used.

References

1. Werr, W. and Lörz, H. (1986) Transient gene expression in a Gramineae cell line. *Mol. Gen. Genet.* **202,** 471–475.
2. Fritze, K., Staiger, D., Czaja, I., Walden, R., Schell, J., and Wing, D. (1991) Developmental and UV light regulation of the snapdragon chalcone synthase promoter. *The Plant Cell* **3,** 893–905.
3. Kaulen, H., Schell, J., and Kreuzaler, F. (1986) Light-induced expression of the chimeric chalcone-synthase-NPTII gene in tobacco cells. *EMBO J.* **5,** 1–8.
4. Rohde, W., Becker, D., and Randles, J. W. (1995) The promoter of coconut foliar decay-associated circular single-stranded DNA directs phloem-specific reporter gene expression in transgenic tobacco. *Plant Mol. Biol.* **27,** 623–628.
5. McCarty, D. R., Hattori, T., Carson, C. B., Vasil, V., and Vasil, I. K. (1991) The *viviparous-1* developmental gene of maize encodes a novel transcriptional activator. *Cell* **66,** 895–905.
6. Skuzeski, J. M., Nichols, L. M., and Gesteland, R. F. (1990) Analysis of leaky viral translation termination codons *in vivo* by transient expression of improved beta-glucuronidase vectors. *Plant Mol. Biol.* **15,** 65–79.

7. Marcotte, W. R., Jr., Bayley, C. C., and Quatrano, R. S. (1988) Regulation of wheat promoter by abscisic acid in rice protoplasts. *Nature* **335,** 454–457.

8. Sheen, J. (1990) Metabolic repression of transcription in higher plants. *Plant Cell* **2,** 1027–1038.

9. Fromm, M., Callis, J., Taylor, L. P., and Walbot, V. (1985) Electroporation of DNA and RNA into plant protoplasts. *Methods Enzymol.* **153,** 351–366.

10. Paszkowski, J., Shillito, R. D., Saul, M., Mandák, V., Hohn, T., Hohn, B., and Potrykus, I. (1984) Direct gene transfer to plants. *EMBO J.* **3,** 2717–2722.

11. Negrutiu, I., Shillito, R., Potrykus, I., Biasini, G., and Sala, F. (1987) Hybrid genes in the analysis of transformation conditions: I. Setting up a simple method for direct gene transfer in plant protoplasts. *Plant Mol. Biol.* **8,** 363–373.

12. Klein, T. M., Wolf, E. D., Wu, R., and Sanford, J. C. (1987) High-velocity microprojectiles for delivering nucleic acids into living cells. *Nature* **327,** 70–73.

13. Klein, T. M., Roth, B. A., and Fromm, M. E. (1989) Regulation of anthocyanin biosynthesis genes introduced into intact maize tissues by microprojectiles. *Proc. Natl. Acad. Sci. USA* **86,** 6681–6685.

14. Voss, A., Niersbach, M., Hain, R., Hirsch, H. J., Liao, Y. C., Kreuzaler, F., and Fischer, R. (1995) Reduced virus infectivity in *N. tabacum* secreting a TMV-specific full-size antibody. *Mol. Breeding* **1,** 39–50.

15. Jefferson, R. A., Kavanagh, T. A., and Bevan, M. W. (1987) GUS fusions: β-glucuronidase as a sensitive and versatile gene fusion marker in higher plants. *EMBO J.* **6,** 3901–3907.

16. Töpfer, R., Pröls, M., Schell, J., and Steinbiβ, H. H. (1988) Transient gene expression in tobacco protoplasts: II. Comparison of the reporter gene systems for CAT, NPTII, and GUS. *Plant Cell Rep.* **7,** 225–228.

17. Watanabe, Y., Meshi, T., and Okada, Y. (1987) Infection of tobacco protoplasts with *in vitro* transcribed tobacco mosaic virus RNA using an improved electroporation method. *FEBS Lett.* **219,** 65–69.

18. Jefferson, R. A. (1989) The GUS reporter system. *Nature* **342,** 837–838.

19. Abel, S. and Theologis, A. (1994) Transient transformation of *Arabidopsis* leaf protoplasts: a versatile experimental system to study gene expression. *Plant J.* **5,** 421–427.

20. Leon, P., Planckaert, F., and Walbot, V. (1991) Transient gene expression in protoplasts of *Phaseolus vulgaris* from suspension culture. *Plant Physiol.* **95,** 968–972.

21. Tovar Torres, J., Block, A., Hahlbrock, K., and Somssich, I. E. (1993) Influence of bacterial strain genotype on transient expression of plasmid DNA in plant protoplasts. *Plant J.* **4,** 587–592.

22. Ballas, N., Wong, L.-M., and Theologis, A. (1993) Identification of the auxin-responsive element, *AuxRE*, in the primary indoleacetic acid-inducible gene, *PS-IAA4/5*, of pea *(Pisum sativum). J. Mol. Biol.* **233,** 580–596.

23. Van der Steege, G., Nieboer, M., Swaving, J., and Tempelaar, M. J. (1992) Potato granule-bound starch synthase promoter-controlled GUS expression: regulation of expression after transient and stable transformation. *Plant Mol. Biol.* **20,** 19–30.

24. Hauptmann, R. M., Ozias-Akins, P., Tabaeizadeh, Z., Rogers, S. G., Horsch, R. B., Vasil, I. K., and Fraley, R. T. (1987) Transient expression of electroporated DNA in monocotyledonous and dicotyledonous species. *Plant Cell Rep.* **6,** 265–270.

25. Rasmussen, J. O. and Rasmussen, O. S. (1993) PEG mediated DNA uptake and transient GUS expression in carrot, rapeseed and soybean protoplasts. *Plant Sci.* **89,** 199–207.

26. Fowke, L. C. and Cutler, A. J. (1994) Plant protoplast techniques, in *Plant Cell Biology* (Harris, N. and Oparka, K. J., eds.). IRL, Oxford, pp. 177–197.

27. Oliveiria, M. M., Barroso, J., and Pais, M. S. S. (1990) Direct gene transfer into *Actinidia deliciosa* protoplasts: analysis of transient expression of the CAT gene using TLC autoradiography and a GC-MS-based method. *Plant Mol. Biol.* **17,** 235–242.

28. Sambrook, J., Fritsch, E. F., and Maniatis, T. (1989) *Molecular Cloning: A Laboratory Manual*, 2nd ed., Cold Spring Harbor Laboratory, Cold Spring Harbor, NY.

14

Stability of Recombinant Proteins in Plants

Dominique Michaud, Thierry C. Vrain, Véronique Gomord, and Loïc Faye

1. Introduction

In almost all living organisms, proteolytic enzymes are involved in a variety of cellular functions not only associated with the control of specific endogenous metabolic reactions, but also with the degradation of abnormal or exogenous ("foreign") proteins (1). Despite the fundamental importance of proteases in cells, studies on these enzymes, for those involved in gene-expression technology, are devised to develop a means of avoiding or minimizing degradation of the recombinant proteins to be produced. Some proteins are rapidly degraded either during or shortly after their synthesis, and others are lost during their extraction from cells or tissues. Although general strategies have been proposed to minimize extraction-related hydrolytic processes in microbial, animal, and plant systems (2,3), in vivo proteolysis still represents one of the most significant barriers to recombinant gene expression in any organism (4). Some exo- and endoproteases from *Escherichia coli* (5) and yeast (6), notably, represent harmful molecules for recombinant ("abnormal") proteins expressed in these systems, and strategies have been developed to counteract potential or actual hydrolytic processes (7,8). Concurrently, the posttranslational ubiquitination of foreign proteins recognized as abnormal in yeast and other eukaryotic cells may lead to their rapid degradation by the multicatalytic complex proteasome via the ubiquitin-mediated proteolysis pathway (9), rendering necessary the study of ubiquitin conjugates when planning to express recombinant proteins in the cytoplasm of yeast or animal cells (10).

In plants, little is known about the actual stability of recombinant proteins, but hydrolytic processes similar to those observed in bacterial and yeast cells

From: *Methods in Biotechnology, Vol. 3:*
Recombinant Proteins from Plants: Production and Isolation of Clinically Useful Compounds
Edited by: C. Cunningham and A. J. R. Porter © Humana Press Inc., Totowa, NJ

are likely to occur. Proteases are known to lurk into the vacuoles of plant cells *(11–13)*, and the major components of the highly conserved ubiquitin system have been identified in plants in the past few years *(12)*. Proteases present in plant cells, especially those active in the mildly acidic pH range and residing in the vacuole, may have a profound influence on stability of the foreign proteins expressed, with negative effects on recombinant protein yields. The identification of potential and actual proteolytic events altering the stability of a protein to be expressed is thus essential to achieve high-yield expression and accumulation of this protein in a given plant species. The present chapter describes simple procedures aimed at predicting and detecting proteolytic processes likely to occur or actually taking place in transgenic plants expressing recombinant proteins.

2. Materials

2.1. Plant Material

Transgenic and nontransgenic plants of a given species grown in greenhouse or field conditions are used for the experiments. The transgenic plants must accumulate a recombinant protein not produced (heterologous protein) or produced in lower amounts (homologous protein) in the wild nontransformed plants.

2.2. Special Laboratory Tools and Materials

1. Liquid nitrogen.
2. A mortar and a pestle.
3. Miracloth (Calbiochem, La Jolla, CA).

2.3. Buffers and Other Solutions

All buffers and solutions are made up as aqueous preparations. Solutions for standard and widely used procedures, such as sodium dodecyl sulfate-polyacrylamide gel electrophoresis (SDS-PAGE), protein blotting, and immunostaining, are not described here. Detailed, step-by-step protocols for SDS-PAGE and protein blotting are described in refs. *14* and *15*, respectively. For immunostaining, simple and efficient procedures involving the use of alkaline phosphatase- or peroxidase-labeled secondary antibodies, and appropriate reagents for chromogenic development, are usually described in detail by the suppliers.

1. Leaf extraction buffer I: 50 mM sodium phosphate, pH 7.5, 20 mM sodium metabisulfite. Prepare fresh.
2. Leaf extraction buffer Ia: Leaf extraction buffer I supplemented with 2 mM EDTA, 1 mM phenylmethylsulfonyl fluoride (PMSF, Sigma, St. Louis, MO),

1 μ*M* pepstatin (Sigma), and 10 μM *trans*-epoxysuccinyl-L-leucylamido (4-guanido) butane (E-64, Sigma). Prepare fresh.

3. Leaf extraction buffer II: 50 m*M* Tris-HCl, pH 7.5, 2 m*M* dithiothreitol (Sigma), 1 m*M* EDTA, 100 μ*M* leupeptin (Sigma), 5 m*M* MgCl$_2$, 5 mM ATP (Sigma), and 10 m*M* phosphocreatine (Sigma). Prepare fresh.

4. Leaf extraction buffer IIa: Leaf extraction buffer II from which MgCl$_2$, phosphocreatine, and ATP are omitted.

5. Proteolysis buffer: 400 m*M* sodium acetate, pH 4.0, 5 m*M* L-cysteine (Sigma). Prepare 400 m*M* sodium acetate and adjust to pH 4.0 with glacial acetic acid. Add L-cysteine before use.

6. Electrophoresis sample buffer (double strength): 62.5 m*M* Tris-HCl, pH 6.8, 2% (w/v) SDS, 20% (w/v) glycerol, 10% (v/v) β-mercaptoethanol, 0.01% (w/v) bromophenol blue. Add β-mercaptoethanol fresh. Nonreduced buffer stable for weeks at 4°C.

7. Protein preparation: The protein of interest is dissolved in a suitable buffer at a concentration of about 1 mg/mL. The protein must be prepared in a low ionic strength buffer (e.g., 25–50 m*M* salt) adjusted to a pH ensuring its stability. Ideally, the pH of the protein-containing buffer should be between 4.0 and 8.0. Store at –20°C until use. Stability depends on each particular protein.

8. Trichloroacetic acid (TCA) solution: 10% (w/v) TCA. Keep at 4°C until use. Stable for months.

9. Acetone solution: 80% (v/v) acetone. Keep at –20°C until use. Stable for months.

10. Ubiquitin preparation: 5 mg/mL ubiquitin in leaf extraction buffer II. Prepare fresh.

11. Creatine phosphokinase solution: 100 units/mL creatine phosphokinase (from rabbit muscle, E.C.2.7.3.2; Boehringer Mannheim, Indianapolis, IN) in 100 m*M* sodium phosphate, pH 7.5. Keep at –20°C until use. Stable for a few weeks.

2.4. Gel Electrophoretic Systems

1. A Mini-Protean II™ slab gel unit (Bio-Rad, Richmond, CA).
2. Gels and solutions for standard SDS-PAGE (*14,16*; Bio-Rad).
3. Gel-staining solution: 0.1% (w/v) Coomassie brilliant blue in 25% (v/v) isopropanol/10% (v/v) acetic acid. Dissolve Coomassie blue in isopropanol before adding water and acetic acid.
4. Gel-destaining solution: 10% (v/v) isopropanol, 10% (v/v) acetic acid.
5. Standard materials for immunoblotting analysis (*see* **ref. *17*** for details).
6. Specific polyclonal antibodies directed against the protein to be analyzed.
7. Specific polyclonal antibodies directed against ubiquitin (Sigma).

3. Methods

3.1. Predicting and Detecting Nonspecific Mildly Acidic Proteolysis

The procedures presented in this section describe a simple way to predict or detect the susceptibility of a given protein to limited or extensive proteolysis by the various "nonspecific" plant leaf proteases active in the mildly acidic pH range (*see* **Note 1**). The protocols were developed to analyze the susceptibility of the protein before or after insertion of its DNA coding sequence into the host plant genome. The protocol involving nontransgenic leaf extracts was devised to predict eventual instability problems, and that involving extracts from transgenic plants was developed to detect degradation actually occurring *in planta*. These protocols are suitable for the analysis of leaf proteases from any species, without the need for important modifications.

3.1.1. Extraction of the Host-Plant Soluble Proteins

1. Grind 5 g of mature leaves from a nontransgenic plant to a fine powder in liquid nitrogen, using a mortar and a pestle (*see* **Note 2**).
2. Transfer the leaf powder to an ice-cold plastic tube, and add 20 mL of cold leaf extraction buffer I supplemented with 0.5% (w/v) polyvinylpyrrolidone (PVP).
3. After extraction of soluble proteins on ice for 30 min, pass the mixture through one layer of Miracloth to remove leaf tissue debris, and centrifuge (15,000*g*) the aqueous solution for 15 min at 4°C. Discard the pellet, and use the supernatant (leaf extract I) as a source of plant endogenous proteases for further analyses. If not used immediately, the extract can be quick-frozen in liquid nitrogen and stored at –80°C for several weeks without loss of protease activity.

3.1.2. Detection of Degradation In Vitro

1. Place 24 µL of leaf extract I in two microcentrifuge tubes. Add 6 µL of the protein preparation and 6 µL of proteolysis buffer in each tube. The resulting pH in the reaction mixture is about 4.5–5.0, as in plant cell vacuoles.
2. Place one tube in a water bath for an appropriate period (e.g., 4 h) at 37°C to allow degradation of the protein by leaf proteases. Keep the other tube on ice (control sample) until TCA precipitation (**step 3**).
3. Stop the reaction by adding 36 µL of cold TCA solution. Keep the tubes standing on ice for 15 min to allow efficient precipitation of the proteins. Centrifuge (12,000*g*) for 10 min at 4°C.
4. Remove supernatant and resuspend the protein-containing pellet in 75 µL of the acetone solution. Centrifuge (12,000*g*) for 10 min at 4°C, and remove the supernatant.
5. Dissolve the pellet in 30 µL of SDS-PAGE sample buffer (double strength). This sample contains the host plant proteins and the protein of interest after its (possible) cleavage by endogenous plant proteases.
6. Perform standard SDS-PAGE *(14,16)* using the Mini-Protean II slab-gel unit from Bio-Rad. Load the samples onto a 0.75-mm thick gel (10 µL per well), and perform migration at 200 V, until the bromophenol blue tracking dye reaches the bottom of the gel.
7. After electrophoresis, incubate the gel for 1 h at 37°C in the gel staining solution, and destain at room temperature in the gel-destaining solution to visualize the

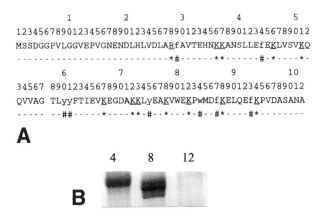

```
         1         2         3         4         5
12345678901234567890123456789012345678901234567890123456789012
MSSDGGPVLGGVEPVGNENDLHLVDLARfAVTEHNKKANSLLEfEKLVSVKQ
- - - - - - - - - - - - - - - - - - - - - - - - - - - -*#- - - - - -**- - - - - - -#- *- - - - -*-
         6         7         8         9        10
34567 890123456789012345678901234567890123456789012
QVVAG TLyyFTIEVKEGDAKKLyEAKVWEKPwMDfKELQEfKPVDASANA
- - - - -  - -##- - - - -*- - - - -**-#- -*- - - -*-#- -#*- - - -#*- - - - - - - -
```

A

```
  4    8   12
```

B

Fig. 1. Degradation of oryzacystatin I (OCI) by strawberry leaf proteases active in the mildly acidic pH range. (**A**) OCI amino acid sequence, showing the occurrence of multiple basic (*) and aromatic (#) residues on each side of the active site, residues 53–57 (sequence from **ref. 26**). Such residues are recognized by various poorly specific endoproteases. (**B**) Coomassie blue-stained SDS-polyacrylamide gel showing limited proteolysis and subsequent complete hydrolysis of OCI by strawberry leaf vacuolar proteases active at pH ~4.5, after 4, 8, and 12 h incubations. OCI was prepared in *E. coli* as described previously (*27*), and diluted to a concentration of 1 mg/mL in a 50 m*M* Tris-HCl buffer, pH 8.0.

proteins resolved (*see* **Note 3**). By comparing protein patterns from test and control samples, estimate the susceptibility of the protein of interest to proteolytic cleavage by the host-plant endogenous proteases active in the mildly acidic pH range. In particular, verify the presence of proteolytic fragments resulting from limited proteolysis of the protein (*see* **Note 4**). As an example, **Fig. 1** shows the effect of strawberry leaf endogenous proteases on the stability of oryzacystatin I, a protein cysteine proteinase inhibitor from rice (*19*), potentially useful for pest control (*20–25*), but containing multiple potential cleavage sites in its primary structure (*see* **Fig. 1A**).

3.1.3. Detection of Degradation In Vivo

1. Extract leaf proteins from a transgenic plant expressing the protein of interest as described previously for nontransgenic plants (*see* **Subheading 3.1.1.**, **steps 1–3**), but use leaf extraction buffer Ia (with protease inhibitors) instead of the leaf extraction buffer I (*see* **Note 5**).
2. Perform standard SDS-PAGE (*14,16*) using the Mini-Protean II slab-gel unit from Bio-Rad. Load the sample on a 0.75-mm-thick gel (20 µL per well), and perform migration at 200 V until the bromophenol blue tracking dye reaches the bottom of the gel. The amount of extract loaded on the gel might have to be calibrated for each particular case.

3. After electrophoretic migration, transfer the proteins resolved onto a nitrocellulose sheet and perform an immunostaining using polyclonal antibodies directed against the protein of interest as primary antibodies. By comparison with a control well containing the pure protein, look for the presence of degradation intermediates recognized by the polyclonal antibodies on the nitrocellulose sheet (*see* **Note 4**). Also look for the presence of multiple bands with molecular masses larger than that of the protein. The occurrence of such bands could indicate degradation of the protein in the transformed cells via the ubiquitin-mediated pathway (*see* **Subheading 3.2**).

3.2. Predicting and Detecting Ubiquitin-Mediated Proteolysis

The protocols presented in this section describe a simple way to predict (in vitro, before transformation) or to detect (in vivo, after transformation) the occurrence of ubiquitin conjugates in recombinant protein-expressing transgenic plant leaf cells. The procedures are based on an intrinsic property of the ubiquitin system, which can be activated after cell breakage only when a suitable source of energy is supplied, allowing in vitro analyses to be performed while avoiding artifactual hydrolytic processes. These protocols are suitable for the analysis of leaf proteins from several species, without the need for major alterations.

3.2.1. Extraction of the Host-Plant Soluble Proteins

Extract leaf proteins from a nontransgenic plant as described previously (*see* **Subheading 3.1.1.**, **steps 1–3),** but use 10 mL of leaf extraction buffer II, instead of 20 mL of leaf extraction buffer I. Use the resulting extract as a source of primary components of the ubiquitin system for predicting the occurrence of ubiquitin conjugates in the plants to be transformed (*see* **Subheading 3.2.2.**). If not used immediately, the extract can be quick-frozen in liquid nitrogen and stored at –80°C for several days without loss of activity.

3.2.2. Detection of Degradation In Vitro

1. Place 24 µL of the leaf extract II in 6 microcentrifuge tubes. Add 3 µL of the protein preparation, 3 µL of the ubiquitin preparation, 3 µL of the creatine phosphokinase solution, and 3 µL water in each tube.
2. To initiate ubiquitin-dependent proteolysis, place the tubes in a water bath at 30°C for 0, 15, 30, 60, 90, or 120 min. Keep the tubes standing on ice until initiating the reaction.
3. Perform a TCA/acetone double precipitation, as described (*see* **Subheading 3.1.2.**, **steps 3–5**), to block the reaction and to recover the proteins from the mixture.
4. Perform standard SDS-PAGE *(16)* using the Mini-Protean II slab-gel unit from Bio-Rad. Load the sample onto two 0.75-mm thick gels (12 µL per well), and

perform migration at 200 V, until the bromophenol blue tracking dye reaches the bottom of the gel.

5. After migration, transfer proteins from the two gels onto nitrocellulose sheets. With one nitrocellulose sheet, perform an immunostaining using polyclonal antibodies, directed against the protein of interest, as primary antibodies. With the other sheet, perform an immunostaining with the antibody directed against ubiquitin.

6. Look for the presence on blots of a series of bands spaced at intervals of about 8 kDa above the molecular mass of the native (nonconjugated) protein. The occurrence of such ladders on both nitrocellulose sheets demonstrates the recognition of the protein by the host plant ubiquitin system components, and its eventual instability in transgenic plant cells if accumulated in the cytoplasm or in the nucleus (*see* **Note 6**).

3.2.3. Detection of Degradation In Vivo

1. Extract leaf proteins from a transgenic plant expressing the protein of interest, as described previously for nontransgenic plants (*see* **Subheading 3.2.1.**), but use leaf-extraction buffer IIa, instead of the leaf extraction buffer II. Use this extract (leaf extract IIa) for detecting ubiquitin conjugates formed *in planta*.

2. Perform standard SDS-PAGE and immunoblotting, as described in **Subheading 3.2.2.**, **steps 4** and **5**.

3. Look for the presence on blots of a series of bands spaced at intervals of 8 kDa above the molecular mass of the native (nonconjugated) protein. Because of the very negligible activity of the ubiquitin system at low temperatures and the rapid depletion of ATP in crude extracts, it appears unlikely that the ubiquitin conjugates detected are the result of a ubiquitination process that occurred during extraction. Rather, the detection of protein ladders on both nitrocellulose sheets strongly suggests ubiquitin-mediated degradation of the protein in vivo, before lysis of the plant cells (*see* **Note 6**).

The minimization of proteolysis in transgenic plants is discussed in **Note 7**.

4. Notes

1. Although confirmation has to be obtained in each specific case by extensive localization studies, the bulk of nonspecific protease activity in plant leaf cells is generally thought to be associated with the vacuole *(11–13)*, especially the activity measured in the mildly acidic pH range. Future progress in identifying specific vacuolar proteases of plant leaf cells will surely help to develop specific tests for each particular group of enzymes, as is notably the case for yeast vacuolar proteases *(7)*.

2. Because the proteolytic status generally varies with leaf development (e.g., during senescence), the analysis of proteases from young and senescent leaves may be useful for comparison purposes.

3. In some cases, it is difficult to discriminate the protein tested from the fragments of this protein, or from other proteins present in the plant extract. In such cases, perform SDS-PAGE *(16)*, transfer the proteins onto a nitrocellulose sheet *(15)*, and identify the protein by immunostaining, using the appropriate primary antibody.

4. According to Linderstrom–Lang *(18)*, there are two extreme types of enzymatic proteolysis: limited (or restricted) and random. Limited proteolysis leads to the formation of relatively stable high-molecular weight intermediate fragments; random proteolysis leads to the breakdown of proteins into small peptidic products without the accumulation of detectable stable intermediates. The majority of protein-protease interactions leads to the accumulation of protein intermediates detectable on gel.

5. Protein degradation is a common phenomenon observed during the extraction of proteins from various plant sources. It occurs as soon as the cells are broken down, potentially leading to a significant alteration of the protein patterns observed on gels, and to artifact-based conclusions about biochemical events taking place in vivo. A simple way to minimize proteolysis during extraction requires the addition of proteinase inhibitors and chelating agents to the extraction buffer, and maintainance of the extraction mixtures at low temperatures *(3)*.

6. The steady-state level of some ubiquitin conjugates is very low, especially for those with a high degradation rate *(28)*. In such cases, the detection of conjugates may require more sensitive approaches (e.g., use of I^{125}-labeled secondary antibodies during immunoblotting and autoradiography of the nitrocellulose sheets, or use of I^{125}-labeled protein and autoradiography of the polyacrylamide gel). Alternatively, ubiquitin-mediated general degradation of the protein may be monitored by immunoblotting (*see* **Subheading 3.2.2.**, **steps 4–5**) carried out after inhibitions with various inhibitors of the ubiquitin-dependent proteolytic pathway (*see* **ref.** *10* for details of the various inhibitors available). For inhibition analyses, the inhibitor solutions diluted to the appropriate concentrations replace the water fraction (3 µL) in the reaction mixture (*see* **Subheading 3.2.2.**, **step 1**).

7. Despite the recent and major advances in the field of plant proteolytic metabolism (reviewed in **refs.** *12* and *13*), little is known about the specific mechanisms underlying the action of plant proteases against either endogenous or exogenous protein substrates. Unlike for bacterial *(5,8)* and yeast *(6,7)* expression systems, the resident proteases of plant cells are not known in detail, and specific mutants lacking proteases potentially damaging to recombinant proteins are not presently available. At this time, targeting recombinant proteins in cellular locations different from the sites of action of proteases and ubiquitin appears as the principal way to effectively overcome, or at least minimize, unwanted proteolysis *in planta*. Results from several experiments suggest that most of the proteases found in vegetative organs of plants are cysteine-type endoproteases active in the mildly acidic pH range, and possibly localized into the vacuole *(13,29,30)*. In parallel, it is now well-established that ubiquitination of damaged and abnormal proteins takes place in the cytoplasm. Fusing a peptide signal to a recombinant protein to

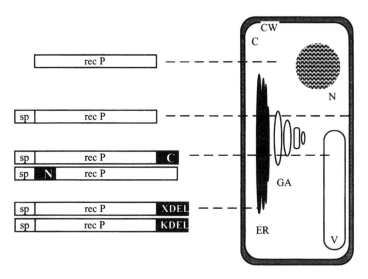

Fig. 2. Targeting recombinant proteins in plant cells. Proteins, including in their primary sequence a signal, like the N-terminal signal peptide (sp) present in several proteins, enter the secretion pathway via the endoplasmic reticulum (ER), while those with no such signal are accumulated in the cytoplasm (C), or subsequently directed to other organelles (e.g., nucleus, mitochondria, chloroplasts, peroxisomes, glyoxisomes) via a cytoplasmic route. Depending on the presence or not of specific additional signals in their primary structure, the sp-containing secreted proteins are retained in the ER (KDEL of HDEL sequence at their C-terminal end [e.g., **refs. *35–37***]) or tranferred to the Golgi apparatus (GA), from which they are directed to the vacuole (V) (C- and N-terminal signal sequences [e.g., **refs. *33,34,38,39***]) or are secreted out of the cell via a default secretory pathway (e.g., **ref. *40***). CW, cell wall; N, nucleus.

direct its accumulation in extracellular compartments, or causing its retention in the endoplasmic reticulum, represents, in this case, an interesting alternative to noncontrolled targeting. Peptide signals directing translocation of proteins in the different cellular compartments of plant cells have been partially elucidated (*see* reviews in **refs. *11*** and ***31***), and their fusion to recombinant proteins using appropriate genetic constructs now appears plausible and even functional (*32–35*) for specifically controlling the final destination of these proteins in transgenic plant cells **(Fig. 2)**. The accumulation of the pea vacuolar storage protein vicilin, for instance, was increased by 100-fold in transgenic mesophyll cells of alfalfa when the tetrapeptidic signal KDEL (Lys-Asp-Glu-Leu) was fused to its carboxy-terminal end (*35*). This increase was associated with a retention of the protein in the endoplasmic reticulum, preventing its normal progression in the secretory pathway and its (eventual) degradation in the vacuole (*31,35*), a demonstration of the high potential of controlled targeting of recombinant proteins in transgenic plants.

Acknowledgments

The authors thank Line Cantin for critical reading of the manuscript. This work was partly supported by a fellowship from the Natural Science and Engineering Research Council of Canada to D. Michaud.

References

1. Wolf, H. (1992) Proteases as biological regulators. *Experientia* **48,** 117–118.
2. North, M. J. (1989) Prevention of unwanted proteolysis, in *Proteolytic Enzymes: A Practical Approach* (Beynon, R. J. and Bond, J. S., eds.), IRL, New York, pp. 105–124.
3. Michaud, D. and Asselin, A. (1995) Review. Application to plant proteins of gel electrophoretic methods. *J. Chromatogr. A* **698,** 263–279.
4. Emr, S. D. (1990) Heterologous gene expression in yeast. *Methods Enzymol.* **185,** 231–233.
5. Maurizi, M. R. (1992) Proteases and protein degradation in *Escherichia coli. Experientia* **48,** 178–201.
6. Jones, E. W. (1991) Minireview. Three proteolytic systems in the yeast *Saccharomyces cerevisiae. J. Biol. Chem.* **266,** 7963–7966.
7. Jones, E. W. (1990) Vacuolar proteases in yeast *Saccharomyces cerevisiae. Methods Enzymol.* **185,** 372–386.
8. Gottesman, S. (1990) Minimizing proteolysis in *Escherichia coli:* genetic solutions. *Methods Enzymol.* **185,** 119–129.
9. Ciechanover, A. (1994) Review. The ubiquitin-proteasome proteolytic pathway. *Cell* **79,** 13–21.
10. Wilkinson, K. D. (1990) Detection and inhibition of ubiquitin-dependent proteolysis. *Methods Enzymol.* **185,** 387–397.
11. Chrispeels, M. J. and Raikhel, N. V. (1992) Minireview. Short peptide domains target proteins to plant vacuoles. *Cell* **68,** 613–616.
12. Vierstra, R. D. (1993) Protein degradation in plants. *Annu. Rev. Plant Physiol. Plant Mol. Biol.* **44,** 385–410.
13. Callis, J. (1995) Regulation of protein degradation. *Plant Cell* **7,** 845–857.
14. Smith, B. J. (1984) SDS polyacrylamide gel electrophoresis of proteins, in *Methods in Molecular Biology,* vol. 1: *Proteins* (Walker, J. M., ed.), Humana, Clifton, NJ, pp. 41–45.
15. Gooderham, K. (1984) Transfer techniques in protein blotting, in *Methods in Molecular Biology,* vol. 1: *Proteins* (Walker, J. M., ed.). Humana, Clifton, NJ, pp. 165–178.
16. Laemmli, U. K. (1970) Cleavage of structural proteins during the assembly of the head of bacteriophage T4. *Nature* **227,** 680–685.
17. Bjerrum, O. J. and Heegaard, N. H. H. (1988) *Handbook of Immunoblotting of Proteins,* vol. I. *Technical Descriptions.* CRC, Boca Raton.
18. Linderstrom-Lang, K. (1952) *Lane Medical Lectures,* vol. 6. Stanford University Press, Stanford, pp. 53–72.

19. Abe, K., Hiroto, K., and Arai, S. (1987) Purification and characterization of a rice cysteine proteinase inhibitor. *Agric. Biol. Chem.* **51,** 2763–2765.
20. Liang, C., Brookhart, G., Feng, G. H., Reeck, G. R., and Kramer, K. J. (1991) Inhibition of digestive proteinase of stored grain Coleoptera by oryzacystatin, a cysteine proteinase inhibitor from rice seeds. *FEBS Lett.* **278,** 139–142.
21. Chen, M.-S., Johnson, B., Wen, L., Muthukrishnan, S., Kramer, K. J., Morgan, T. D. and Reeck, G. R. (1992) Rice cystatin: bacterial expression, purification, cysteine proteinase inhibitory activity, and insect growth suppressing activity of a truncated form of the protein. *Protein Expres. Purif.* **3,** 41–49.
22. Bonadé-Bottino, M. (1993) Défense du colza contre les insectes phytophages déprédateurs: étude d'une stratégie basée sur l'expression d'inhibiteurs de protéases dans la plante. Ph.D. Thesis, Université de Paris-Sud, Centre d'Orsay.
23. Michaud, D., Nguyen-Quoc, B., and Yelle, S. (1993) Selective inactivation of Colorado potato beetle cathepsin H by oryzacystatins I and II. *FEBS Lett.* **331,** 173–176.
24. Michaud, D., Bernier-Vadnais, N., Overney, S., and Yelle, S. (1995) Constitutive expression of digestive cysteine proteinase forms during development of the Colorado potato beetle, *Leptinotarsa decemLineata* Say (Coleoptera: Chrysomelidae). *Insect Biochem. Mol. Biol.* **25,** 1041–1048.
25. Leplé, J.-C., Bonade Bottino, M., Augustin, S., Delplanque, A., Dumanois, V., Pilate, G., Cornu, D., and Jouanin, L. (1995) Toxicity to *Chrysomela tremulae* (Coleoptera: Chrysomelidae) of transgenic poplars expressing a cysteine proteinase inhibitor. *Mol. Breed.* **1,** 319–328.
26. Abe, K., Emori, Y., Kondo, H., Arai, S., and Suzuki, K. (1988) The NH_2-terminal 21 amino acid residues are not essential for the papain-inhibitory activity of oryzacystatin, a member of the cystatin superfamily. Expression of oryzacystatin cDNA and its truncated fragments in *Escherichia coli. J. Biol. Chem.* **263,** 7655–7659.
27. Michaud, D., Nguyen-Quoc, B., and Yelle, S. (1994) Production of oryzacystatins I and II in *Escherichia coli* using the glutathione *S*-transferase gene fusion system. *Biotechnol. Prog.* **10,** 155–159.
28. Bachmair, A., Finley, D., and Varshavsky, A. (1986) *In vivo* half-life of a protein is a function of its amino-terminal residue. *Science* **234,** 179–186.
29. Storey, R.D. (1986) Plant endopeptidases, in *Plant Proteolytic Enzymes* (Dalling, M., ed.), CRC, Boca Raton, pp. 119–135.
30. Canut, H., Dupré, M., Carrasco, A., and Boudet, A. M. (1987) Proteases of *Melilotus alba* mesophyll protoplasts. *Planta* **170,** 541–549.
31. Chrispeels, M. J. (1991) Sorting of proteins in the secretory system. *Annu. Rev. Plant Physiol. Plant Mol. Biol.* **42,** 21–53.
32. Bednarek, S. Y., Wilkins, T. A., Dombrowski, J. E., and Raikhel, N. V. (1990) A carboxy-terminal propeptide is necessary for proper sorting of barley lectin to vacuoles of tobacco. *Plant Cell* **2,** 1145–1155.
33. Wilkins, T. A., Bednarek, S. Y., and Raikhel, N. V. (1990) Role of propeptide glycan in post-translational processing and transport of barley lectin to vacuoles in transgenic tobacco. *Plant Cell* **2,** 301–313.

34. Neuhaus, J.-M., Sticher, L., Meins, F., Jr., and Boller, T. (1991) A short C-terminal sequence is necessary and sufficient for the targeting of chitinase to the plant vacuole. *Proc. Natl. Acad. Sci. USA* **88,** 10362–10366.

35. Wandelt, C. I., Khan, M. R. I., Craig, S., Schroeder, H. E., Spencer, D., and Higgins, T. J. V. (1992) Vicilin with carboxy-terminal KDEL is retained in the endoplasmic-reticulum and accumulates to high-levels in the leaves of transgenic plants. *Plant J.* **2,** 181–192.

36. Denecke, J., Goldman, M. H. S., Demolder, J., Seurinck, J. and Bottermann, J. (1991) The tobacco luminal binding protein is encoded by a multigene family. *Plant Cell* **3,** 1025–1035.

37. Herman, E. M., Tague, B. W., Hoffman, L. M., Kjemtrup, S. E., and Chrispeels, M. J. (1990) Retention of phytohemagglutinin with carboxyterminal tetrapeptide KDEL in the nuclear envelope and the endoplasmic reticulum. *Planta* **182,** 305–312.

38. Matsuoka, M. and Nakamura, K. (1991) Propeptide of a precursor to a plant vacuolar protein required for vacuolar targeting. *Proc. Natl. Acad. Sci. USA* **88,** 834–838.

39. Holwerda, B. C., Galvin, N. J., Baranski, T. J., and Rogers, J. C. (1990) *In vitro* processing of aleurain, a barley vacuolar thiol protease. *Plant Cell* **2,** 1091–1106.

40. Denecke, J., Botterman, J., and Deblaere, R. (1990) Protein secretion in plant cells can occur via a default pathway. *Plant Cell* **2,** 51–59.

15

Trafficking and Stability of Heterologous Proteins in Transgenic Plants

Johnathan A. Napier, Gaelle Richard, and Peter R. Shewry

1. Introduction

The ability to introduce foreign genes into plant species by techniques such as *Agrobacterium*-mediated transformation or by direct gene transfer (e.g., biolistics) has opened up the possibility of using transgenic plants as host organisms for the production of heterologous proteins (*see* **ref.** *1* for an excellent review). Although a number of host organisms (such as *Escherichia coli*, yeasts, and mammalian cell cultures) are already used for the production of recombinant proteins, plants have several advantages that make them highly attractive for this purpose. For example, the large-scale production and processing of plant material is a routine, cost-effective process carried out by every nation with any form of agricultural industry, thus reducing the need for specialized growth and harvesting equipment. Plants offer additional advantages as potential producers of recombinant proteins. Not only do they modify proteins (in terms of glycosylation, prenylation, and so on) in the same way as other higher eukaryotes *(2)*, but also, since they transmit the transgene encoding the heterologous protein in a Mendelian fashion, the seed of a transgenic plant can be used as the source of further producing lines. This seed also serves as a convenient way of storing biologically viable transgenic material without the need for either cryopreservation or continual passaging/subculturing. The potential offered by transgenic plants has been widely appreciated and anticipated, and a range of innovative strategies have been taken to either alter the endogenous compositions of plant products, or use plants as bio-reactors for the synthesis and accumulation of heterologous proteins. However, as a result of these studies, it has become apparent that there are a number of constraints

From: *Methods in Biotechnology, Vol. 3:*
Recombinant Proteins from Plants: Production and Isolation of Clinically Useful Compounds
Edited by: C. Cunningham and A. J. R. Porter © Humana Press Inc., Totowa, NJ

on the expression and stability of foreign proteins in transgenic plants. Examples were first noted when researchers tried to improve the amino acid contents of various plant species by the constitutive (and hence ectopic) expression of seed-storage proteins. It was observed that proteins, such as soybean conglycinin *(3)* and pea vicilin *(4)*, accumulated in nonseed tissues, such as leaves, at very low levels (less than 0.01%), although the mRNA encoding the transgene was present at a high level in all tissues. A similar situation was observed when an engineered form of maize zein (containing a 45-bp insert encoding six methionines) was expressed in tobacco *(5)*, though this situation is not unique to the ectopic expression of seed-storage proteins, since a human serum albumin gene introduced into tobacco under the control of the same constituitive promoter resulted in only 0.02% total protein accumulation *(6)*. Therefore, it is clear that protein stability is a key factor in regulating the accumulation of an heterologous protein in transgenic plants. In all the above examples (and the example to be described in this chapter), the heterologous proteins were synthesized with endogenous signal peptides that would target them into the secretory pathway *(7)*. Since one of the major compartments of the secretory pathway in plant cells is the vacuole, which is the location of a wide range of proteolytic enzymes, it seems likely that the fate of these foreign proteins was transport to, and degradation in, this organelle *(8)*.

Another problem associated with the expression of foreign proteins in transgenic plants may arise when the protein of interest requires specific post-translational processing or modification, for example, proteins, such as a number of seed-storage proteins, containing disulphide bonds, whose formation is assisted catalytically, and must occur in a specific order to produce a correctly folded protein. Disulphide bond formation and rearrangement are carried out by enzymes, such as the ER-lumenal protein disulphide isomerase (PDI), and by thioredoxins localized in other subcellular compartments *(9)*. Catalytically assisted disulphide bond formation has been described as a chaperone-mediated process and is not spontaneous *(2,7)*. Consequentially, if a foreign protein is expressed and targeted to the wrong subcellular compartment, i.e., one that does not contain the folding chaperone, then it is unlikely to be synthesized as an active protein. Attempts have been made in other expression systems to co-express chaperones, such as BiP, and enzymes, such as thioredoxin and PDI, with the foreign protein of interest *(10)*. It must also be remembered that the disulfide-forming enzymes will only be able to function when in the correct oxo-reductive environment. It is therefore clear that some plant subcellular domains are more suitable as target compartments for the expression of foreign proteins. Such compartmentalization, as a method for increasing the stability of heterologous protein in transgenic plants, has been exploited by other workers who have targeted transgene products to either the chloroplast *(11)* or the oil body *(12)*.

It is also clear from a number of studies that the lumen of the ER is a highly benign environment for proteins *(13)*, and it also contains a wide range of chaperones and folding enzymes, such as PDI, BiP, and s-cyclophilin *(7)*. Entry of a protein into the ER lumen is mediated by a signal sequence of about 20 residues, and this process is a co-translational event requiring the signal recognition particle (SRP) *(14)*. However, the ER lumen is only the starting point of the secretory pathway, and it has been demonstrated that trafficking through this network is by a default system *(15)*, i.e., if no further signaling determinants are present in the protein loaded into the secretory system, then the protein will be exported to the default destination. In mammalian cells, the default destination of the secretory pathway is the cell surface *(16)*, but in yeast it appears that the default destination can be the vacuole *(2)*, and a similar situation may occur in some plant cell types. In other cases, transport to the vacuole requires specific targeting sequences *(2)*. Therefore, to exploit the hospitality of the ER lumen as a compartment for the expression of a heterologous protein, it is vital to prevent the same protein being trafficked by default out of the ER and into the protease-rich environment of the vacuole. Fortunately, the mechanism by which ER-lumenal proteins themselves are retained in the ER is well-characterized and it has been shown that a tetrapeptide motif, usually KDEL or HDEL, present at the extreme C-terminus of the protein, is necessary and sufficient to cause ER retention, or at least recycling back to the ER *(17)*. In the example given in this chapter, we wished to ectopically express a cereal seed-storage protein in transgenic tobacco. All seed-storage proteins are synthesized as precursors containing signal sequences to direct them into the ER lumen; therefore, we wished to compare the stability of a wild-type protein with that of a protein that had either an KDEL or HDEL ER-retention signal added to it. From our results, it was clear that the accumulation of the wheat seed-storage protein in the leaves of transgenic tobacco plants was greatly increased by the presence of the C-terminal retention signal and that this resulted from an increased half-life of the transgene product. In the absence of a retention signal, the ectopically expressed protein was co-translationally loaded into the secretory system by translocation into the ER, but was then very rapidly degraded in a post-ER compartment (probably the vacuole). These experiments demonstrate that the ER lumen may indeed be a suitable environment to allow the accumulation of heterologous proteins and that the simple addition of an ER-retention tetrapeptide to proteins destined to enter the secretory pathway, e.g., storage proteins, results in increased stability. This obviously has important implications in a number of situations; for example, the expression and stability of a 2S albumin seed-storage protein, with a biologically active peptide (leu-enkephalin) inserted into its coding sequence *(18)*, might be predicted to be increased by the addition of an ER-retention signal. It may also be used to improve the amino acid content of fodder crops by allowing the accumulation in the leaf of proteins (such as a sunflower storage protein) with high contents of sulphur amino acids *(7,13)*. The described system used proteins with their own endogenous N-terminal signal sequences, to target them into the ER lumen, but clearly the principle of exploiting the ER lumen as a resting place for heterologous proteins can be extended beyond secretory proteins by the simple addition of the above-mentioned N-terminal SRP signal sequence *(2,14)*. Thus, the ER lumen could be exploited for its ability to fold and assemble proteins and for its low level of proteolytic activity. One other advantage of the addition of an ER retention signal to a protein of interest is that highly specific monoclonal antibodies (2E7/MAC 256) are available to these epitopes *(19)*, facilitating the monitoring of the transgene product.

2. Materials

2.1. DNA Material

1. The DNA encoding the protein of interest, which is to be expressed in transgenic plant, should ideally be a cDNA with as short a 5' leader sequence prior to the initiating methionine as possible.
2. The cloning vector of choice in this study, the pGEM3Zf(-) vector (Promega, Madison, WI), was used to clone the PCR-modified cDNA. The choice of plant-transformation vector will depend on the plant-transformation system to be used and the plant species employed. We used the Ω–enhanced 35S promoter cassette pJD330 *(20)* and the binary vector pBIN19 *(21)*.

2.2. Bacterial Strains

1. Standard *E.coli* cells for propagating cloned DNA, e.g., JM109, MC1022, SURE™.
2. Appropriate *Agrobacterium* strain relevant to the transformation vector used. In this study, the strain LBA4404 was used.

2.3. Plant Material

Tobacco plants (*Nicotiana tabacum* L. cv. Samsun) were grown under sterile conditions on agar and were used as a source of leaf disks for transformation.

2.4. Laboratory Equipment

1. PCR thermocycler (Cetus/Perkin-Elmer 480).
2. Gene-Pulser Electroporator (Bio-Rad).

2.5. Buffers and Other Solutions

2.5.1. PCR Reaction

1. Thermo-reaction buffer: 500 mM KCl, 100 mM Tris-HCl, pH 9.0, 1% Triton X-100 (usually provided with the commercially available *Taq*-polymerases).
2. 25 mM MgCl$_2$.
3. dNTP mix: A 2.5 mM solution containing dATP, dCTP, dGTP, and dTTP in sterile distilled water (SDW).

2.5.2. Protein Extraction and SDS-PAGE

1. Protein extraction buffer: 100 mM Tris-HCl, pH 8.0, 10 mM EDTA, 5% (w/v) SDS, 20% (w/v) sucrose, 5% (v/v) 2-mercaptoethanol.
2. SDS-PAGE gel buffer: 1.5M Tris-HCl, pH 8.8.
3. Acrylamide stock solution: 30 g (w/v) acrylamide, 0.8 g (w/v) *bis*-acrylamide dissolved in 100 mL of SDW.
4. 10% (w/v) SDS: 10 g sodium dodecyl sulphate dissolved in 100 mL SDW.
5. Stacking gel buffer: 0.5M Tris-HCl, pH 6.8.
6. SDS-PAGE running buffer: 25 mM Tris, 192 mM glycine, 0.1% (w/v) SDS.
7. Coomassie staining solution: 30 mL methanol, 10 mL glacial acetic acid, 70 mL water, 0.1g (w/v) Coomassie BBR250 blue.
8. Destain solution: 10% (v/v) methanol, 10% (v/v) glacial acetic acid in water.

2.5.3. Western Blotting

1. Transfer buffer: 25 mM Tris, 192 mM glycine, 0.02% (w/v) SDS, 20% (v/v) methanol.
2. TBS: 20 mM Tris-HCl, 0.5M NaCl, pH 7.5.
3. Blocking buffer: 3% (w/v) BSA (bovine serum albumin) in TBS.
4. Binding buffer: 1% (w/v) BSA in TBS.
5. Carbonate buffer: 0.1M NaHCO$_3$, 1 mM MgCl$_2$, pH 9.8 (adjusted with NaOH).
6. BCIP solution: 25 mg/mL 5-bromo-4-chloro-3-indolyl phosphate dissolved in 100% N,N-dimethylformamide (DMF).
7. NBT solution: 25 mg/mL nitro blue tetrazolium dissolved in 70% (v/v) DMF.

2.5.4. Protoplast Isolation and Immunoprecipitation

1. Protoplast incubation (PIN) buffer: 0.4M sucrose, 6 mM CaCl$_2$, 25 mM KNO$_3$, 5 mM MES-pH 5.7, 3 mM NH$_4$NO$_3$, 1 mM MgSO$_4$, 1 mM (NH$_4$)$_2$SO$_4$, 600 μM Xylose, 367 μM CaHPO$_4$, 100 μM Na$_2$-EDTA, 100 μM FeSO$_4$, 55 μM m-inositol, 30 μM thiamine-HCl, 25 μM H$_3$BO$_3$, 3 μM MnSO$_4$, 3 mM ZnSO$_4$, 2 μM KI, 1 μM nicotinic acid, 500 nM Na$_2$MoO$_4$, 500 nM pyridoxine HCl (*see* **Note 1**).
2. Protoplast isolation buffer: 1% (w/v) cellulase (R10), 0.1% (w/v) macerozyme (R10) dissolved in PIN buffer. This solution should be filter-sterilized.

3. Immunoprecipitation buffer: 50 mM Tris-HCl, pH 7.5, 150 mM NaCl, 1% (v/v) Nonidet P-40 (NP-40), 0.5% (w/v) sodium deoxycholate, 0.1% (w/v) SDS.
4. Washing buffer: 10 mM Tris-HCl, pH 7.5, 0.1% (v/v) NP-40.

3. Methods

3.1. Addition of an ER-Retention Signal by PCR

The aim of this procedure is the addition of a C-terminal tetrapeptide ER-retention signal to the DNA encoding the protein of interest (*see* **Note 2**). The ER-retention signal may take the form of HDEL or KDEL, since both work efficiently in plants. The simplest method for the introduction of this tetrapeptide is by the use of the polymerase chain reaction (PCR). A complementary oligonucleotide is designed to anneal to the region of the DNA encoding the protein of interest, so that it spans the last six amino acids, but excludes the stop codon. The ER-retention signal and stop codon are then introduced into the oligo as an in-frame fusion; although this region of the oligo will be noncomplementary to the target template, this mismatch will be tolerated for the initial cycle of amplification, since the nonannealed extension will be present only at the 5' end of the resulting duplex. In this respect, it is also possible to introduce a restriction site to aid the subsequent cloning of the amplified product. Thus, the PCR will require the synthesis of a mutagenic oligonucleotide primer of about 45 bases, including the restriction site of choice and the addition of two or three spacer bases at the 5' end to enhance the restriction digestion of the template. A second primer is required for the amplification of the DNA by the PCR, but usually it is convenient to use the M13 universal forward or reverse sites, since they are present in virtually all standard cloning vectors (it is assumed that the investigator already has a cloned template encoding the protein of interest; if this is not the case, then a primer will be required that anneals to the region surrounding the intiating methionine of the coding sequence). It is not the aim of this chapter to describe PCR amplification in detail, since a number of technical bulletins are available (Promega) which discuss some of the practical problems, but the basic protocol is given below. It is also assumed that the investigator is familar with standard recombinant DNA techniques; however, if not, then the reader is referred to ref. *22* as an excellent short introductory text.

1. Make up a 100 μL master mix on ice containing 10 μL 10X Thermo-reaction buffer, 10 μL 25 mM MgCl$_2$, 8 μL 2.5 mM dNTP mix, 1 μL each of the two oligonucleotide primers (50 ng/μL), and 70 μL of sterile distilled water. Finally, add 0.1 μL of *Taq*-polymerase (@5 U/μL) and mix well using a pipet; this mixture can be scaled-up in these proportions.

2. Aliquot 75 µL of this mix into a 0.5 mL Eppendorf tube containing approx 20 ng of template DNA; use the other 25 µL as a negative (no template) control reaction. Overlay the samples with mineral oil.

3. Using a suitable thermocycler (such as the Cetus/Perkin-Elmer 480), carry out the following amplification reaction. Initially, heat the samples at 94°C for 2 min, then perform 35 cycles at 94°C, 1 min; 50°C, 1 min; and 72°C, 1 min.

4. Check that the PCR amplification has been successful by analysis of a fraction of the sample and the negative control on an agarose gel. If the resulting product is of the predicted size and the negative control is blank, then the remaining portion of the experimental sample should be phenol-extracted, ethanol-precipitated, restricted with the appropriate enzyme(s) overnight, and then gel-purified using a method such as GeneClean (Bio 101). The recovery of this fragment should be checked by running one-tenth of the sample on an agarose gel, as above.

5. The purified template should then be cloned into an appropriate vector; this could be either a standard cloning vector, such as pGEM3 (Promega), or into a promoter/terminator cassette (in the given example, we used pJD330, a 35S/Ω-enhancer/nos terminator cassette present in pUC18) suitable for direct subcloning into a transformation vector such as pBIN19. In the latter case this obviously reduces the number of cloning steps, but since the vector already contains an insert, there is no blue/white selection to aid the identification of transformants. It is also easier to sequence the cloned PCR product in the smaller cloning vectors (*see* **Note 3**).

6. Once it has been established that the modified gene is correctly inserted into the required promoter/terminator cassette, then this entire fragment (i.e., promoter/PCR-modified coding sequence/terminator) can be introduced into a plant transformation vector (in the given example, the *Agrobacterium* binary vector pBIN19 was used); the choice of transformation vector and method used will be made by the individual investigator, since different plant species require different methods. In this study, the resulting plasmid construct was transformed into *Agrobacterium* by electroporation, and recombinant cells were then used to infect tobacco leaf disks using the method described by Horsch et al. *(23)*. Briefly, leaf disks were placed on selective medium after 2 d of *Agrobacterium* infection, and resulting callus tissue was allowed to form leaves, which were transferred to rooting media. Rooted plantlets were then transferred to the glasshouse and grown under controlled and confined conditions.

3.2. Analysis of Transgenic Plants

When a transgenic plant or cell line has been produced, with the goal of expressing a heterologous protein at high levels, the easiest and most convenient method of screening the population of primary transformants is by Western blotting; this involves the extraction of total proteins from the appropriate plant tissue, analysis by SDS-PAGE, transfer to nitrocellulose, and detection of transgene product by the use of an antibody. If no specific antibody is available, then, as mentioned above, the ER-retention tetrapeptide can serve as an epitiope tag. However, it is probable that most investigators who have progressed to the stage of producing recombinant protein in transgenic plants will have a suitable antibody available to detect the protein of interest. In the given example, a rabbit polyclonal antiserum was used, though a similar protocol can be used for monoclonal antibodies. Detection of the antibody/antigen complex is carried out by using a commercially available secondary antibody that will recognize total rabbit IgG; this antibody is also linked to a detection system, in this case, alkaline phosphatase. The resulting sandwich of antigen/1st antibody/ 2nd antibody-conjugate can be detected by the addition of a chlorometric substrate. In the given example, ectopically expressed wheat storage proteins are detected in the leaves of transgenic tobacco plants, but only if the wheat proteins have been modified by the addition of a ER-retention signal (**Fig 1.**).

1. Homogenize 0.1 g of leaf tissue in 1 mL of protein extraction buffer; this can be performed in an Eppendorf tube with a teflon Eppendorf homogenizer (Scotlab). Denature the sample by heating to 95°C for 5 min and then centrifuge for 5 min at 13,000g. Remove clarified supernatant into a fresh tube and retain.

2. Pour two 12% (w/v) polyacrylamide gels (the acrylamide concentration of the gel may be varied, depending on the size of the protein of interest); we routinely use a minigel system (Bio-Rad). For two gels, mix 2.5 mL of SDS-PAGE gel buffer with 4 mL acrylamide stock solution, 3.2 mL SDW, 200 µL 10% (w/v) SDS, 100 µL 10% (w/v) ammonium persulphate (fresh), and 10 µL of TEMED. The resulting solution should be mixed well and poured into the assembled gel apparatus, then overlayed with water-saturated butanol and allowed to polymerize.

3. After the separating gel has set, the butanol is removed and a 4% (w/v) polyacrylamide stacking gel poured. This consists of 1.25 mL stacking gel buffer, 1.0 mL acrylamide stock solution, 2.5 mL SDW, 100 µL 10% (w/v) SDS, 100 µL 10% (w/v) ammonium persulphate (fresh), and 10 µL of TEMED.

4. Two replicate gels are run; one for staining, to determine the amount of protein per sample, and so allow a rough estimate of the efficiency of extraction for each sample, and the other for Western blotting and immunodetection of the expressed protein. Load about 20 µL of protein extract/sample (it may be necessary to denature the samples again prior to loading) and run the two gels together at a constant current of 80 mA using SDS-PAGE running buffer. Do not forget to load positive and negative controls and also prestained markers that will be visible after the transfer to nitrocellulose.

5. When the dye front has run off the bottom of the gel (usually after about 40 min), switch off the power supply and dismantle the gel rigs. Put one gel in Coomassie blue staining solution; place the second gel in transfer buffer and allow to equilibrate for 15 min. Cut a piece of nitrocellulose to the same size as the gel and

1 2 3 4 5

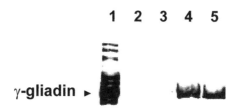

γ-gliadin ►

Fig 1. Expression of wild-type and ER-retention signal-containing forms of the wheat γ-gliadin protein in transgenic tobacco leaves. Equal amounts of total protein extracts were analyzed by SDS-PAGE and transferred to nitrocellulose. The presence of the γ-gliadin protein was detected by Western blotting with a rabbit polyclonal antisera (γ-326). It was observed that only the mutated forms of the γ-gliadin, containing KDEL or HDEL C-terminal extensions, accumulated in the tobacco leaf. Lane 1, positive control wheat endosperm sample; lane 2, untransformed tobacco leaf; lane 3, transgenic tobacco leaf expressing wild-type γ-gliadin; lane 4, transgenic tobacco leaf expressing γ-gliadin-KDEL; lane 5, transgenic tobacco leaf expressing γ-gliadin-HDEL.

   allow to soak in transfer buffer; also cut six pieces of Whatman 3MM filter paper to a size slightly larger than the gel. Assemble the sandwich so that the gel is on top of three pieces of 3MM paper, followed by the nitrocellulose and the three other pices of 3MM paper. Make sure no air bubbles are trapped between layers. Assemble the blotting apparatus containing the blot sandwich according to the manufacturer's instructions and transfer for 1 h at a constant voltage of 70 V.

6. While one gel is blotting, the second gel can be placed in destain solution; the addition of a piece of foam rubber to the destain will speed up the process by absorbing the Coomasie Blue. When the gel is destained, it can be photographed.

7. When the transfer of the total protein extracts to the nitrocellulose membrane is complete, dismantle the apparatus and put the membrane into 50 mL of blocking buffer for 30 min; this will reduce the nonspecific binding of the antibodies to the membrane.

8. Incubate the membrane in binding buffer containing the primary antibody (i.e., the antibody against the protein of interest) at the appropriate dilution; the volume of this reaction can be varied, depending on the dilution of the antibody. Incubate either at room temperature for at least 1 h or overnight in the cold room.

9. After incubation with the primary antibody, wash the membrane once with TBS and then incubate for 1 h in binding buffer containing the secondary antibody/ conjugate at the recommended dilution. After incubation, wash the membrane three times in TBS.

10. To detect the presence of an antigen/antibody reaction, add 40 μL BCIP solution and 40 μL NBT solution to 15 mL carbonate buffer and mix well; pour the solution over the membrane (make sure you have it the right side up) and wait for the bands to appear. Stop the reaction, when the optimum intensity is obtained, by placing the membrane in SDW.

11. Estimate protein expression levels present in the different samples by comparing the Western blot with the stained protein gel. If the transgene product is absent or present at very low levels (and the positive control is present at the expected level), it is probable either that the transgene is not being transcribed (check by Northern blotting; *see* **ref. 22**) or that the stability of the introduced protein is very low.

3.3. Analysis of Protein Stability in Transgenic Plants

The analysis of protein stability is carried out by determining the turnover time of the protein of interest; this is done by pulse-chase in vivo labeling of total proteins and then monitoring the abundance of the target protein by immunoprecipitation and SDS-PAGE. Protoplasts are prepared from the plant tissue and are then incubated in the presence of ^{35}S-methionine (Amersham) for several hours (this is the "pulse"). The protoplasts are then incubated in the presence of an excess of nonradioactive L-methionine (the "chase") for up to 24–48 h, with aliquots being removed for immunoprecipitation of the target protein at suitable time-points. In this study, protoplasts were prepared from sterile tobacco leaves (*see* **Note 4**) and the cells were chased for up to 24 h after a 6-h pulse.

3.3.1. Isolation of Tobacco Leaf Protoplasts

Protoplasts are prepared and maintained under sterile conditions; this means that it is best to use plant material that has been propagated in vitro and is therefore axionic. This is important, since the protoplasts must remain viable for several days and the isolation medium is a rich source of carbon for micro-organisms.

1. Eight leaves are cut from sterile tobacco plants at 3–5 wk old and their midribs removed; the leaves are then cut into thin strips and placed in 20 mL of protoplast isolation buffer in a Petri dish. The Petri dish is wrapped in foil and incubated overnight at 25°C.
2. The following morning, gently shake the Petri dishes to aid the release of the protoplasts. Carefully remove the dark green protoplast-containing solution (use a 5-mL Gilson pipet) and pass through a 100-micron mesh filter to remove cell debris and undigested leaf material.
3. Spin the filtrate at 50g for 5 min. Intact protoplasts remain at the top of the solution and can be carefully removed by the use of a 5-mL Gilson pipet. Add the recovered protoplasts to a tube containing 20 mL protoplast incubation medium (i.e., containing no cell wall-digesting enzymes). Mix very gently and repeat the centrifugation step.
4. Repeat **step 3** and finally recover the protoplasts in as small a volume of incubation medium as possible. Count the number of protoplasts produced using a hemocytometer; a yield of 1×10^6/Petri dish should be expected.
5. The protoplasts should be used immediately for the pulse-chase analysis.

3.3.2. Pulse-Chase In Vivo Labeling and Immunoprecipitation

This procedure is carried out using a modification to the protocol described by Wilkens et al. *(24)*. In the given example, we wished to check that our transgene product was not being secreted by a default mechanism, and so examined the proteins present in the protoplast incubation medium as well as those within the cell.

1. For each transgenic plant being analyzed, two replicates of 5×10^6 protoplasts should be resuspended in 2 mL of protoplast incubation medium in the presence of 75 µCi ^{35}S L-methionine (Amersham). The pulse incubation is carried out for 6 h in a suitably-sized microtiter plate at 25°C in the dark.
2. After this time, add L-methionine to a final concentration of 1 m*M* to commence the chase. Remove an aliquot (e.g., 200 µL) of the medium and recover the protoplasts by centrifugation at 100*g*. Remove the supernatant and keep on ice until ready to proceed with the immunoprecipitation. Meanwhile, lyse the protoplast pellet by resuspension in 500 mL of immunopreciptation (IP) buffer; lysis can be aided by freeze-thawing or by the use of a narrow-bore pipet tip. Keep the lysed cells on ice (*see* **Note 5**).
3. Resuspend an appropriate volume of Protein A-Sepharose (Sigma) in IP buffer to give a final concentration of 62.5 mg/mL; allow 70 µL of this solution for each sample to be immunoprecipitated. Make up the resuspended pellet sample to 930 µL with IP buffer and add 70 µL of Protein A-Sepharose solution. Add 730 µL of IP buffer to the recovered medium sample and also add 70 µL of Protein A-Sepharose solution. Add an appropriate amount of antibody to the sample (usually a few microliters of polyclonal antisera is sufficient) and incubate overnight on a rotating wheel at 4°C (*see* **Note 6**).
4. Collect samples at appropriate time-points over a period of 24–48 h of the chase, and process in exactly the same way. Obviously, if you wish to examine a large number of time-points, then the aliquots removed each time will need to be smaller. It may be of interest to take several time-points close to the start of the chase, if it is suspected that your protein of interest is being rapidly degraded.
5. After the samples have been immunoprecipitated overnight, collect the antigen/ Protein A-Sepharose by centrifugation at 6,000*g* for 5 min. Discard the supernatant (remember that it is likely to be radioactive) and wash the pellet by resuspension in 1 mL of IP buffer.
6. Repeat step 5. Carry out one final wash by resuspension in washing buffer.
7. Recover the pellet by centrifugation and carefully remove all the supernatant, so that the pellet appears granular. Add 20 µL of protein-extraction buffer and incubate at 65°C for 10 min. The sample can then be stored at –20°C or analyzed immediately by SDS-PAGE gel.
8. For analysis by SDS-PAGE, load 10 µL of the sample and separate as described. The gel can then be fixed, dried, and exposed to X-ray film, or soaked in a fluorographic reagent, such as Amplify (Amersham), prior to drying to enhance the signal. An example is shown in **Fig. 2**.

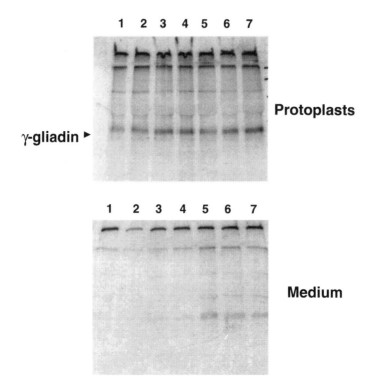

Fig 2. Pulse-chase analysis of γ-gliadin-KDEL in transgenic tobacco protoplasts. Isolated protoplasts were in vivo labeled with ^{35}S-methionine for 6 h and then chased with an excess of 1 mM methionine for different time points. At each time-point, the samples were separated into protoplast and media fractions by centrifugation, to allow the analysis of retained and secreted proteins. Samples were immunoprecipitated and analyzed by SDS-PAGE; it can be seen that the γ-gliadin-KDEL mutant is highly stable and is found almost exclusively in the protoplast fraction. A small amount of protein is seen in the media, but this may result from cell lysis rather than secretion. Protoplasts expressing the wild-type form of the γ-gliadin completely degraded the protein in under 1 h (data not shown). Time-points for the chase are: lane 1, 0 h (i.e., immediately after the 6 h pulse); lane 2, 1 h; lane 3, 3 h; lane 4, 5 h; lane 5, 7 h; lane 6, 17 h; lane 7, 24 h.

4. Notes

1. Fortunately, the majority of components of the PIN buffer can be purchased in the form of concentrated mixtures, such as macronutrients, micronutrients, and Gambourg's salts, from Sigma.
2. It must be emphasized that the proteins studied above all have endogenous signal sequences to target them into the ER lumen. If the protein of interest does not contain such a signal, it will have to be added, in addition to the C-terminal retention motif.

3. As with any construct generated by PCR, it is very important to verify it by sequencing, since the PCR reaction can introduce mutations.
4. The sterility of the plant material is very important for in vivo labeling, but it has also been shown that material derived from tissue-cultured plants expressed higher levels of heterologous protein than those grown in the glasshouse *(13)*.
5. The immunoprecipitation protocol is for an antibody that recognizes the denatured form of the antigen; if you wish to use an antibody that recognizes the native form of the protein, then a different method should be used (*see* **ref.** *22* for details).
6. Not all antibodies bind well to Protein A; particular examples include IgG_1 subclasses. If this is likely to be a problem, then it may be possible to use a secondary antibody coupled to agarose as the affinity matrix.

Acknowledgments

Institute of Arable Crops Research receives grant-aided support from the Biotechnology and Biological Sciences Research Council (UK).

References

1. Whitelam, G. C., Cockburn, B., Gandecha, A. R., and Owen, M. R. L. (1993) Heterologous protein production in transgenic plants. *Biotechnol. Gene. Eng. Rev.* **11,** 1–29.
2. Chrispeels, M. J. and Tague, B. W. (1991) Protein sorting in the secretory system of plant cells. *Int. Rev. Cytol.* **125,** 1–45.
3. Lawton, M. A., Tierney, M. A., Nakamura, I., Anderson, E., Komeda, Y., Dube, P., Hoffman, N., Fraley, R. T., and Beachy, R. N. (1987) Expression of a soybean β-conglycinin gene under the control of the cauliflower mosaic virus 35S and 19S promoters in transformed petunia tissues. *Plant Mol. Biol.* **9,** 315–324.
4. Higgins, T. J. V. and Spencer, D. (1991) The expression of a chimeric cauliflower mosaic virus (CaMV 35S)-pea vicilin gene in tobacco. *Plant Sci.* **74,** 89–98.
5. Hoffman, L. M., Donaldson, D. D., and Herman, E. M. (1988) A modified storage protein is synthesised, processed and degraded in the seeds of transgenic plants. *Plant Mol. Biol.* **11,** 717–729.
6. Sijmons, P. C., Dekker, B. M. M., Schrammeijer, B., Verwoerd, T. C., Van Den Elzen, P. J. M., and Hoekema, A. (1990) Production of correctly processed human serum albumin in transgenic plants. *Bio/Technology* **8,** 217–222.
7. Shewry, P. R., Napier, J. A., and Tatham, A. S. (1995) Seed storage proteins: structures and biosynthesis. *Plant Cell* **7,** 945–956.
8. Van der Valk, H. C. P. M. and Van Loon, L. C. (1988) Subcellular localization of proteases in developing leaves of oats (*Avena sativa* L.). *Plant. Physiol.* **87,** 536–541.
9. Freedman, R. B. (1984) Native disulphide bond formation in protein biosynthesis: evidence for the role of protein disulphide isomerase. *Trends Biochem. Sci.* **9,** 438–441.
10. LaVallie, E. R., DiBlasio, E. A., Kovacic, S., Grant, K. L., Schendel, P. F., and McCoy, J. M. (1993) A thioredoxin gene fusion expression system that circumvents inclusion body formation in *E. coli* cytoplasm. *Bio/Technology* **11,** 187–193.

11. McBride, K. E., Svab, Z., Schaaf, D. J., Hogan, P. S., Stalker, D. M., and Maliga, P. (1995) Amplification of a chimeric *Bacillus* gene in chloroplasts leads to an extraordinary level of an insecticidal protein in tobacco. *Bio/Technology* **13,** 362–367.

12. van Rooijen, G. J. H. and Moloney, M. M. (1995) Plant seed oil-bodies as carriers for foreign proteins. *Bio/Technology* **13,** 72–77.

13. Wandelt, C. I., Khan, M. R. I., Craig, S., Schroeder, H. H., Spencer, D., and Higgins, T. J. V. (1992) Vicilin with carboxy-terminal KDEL is retained in the endoplasmic reticulum and accumulates to high levels in the leaves of transgenic plants. *Plant J.* **2,** 181–192.

14. Lutcke, H. (1995) Signal recognition particle (SRP), a ubiquitous initiator of protein translocation. *Eur. J. Biochem.* **228,** 531–550.

15. Denecke, J., Botterman, J., and Deblaere, R. (1990) Protein secretion in plant cells can occur via a default pathway. *Plant Cell* **2,** 51–59.

16. Pfeffer, S. R. and Rothman, J. E. (1987) Biosynthetic protein transport and sorting by the endoplasmic reticulum and golgi. *Ann. Rev. Biochem.* **56,** 829–852.

17. Munro, S. and Pelham H. R. B. (1987) A C-terminal signal prevents secretion of luminal ER proteins. *Cell* **48,** 899–907.

18. Vandekerckhove, J., Van Damme, J., Van Lijsebettens, M., Botterman, J., De Block, M., Vandewiele, M., De Clercq, A., Leemans, J., Van Montague, M., and Krebbers, E. (1989) Enkephalins produced in transgenic plants using modified 2S seed storage proteins. *Bio/Technology* **7,** 929–932.

19. Napier, R. M., Fowke, L. C., Hawes, C., Lewis, M., and Pelham, H. R. B. Immunological evidence that plants use both HDEL and KDEL for targeting proteins to the endoplasmic reticulum. *J. Cell Sci.* **102,** 261–271.

20. Zakai, N., Ballas, N., Hershkovitz, M., Broido, S., Ram, R., and Loyter, A. (1993) Transient gene expression of foreign genes in preheated protoplasts: stimulation of expression of transfected genes lacking heat shock elements. *Plant Mol. Biol.* **21,** 823–834.

21. Frisch, D. A., Harris-Haller, L. W., Yokubaitis, N. T., Thomas, T. L., Hardin, S. H., and Hall, T. C. (1995) Complete sequence of the binary vector Bin 19. *Plant Mol. Biol.* **27,** 405–409.

22. Sambrook, J., Fritsch, E. F., and Maniatis, T. (1989) *Molecular Cloning: A Laboratory Manual*, 2nd ed., Cold Spring Harbor Laboratory, Cold Spring Harbor, NY.

23. Horsch, R. B., Fry, J. E., Hoffman, N. E., Eichholtz, D., Rogers, S. G., and Frayley, R. T. (1985) A simple and general method for transferring genes into plants. *Science* **227,** 1229–1231.

24. Wilkins, T. E., Bednarek, S. Y., and Raikhel, N. V. (1990) Role of propeptide glycan in post-translational processing and transport of barley lectin to vacuoles in transgenic tobacco. *Plant Cell* **2,** 301–313.

16

Screening for Transgenic Lines with Stable and Suitable Accumulation Levels of a Heterologous Protein

Myriam De Neve, Helena Van Houdt, Anne-Marie Bruyns, Marc Van Montagu, and Ann Depicker

1. Introduction

Genes encoding heterologous proteins are introduced into the plant genome for several purposes. First, the plant-made protein can be used as a tool in fundamental research. Reporter proteins can be used to characterize promoter sequences and other *cis*-acting sequences; the overproduction of proteins in plants can help to elucidate their function. Second, the synthesis of heterologous proteins in plants can have biotechnological applications. The introduced gene can confer new properties to the plant, or transgenic plants can be used as a source of important heterologous proteins. For all these purposes, the expression profile and accumulation level of the heterologous protein in plants should be stable, reproducible, and suitable. This chapter deals with the following problems: Most primary transformants show relatively low expression levels of the transgene *(1)*; and expression levels are not always stably transmitted to the progeny.

Different systems have been developed to introduce foreign DNA into plant cells *(2)*. One of these systems is based on the natural gene transfer system of *Agrobacterium*. This *Agrobacterium*-mediated transformation system has been widely used for dicotyledonous species *(3)*. Recently, successful transformation of some monocotyledonous plants by *Agrobacterium* has been reported *(4–6)*. In the *Agrobacterium*-mediated transformation system, the gene of interest is cloned into *Escherichia coli* between the borders of the transferred DNA (T-DNA). Subsequently, the T-DNA construct is introduced into

From: *Methods in Biotechnology, Vol. 3:*
Recombinant Proteins from Plants: Production and Isolation of Clinically Useful Compounds
Edited by: C. Cunningham and A. J. R. Porter © Humana Press Inc., Totowa, NJ

Agrobacterium. The recombinant *Agrobacterium* transfers the T-DNA into the plant cell during co-cultivation. The transferred DNA is targeted to the nucleus and is found to integrate at random positions into plant chromosomes *(7,8)*. Transformed cells are selected and regenerated into flowering plants. This last step remains a major bottleneck for some plant species.

Primary transformants can differ substantially in the expression levels of introduced genes. Differences in heterologous protein levels, up to 100-fold or more, have been reported *(1,9,10)*. This intertransformant variability has been shown for different constructs and plant species *(1,11)*. In some studies, a reduced intertransformant variability has been reported using particular elements at the T-DNA borders, such as a 3'-untranslated region *(12)*, matrix-associated regions (MAR) *(13)*, and scaffold-associated regions (SAR) *(14)*, or placing the gene between two different selection markers in the T-DNA and selecting for both markers *(15)*.

The high intertransformant variability of expression levels in transgenic populations has been explained in several ways *(1,16)*. Probably all of the available explanations contribute to a different extent to the observed variation. Integration of foreign DNA appears to occur randomly in the plant genome. Consequently, independent transformants contain the introduced DNA at different positions of the genome. Gene expression can be influenced by the properties of the surrounding plant DNA: the presence of flanking promoters, the presence of enhancing or silencing DNA sequences, and the chromatin structure or methylation status of the surrounding plant DNA. The relative position and/or orientation of the gene of interest in the T-DNA can therefore also influence the intertransformant variability *(12)*. In addition to these positional effects, the number of inserted T-DNA copies can modulate the expression level. Indeed, positive *(15,17,18)*, indeterminate *(19–21)*, and negative correlations *(18,22)* between copy number and gene expression have been reported in different studies. Additionally, somaclonal variation *(23)*, as well as mutations *(24)* and rearrangements in the transgene, can be responsible for variation. Finally, along with the transgene position and its genotype structure, the developmental and physiological state of the analyzed tissue can also influence the expression level of the transgene.

In some transformants, the accumulation of the heterologous protein is very low or even undetectable, even though an intact copy of the introduced gene is present *(20,25,26)*. This can be a result of a phenomenon called gene silencing. Many cases of gene silencing have been described in plants over the past 5 yr *(27–29)*. Transgenes introduced by *Agrobacterium*-mediated transformation, as well as DNA fragments introduced by direct gene transfer, are prone to silencing. Critical for the triggering of the silencing phenomenon is the presence of homologous sequences residing within two or more introduced transgenes or within an introduced transgene and a homologous endogenous gene.

For most applications of transgene expression in plants, these silenced plants are undesirable. Gene silencing occurs to a different extent in independent transformations, even when the same construct and the same transformation conditions are used *(16)*. Silenced primary transformants are easily eliminated on the basis of low or absent expression of the transgene. However, it is important to keep in mind that silencing sometimes only occurs in progeny plants *(26,30–36)*. In some cases, homozygous progeny plants show a reduced activity of the gene, when compared to hemizygous plants, and this as soon as the homozygous state is reached. In other cases, the transgene silencing is progressively built up in subsequent generations. Clearly, it is very important to follow the transgene expression over several generations in selected transgenic lines.

Finnegan and McElroy *(9)* discussed the mechanisms that might cause gene expression instability following sexual propagation, and suggested ways to stabilize it. Because transgene silencing is more commonly observed in plants with multiple (linked or unlinked) copies than in plants with a single copy of the transgene, problems with expression instability can be minimized by selecting transformants with one copy of the introduced transgene. Even then, the progeny should be analyzed, because gene inactivation of single copies has been observed following sexual propagation *(30)*.

We propose a protocol that should allow identification of lines with suitable accumulation levels of the heterologous protein and stable transmission of their expression level in a homozygous condition through subsequent generations. The protocol starts with the assumption that a plant species is used that can be self-fertilized. Moreover, methods are described for the screening of a transgenic *Arabidopsis* or tobacco population obtained by *Agrobacterium*-mediated transformation. Obviously, the general principles of the proposed method also apply to transformants of other plant species or transformants obtained by other transformation methods. However, adaptations might have to be made at different levels in order to meet the specific characteristics of the species or of the transformation method used.

The proposed scheme has three steps (*see* **Note 1**). First, many primary transformants are screened by a fast and easy assay for transgene expression in order to select a limited amount of interesting transformants. Second, the number of T-DNA copies integrated into each of the selected transformants is determined and transformants with only one T-DNA copy are preferentially retained. Finally, these transformed lines are sexually propagated, and progeny plants are analyzed for the stability of their transgene expression.

2. Materials

2.1. DNA Material

1. λ DNA (Gibco BRL, Gaithersburg, MD).
2. Salmon sperm DNA (D-1626; Sigma, St. Louis, MO).

2.2. Special Laboratory Tools and Materials

1. Corex® tubes (DuPont, Wilmington, DE) siliconized, wrapped in aluminium foil, and sterilized by incubation overnight at 180°C.
2. Parafilm® (American National Can, Neenah, WI).
3. Wide-bore sterile 10-mL pipets (Falcon® 7543; Becton Dickinson, Bedford, MA).
4. 2.0-mL tubes (Eppendorf, Hamburg, Germany).
5. Miracloth (Calbiochem, La Jolla, CA).
6. Filter paper sheets (3MM; Whatman, Clifton, NJ).
7. 15-cm Optilux Petri dishes (Falcon® 1013).
8. Sealing tape (Urgopore, Chenoves, France).

2.3. Buffers and Other Solutions

The solutions below are not sterilized unless specified.

1. NTES buffer: 10 mM Tris-HCl, pH 7.5, 0.1M NaCl, 1 mM ethylene-diaminetetraacetic acid (EDTA) and 1% (w/v) sodium dodecyl sulfate (SDS).
2. Phenol: Molten phenol saturated with 0.1M Tris-HCl, pH 8.0. The Tris-HCl solution used is sterilized by autoclaving.
3. Chloroform solution; chloroform:isoamyl alcohol, 24:1.
4. Extraction buffer: 0.1M Tris-HCl, pH 8.0, 0.5M NaCl, 50 mM EDTA, and 10 mM β-mercaptoethanol. The buffer is prepared from autoclave-sterilized stock solutions of Tris-HCl, NaCl, and EDTA. Sterile water is used for further dilution to the required volume.
5. 3M KAc solution: prepared by mixing 60 mL 5M KAc, 11.5 mL glacial acetic acid, and 28.5 mL sterile H$_2$O. The KAc solution is sterilized by autoclaving.
6. TE: 10 mM Tris-HCl, pH 8.0, and 1 mM EDTA. The solution is made with auto-clave-sterilized stock solutions.
7. CTAB buffer: 0.2M Tris-HCl, pH 7.5, 2M NaCl, 50 mM EDTA, and 2% (w/v) cetyltrimethylammonium bromide (CTAB). Sterilize by autoclaving. Store at room temperature. The buffer should not contain any precipitate. If this is the case, heat the CTAB buffer at 65°C to dissolve the precipitate prior to using it for the purification of DNA preparations.
8. DNA loading buffer: 25% Ficoll (type 400), 0.25% bromophenol blue, 0.25% xylene cyanol.
9. Denaturation buffer: 0.5M NaOH, 1.5M NaCl.
10. Neutralization buffer: 1M Tris-HCl, pH 7.5, 1.5M NaCl.
11. SSC: 3M NaCl, 0.3M sodium citrate, pH 7.0.

12. Prehybridization solution: 3X SSC, 0.25% (w/v) milk powder.
13. Hybridization solution: 3X SSC, 0.5% (w/v) SDS, 1 m*M* EDTA, 0.25% (w/v) milk powder.
14. Washing solution 1: 3X SSC, 0.1% (w/v) SDS.
15. Washing solution 2: 1X SSC, 0.1% (w/v) SDS.
16. Germination medium (*Arabidopsis*): 1X Murashige and Skoog salt mixture (Gibco), 3% (w/v) sucrose, 100 mg/L inositol, 1.0 mg/L thiamine, 0.5 mg/L pyridoxine, 0.5 mg/L nicotinic acid, 0.5 g/L 2-(*N*-morpholino)ethanesulfonic acid (MES), pH 5.7, and 0.7% agar. 1X Murashige and Skoog salt mixture, sucrose, MES, and agar are sterilized by autoclaving. The other components are added from filter-sterilized stock solutions.
17. Germination medium (tobacco): 1X Murashige and Skoog salt mixture (Gibco), 10 g/L sucrose, pH 5.6, and 0.7% agar. Sterilize by autoclaving.
18. Kanamycin stock solution: 100 mg kanamycin (Duchefa, Haarlem, The Netherlands)/mL distilled H_2O; filter-sterilize and store at $-20°C$ in aliquots; use only freshly prepared, because the solution is not stable.
19. Hygromycin stock solution: 50 mg hygromycin (Calbiochem)/mL distilled H_2O; filter-sterilize; the solution is very stable and is stored at $-20°C$ in aliquots.
20. Bleach solution (*Arabidopsis*): NaOCl, 13% active chlorine (Acros, Geel, Belgium) diluted in the same volume of sterile H_2O (final concentration of 6.5% active chlorine), and 0.05% Tween-20.
21. Bleach solution (tobacco): NaOCl, 13% active chlorine (Acros).

2.4. Gel Systems

1. 10X TAE: 48.4 g Tris base, 11.4 mL glacial acetic acid, 20 mL 0.5*M* EDTA, pH 8.0 per liter.
2. 0.8% Agarose gel in 1X TAE.
3. Running buffer: 1X TAE, prepared from 10X TAE with distilled H_2O.

3. Methods
3.1. First Step: Rapid Screening of Primary Transformants

The described procedure assumes that an easy and fast screening test is available. A test for transgene expression can be based on quantification of the transgene product at the protein level or at the RNA level, of which analysis at the protein level is to be preferred (*see* **Note 2**). Accumulation levels of the heterologous protein can be determined by different assays, depending on the nature of the protein. Enzymes can be quantified on the basis of their activity. Enzymatic assays have been developed for reporter proteins such as β-glucuronidase (GUS) *(37)*, chloramphenicol acetyl transferase (CAT) *(38,39)*, neomycin phosphotransferase II (NPTII) *(40)*, phosphinothricin acetyltransferase (PAT) *(41)*, and firefly luciferase *(42)*. Enzymes and heterologous proteins for which no functional assay is available can be quantified by sandwich ELISA. This method is easy and fast and gives quantitative data. It can be set up for any heterologous protein, as described by Bruyns et al. in Chapter 18. Sandwich ELISAs have been developed for the reporter enzymes CAT and NPTII and are commercially available.

1. Prepare protein extracts from the appropriate tissue (*see* **Note 3**) of 100 independent transformants.
2. Determine the concentration of the total soluble protein (*see* Chapter 18, Note 14) and of heterologous protein in each protein extract, using the selected protein assay (*see* **Notes 4** and **5**).
3. Retain 25 shoots with the highest accumulation levels (tobacco) or regenerate 25 shoots from the callus tissues with the highest accumulation levels (*Arabidopsis*). Grow these shoots into mature plants.
4. Prepare protein extracts from leaf tissue of the developed tobacco plant (*see* **Note 6**).
5. Determine the accumulation of the total soluble protein and of the heterologous protein in each protein extract, using the selected protein assay.
6. Select 10 primary tobacco transformants with the highest accumulation levels.

3.2. Second Step: Southern Analysis

A Southern analysis is performed in order to determine the number of T-DNA copies in each retained transformant (*see* **Subheading 3.1.** and **Note 7**). This step allows the selection of single-copy plants, which have been found to be less prone to silencing *(18,26,30,32)*, or plants with unlinked T-DNAs, from which single-copy plants can be obtained in the next generation. Southern analysis will reduce the number of candidates significantly and, although it can seem time-consuming, the correct choice of transformed lines can prevent many problems in later generations.

3.2.1. Isolation of Genomic DNA

Genomic DNA is isolated from the selected transformants (*see* **Subheading 3.1.**), and from an untransformed plant, for use as a negative control in the Southern analysis. Two different protocols are described. The first protocol (**Subheading 3.2.1.1.**) was adapted from **ref. *43***, and is routinely used in our laboratory for DNA isolation from tobacco leaves. The second protocol (**Subheading 3.2.1.2.**) was adapted from **ref. *44*** and is routinely used in our laboratory for isolation of DNA from *Arabidopsis* seedlings or leaves of greenhouse plants. Because primary *Arabidopsis* transformants tend to have very few leaves, the Southern analysis will have to be performed on plants of the next generation. All plant material is harvested, weighed, wrapped in aluminium foil and put in liquid N_2. The packages are transferred from the liquid N_2 to the –80°C freezer for storage. Samples can then be processed from frozen plant material.

3.2.1.1. PROTOCOL FOR ISOLATION OF GENOMIC DNA FROM TOBACCO LEAVES

This method involves lysis of plant cells, followed by separation of the DNA from protein and RNA. Routinely, about 2 g of leaf material are used for each DNA preparation (*see* **Note 8**). If the weight of the material exceeds 4 g, it should be divided into two.

1. Fill a Corex tube with 9 mL NTES buffer, 3 mL phenol, and 3 mL chloroform solution. Prepare one tube for each DNA preparation.
2. Prechill the mortar and pestle with liquid N_2. Put frozen leaf material in the mortar and pour liquid N_2 over it. Carefully break the leave into large pieces with the pestle. When most of the liquid N_2 has evaporated, grind to a fine powder.
3. Transfer the fine powder into a Corex tube with a weighing spoon, chilled by putting it for some seconds in liquid N_2. Close the tube twice with parafilm and vortex immediately at the highest speed for 1 min. Store the Corex tube on ice.
4. Repeat steps 2 and 3 for the homogenization of leaf material of other transformants.
5. Vortex all samples again for 20 s. Centrifuge at 10,400g for 25 min in a HB4 rotor, 4°C.
6. Handle the solutions containing DNA gently from this point on, in order to reduce shearing of the DNA. Also use widebore pipets. Do not vortex or vigorously mix the solutions.
7. Carefully transfer most of the top (aqueous) phase to a fresh Corex tube. The transferred solution should be limpid. If this is not the case, centrifuge again for 10 min at 10,400g in a HB4 rotor, 4°C, after extraction of the solution with 3 mL phenol and 3 mL chloroform solution.
8. Precipitate the nucleic acids by adding one-tenth vol 2M sodium acetate, pH 4.8, and 1 vol isopropanol. Gently mix by inverting the tube several times. A nucleic acid precipitate should be visible (*see* **Note 9**). Incubate for at least 1 h at –20°C.
9. Pellet the nucleic acids by centrifugation at 10,400g for 10 min in a HB4 rotor, 4°C.
10. Discard the supernatant and wash the pellet with 70% ice-cold EtOH.
11. Centrifuge again at 10,400g for 10 min in a HB4 rotor, 4°C.
12. Discard the supernatant and invert the tube briefly on a paper towel. Dry the pellet in a desiccator under vacuum until the smell of evaporating isopropanol has disappeared. Do not dry the pellet longer than necessary or it will become very difficult to dissolve.
13. Dissolve the pellet in 2.5 mL sterile H_2O by incubating the tube on ice, with gentle mixing from time to time. Do not dissolve by pipeting, because this might shear the DNA.
14. Add 2.5 mL 4M lithium acetate and mix by inverting the tube several times. The LiAc will cause precipitation of single-stranded nucleic acids. Incubate overnight at 4°C.
15. Centrifuge at 10,400g in a HB4 rotor for 10 min, 4°C.
16. Transfer the DNA-containing supernatant to a fresh Corex tube.

17. Add 10 mL 99% ethanol, mix by inverting the tube, and incubate at –20°C for 1 h.
18. Centrifuge at 10,400g in a HB4 rotor for 10 min, 4°C.
19. Discard the supernatant and dry the pellet until the smell of evaporating ethanol has disappeared.
20. Gently resuspend the pellet in 450 µL H_2O. Transfer the DNA solution to an Eppendorf tube, pipeting slowly to avoid DNA breakage.
21. Add 45 µL 2M sodium acetate, pH 4.8, and 1 mL 99% ethanol, and mix well. Incubate at –20°C for 1 h.
22. Centrifuge in a minicentrifuge at maximal speed (14,900g) for 5 min.
23. Discard the supernatant and wash the pellet with ice-cold 70% EtOH.
24. Centrifuge again for 5 min at maximal speed.
25. Remove the supernatant with a pipet and briefly air-dry the pellet.
26. Gently resuspend the DNA in 200 µL sterile H_2O.
27. Determine the DNA concentration in each preparation by measuring the absorption of a 1:100 dilution in H_2O at 260 nm and 280 nm in a spectrophotometer. The figure of the ratio between the readings at 260 nm and 280 nm should be approx 1.8 for pure DNA solutions. The reading at 260 nm allows calculation of the DNA concentration, taking into account that an OD = 1 corresponds to approx 50 µg/mL double-stranded DNA. Normally, DNA concentrations of 20–195 µg/g leaf are obtained for axenically grown tobacco plants. It is advisable to check the quality of the isolated DNA by running an undigested sample on gel.

3.2.1.2. Protocol for Isolation of Genomic DNA from *Arabidopsis*

The method involves the lysis of plant cells, followed by separation of the DNA from proteins, RNA, and polysaccharides. The protocol can be used for the isolation of genomic DNA from 0.2 g to 2 g plant material. No more than 2 g plant material should be used for a single DNA preparation.

1. Pipet 6 mL ice-cold extraction buffer in a 14-mL Falcon tube. Prepare one tube for each DNA preparation.
2. Grind the plant tissue to a fine powder, as described in **step 2** of **Subheading 3.2.1.1.**
3. Transfer the fine powder to the 14-mL Falcon tube using a cold weighing spoon. Mix immediately by inversion. Store on ice.
4. Repeat **steps 2** and **3** for each transformant.
5. Add 0.8 mL 10% (w/v) SDS solution to each tube and gently mix by inverting the tube several times. Incubate for 30 min at 65°C, swirling the solution periodically.
6. Handle solutions containing DNA gently from this point on, in order to reduce shearing of the DNA. Use wide-bore pipets and do not vortex or vigorously mix the solutions.
7. Add 2 mL of an ice-cold 3M KAc solution, mix by inverting the tube, and incubate on ice for 30 min.
8. Centrifuge for 10 min at 2000g in a Sorvall SM24 rotor, 4°C.

9. Gently take off the supernatant and filter it through a Miracloth filter into a fresh 14-mL Falcon tube.
10. Add 6 mL isopropanol to each tube, mix well by inversion, and incubate at room temperature for 20 min.
11. Centrifuge for 10 min at 2000g in a SM24 rotor at room temperature.
12. Discard the supernatant. Recentrifuge for a few seconds and remove the remainder of the supernatant with a pipet, leaving the pellet as dry as possible.
13. Dissolve the pellet, which can have a light yellow-brown color, in 400 µL TE; incubate the tubes on ice, flicking the tube with a finger from time to time. Do not vortex or pipet vigorously, to avoid DNA breakage. Transfer the solution, which may be turbid, to a 2.0-mL Eppendorf tube.
14. Add 2 µL RNase solution (10 mg/mL) to each tube and mix. Incubate for 20 min at 37°C.
15. Add 400 µL CTAB buffer and mix well by inversion. Incubate for 15 min at 65°C, mixing the solution periodically.
16. Add 800 µL chloroform solution to each tube and mix well by inverting the tube several times.
17. Centrifuge in a minicentrifuge for 5 min.
18. Transfer the top (aqueous) phase to a fresh 2.0-mL Eppendorf tube. If the solution is still turbid, repeat **steps 16** and **17.**
19. Add 1.4 mL 99% ethanol and mix. Incubate at room temperature for 15 min.
20. Centrifuge for 10 min at maximum speed in a minicentrifuge.
21. Discard the supernatant and wash the pellet with 70% ethanol.
22. Recentrifuge for 10 min at maximum speed.
23. Remove the supernatant with a pipet. Centrifuge the tube again briefly. Remove the remainder of the supernatant with a pipet, leaving the pellet as dry as possible. Briefly air-dry the pellet until the smell of evaporating ethanol has disappeared.
24. Gently dissolve the pellet in 200 µL–1 mL sterile H_2O—the volume depending on the amount of plant material from which the DNA preparation was started.
25. Determine the DNA concentration by measuring the absorption of a 1:100 dilution in H_2O at 260 nm and 280 nm in a spectrophotometer (*see* **Subheading 3.2.1.1., step 27**).
26. Check the DNA concentration on gel by loading 2 µg DNA for each DNA preparation, as determined in **step 25**, on a 1% (w/v) agarose gel. Load also 2.5-fold serial dilutions of phage λ DNA in a range of 0.05 µg–2 µg. Stain the gel in an ethidium bromide solution (0.5 µg/mL in H_2O) for about 20 min. Estimate the DNA concentration in each DNA preparation by comparing the intensity of the DNA band with the intensities of the bands of the λ DNA dilution series. Estimation of the DNA concentration on gel is essential for DNA prepared by this CTAB method, because we have found that for DNA preparations from *Arabidopsis* seedlings the DNA concentration estimated on gel is about five- to 10-fold lower than the spectrophotometrically determined DNA concentration. Normally, about 10–40 µg DNA/g fresh weight of *Arabidopsis* seedlings are obtained, estimated on gel.

3.2.2. Southern Blotting

The Southern blotting described here is the method described in **ref. 45**, with some modifications.

1. Choose the appropriate restriction enzymes for the Southern analysis, based on the following stipulations. The enzymes should not be prone to cytosine methylation (**46**) and should cut at least 1000–2000 base pairs away from the T-DNA borders, allowing the generation of border fragments containing part of the T-DNA and flanking plant DNA. Choose two different restriction enzymes.
2. Check whether the chosen restriction sites are present at the predicted places in the T-DNA by performing restriction digests on T-DNA-containing plasmid DNA. Perform digests on plasmid DNA with the enzyme stock and buffer solution, which will be used for the Southern analysis, to check whether the enzyme cuts with a 100% efficiency. Select the optimal restriction buffer for each digest.
3. Perform the two different restriction digests on the genomic DNA isolated from the selected primary transformants and from an untransformed plant (*see* **Subheading 3.2.1.**). Digest 10 µg of tobacco DNA or 1 µg of *Arabidopsis* DNA (as estimated on gel) with 2 U restriction enzyme/µg DNA (tobacco) or 4 U restriction enzyme/µg DNA (*Arabidopsis*) for 3 h. Add a further 2 U/µg DNA (tobacco) or 4 U/µg DNA (*Arabidopsis*) and incubate again for 3 h. Make sure that the enzyme is not added at more than 1/10 of the total volume. If the total volume of the restriction digest exceeds the volume that can be loaded on gel, reduce the volume of the digests after digestion by evaporation in a speed-vac to about 15–20 µL. Add one-tenth vol of DNA loading buffer and put the Eppendorf tubes in an Eppendorf shaker for at least 10 min.
4. Pour a 0.8% (w/v) agarose gel, making sure that the thickness does not exceed 7 mm. Load the digests, including at least one lane with DNA mol wt markers (*see* **Note 10**), and, preferably, also a positive control (*see* **Note 11**).
5. Run the gel in running buffer at about 6 V/cm, until the markers in the loading buffer have migrated into the gel. Lower the voltage to approx 1.5 V/cm and run overnight.
6. Turn off the power supply when the bromophenol dye from the loading buffer has migrated about two-thirds of the length of the gel. Remove the slots, and make sure that the orientation of the gel is still recognizable by cutting a little piece of the gel at the upper left corner.
7. Stain the gel with an ethidium bromide solution for 30 min.
8. Photograph the gel with a ruler laid beside it, so that the positions of the bands can be later identified on the membrane (film). Check whether the genomic DNA have been digested by the restriction enzymes; fully digested genomic DNA is visible as a smear from the top to the middle or bottom of the gel.
9. Place the gel in a clean glass dish.

10. Cover the gel with a 0.25M HCl solution for 10 min, shaking slowly on a platform shaker. Check the color of the xylene cyanol and bromophenol blue dyes. These will change to green and yellow, respectively.
11. Immediately remove the HCl solution and rinse the gel three times with distilled H$_2$O.
12. Cover the gel with denaturation buffer for 15 min and put the tray on a platform shaker, shaking slowly.
13. Pour off the solution, cover the gel with fresh denaturation buffer, and shake again for 15 min.
14. Rinse the gel three times with distilled H$_2$O.
15. Cover the gel with neutralization buffer and put the tray on a platform shaker for 30 min, shaking slowly.
16. Pour off the solution, add fresh neutralization buffer, and shake for 15 min.
17. Fill a clean glass dish with 20X SSC, so that the level of the solution is at least 1 cm high. Lay a glass plate on the glass dish.
18. Cut two wicks of the appropriate dimensions out of a Whatman 3MM filter paper sheet. Their width should be at least 2 cm larger than the gel. Their length should be such that both edges submerge in the 20X SSC solution in the glass tray once the wicks have been put on the glass plate after wetting in 20X SSC. Squeeze out air bubbles by rolling a pipet over the surface.
19. Place the gel on the wicks and squeeze out air bubbles carefully.
20. Cut a piece of nylon membrane, just large enough to cover the surface of the gel, and mark one corner with a pencil. Follow the instructions of the supplier for the treatment of the membrane before putting it on the gel.
21. Place the membrane on the surface of the gel, putting the marked corner of the membrane on the left upper corner of the gel. Lay the membrane down precisely. Never remove the filter once it has been in contact with the gel, because detectable transfer can take place very quickly. Squeeze out air bubbles carefully.
22. Cut two pieces of Whatman 3MM paper to the same size of the gel.
23. Wet the pieces with 20X SSC and put them on the filter one by one. Squeeze out air bubbles carefully.
24. Cut four strips of plastic wrap and place them carefully around the gel in order to avoid flow of the 20X SSC buffer around the gel, instead of through the gel.
25. On top of the Whatman 3MM paper, stack napkins to a height of about 6 cm.
26. Lay a glass plate on top of the structure, put on a weight of approx 0.5 kg, and leave overnight.
27. Disassemble the structure.
28. Recover the membrane and air-dry it for 1 h.
29. Recover the gel from the structure and stain it in an ethidium bromide solution. Check whether the DNA fragments have been transferred to the membrane.
30. Immobilize the DNA on the membrane by UV crosslinking in a UV-light box or by baking between two sheets of Whatman 3MM paper for 2 h at 80°C (*see* **Note 12**).

31. Wrap the filters in plastic wrap and store them dry between sheets of Whatman 3MM paper at room temperature.

3.2.3. Hybridizations of the Membranes

The method described here is a protocol that is routinely used in our laboratory involving the use of radioactivity. We propose a method in which two different probes are used in succession on the same membrane. Thus, both left and right border-plant DNA junctions are detected. We strongly favor the use of two different probes instead of only one (*see* **Note 13**), because this facilitates the interpretation of the data (*see* **Subheading 3.2.4.**).

1. Determine which DNA fragments have to be isolated to prepare the two different probes (*see* **Note 14**). The right border (RB) probe and left border (LB) probe should specifically detect right and left border T-DNA-plant DNA junctions, respectively. Therefore, the DNA fragment for the RB probe should be specifically located between the right border of the T-DNA and the restriction site closest to this border (*see* **Fig. 1**). The DNA fragment for the LB probe should be specifically located between the left border of the T-DNA and the restriction site closest to this border (*see* **Fig. 1**). Choose, if possible, a DNA fragment that is located far away from the LB of the T-DNA, because substantial T-DNA truncations at that border have been reported.

2. Shear a solution of 1 mg/mL sperm DNA by punching a hole in the cap of the Eppendorf tube with a needle, attached to a 2-mL syringe, and by passing the solution vigorously through the needle 20 times. Heat-denature the sheared sperm DNA solution by boiling at 95–100°C for 5 min, followed by quick cooling on ice prior to use. Store the solution on ice.

3. Pour prewarmed (65°C) prehybridization solution in a hybridization box and add 200 µL denatured (1 mg/mL) sperm DNA/10 mL. We routinely use at least 35 mL prehybridization solution for one membrane and at least 40 mL for two membranes for a box of 20.5 × 14.5 cm. Bring the membrane in contact with the solution at one edge and gently put the whole membrane into the solution, avoiding bubbles between the surface of the solution and the membrane. Prehybridize the membranes for at least 2 h in a 65°C water bath, shaking at 70 rpm, to block nonspecific DNA binding sites on the surface. Prepare the probe in the meantime.

4. Prepare a RB probe, using 50 ng of the DNA fragment isolated in **step 1**.

5. Remove unincorporated nucleotides, using spin columns (*see* **Note 15**).

6. Punch several holes in the cap of the probe-containing Eppendorf tube. Heat-denature the probe by incubation at 95–100°C in a water bath or incubator, followed by quick cooling on ice for at least 5 min.

7. Pour off the prehybridization solution and pipet at least 25 mL (one membrane) or at least 30 mL (two membranes) prewarmed (65°C) hybridization solution in the box (20.5 × 14.5 cm). Add 200 µL denatured sperm DNA (1 mg/mL)/10 mL hybridization solution.

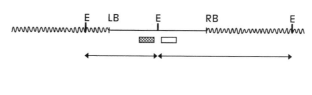

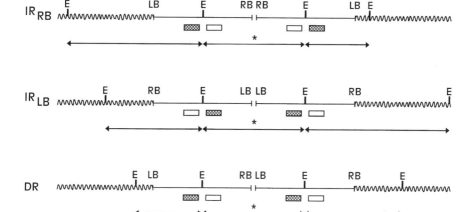

Fig. 1. Schematic representation of the fragments expected in the Southern analysis for physically unlinked or linked T-DNAs. Symbols and abbreviations: wavy line, flanking plant DNA; open box, RB probe; cross-hatched box, LB probe; ($\overset{*}{\leftrightarrow}$), fragment, revealed by the indicated probes, of which the size can be predicted if the T-DNAs are precisely physically linked; ($\leftrightarrow$), fragment, revealed by the indicated probes, of which the size cannot be predicted. Indeed, the size of the fragment depends on the distance from the recognition site in the flanking plant DNA to the T-DNA border; E, recognition site of the restriction enzyme; LB, left T-DNA border; RB, right T-DNA border; IR_{RB}, inverted repeat about the right border; IR_{LB}, inverted repeat about the left border; DR, direct repeat.

8. Add the denatured probe and incubate in a 65°C water bath overnight, shaking at 60 rpm.
9. Wash the membranes twice for 15 min with prewarmed (65°C) washing solution 1 in a shaking (70 rpm) water bath at 65°C.
10. Wash the membranes twice for 15 min with prewarmed (65°C) washing solution 2 in a shaking (70 rpm) water bath at 65°C.
11. Check the membranes for background signal in a corner. If the background signal is still very strong, wash the membranes under more stringent conditions (e.g., 0.1X SSC, 0.1% [w/v] SDS, 50–55°C). If the background signal is negligible, put

the membranes on a paper towel to remove excess fluid. Do not let them dry, because otherwise the removal of the probe can become impossible. Wrap the membranes in plastic film.

12. Expose the membranes to an X-ray film for 2–3 d (*Arabidopsis*) or 1 wk (tobacco) at –70°C.

13. Strip the RB probe from the membranes by pouring 400 mL of boiling 0.1% (w/v) SDS onto the membranes in a glass tray. Gently shake on a platform shaker until the solution has cooled down to room temperature.

14. Put the membranes on a paper towel to remove excess fluid and wrap them in plastic film.

15. Expose the stripped membranes to an X-ray film to make sure that the RB probe was removed from the membranes.

16. Hybridize the stripped membranes with the LB probe, as described in **steps 2–11**, and expose to an X-ray film.

3.2.4. Interpretation of the Data

Single-copy plants or plants with physically unlinked T-DNA copies are selected based on the data of the Southern analysis. Special care is taken to distinguish these plants from plants with T-DNAs linked in repeat structures, because the presence of repeat structures has been correlated with silencing. Physically linked T-DNA copies can be arranged in three different orientations (*see* **Fig. 1**), namely, as an inverted repeat about the right border (IR_{RB}), an inverted repeat about the left border (IR_{LB}), or a direct repeat (DR). **Subheadings 3.2.4.1.** and **3.2.4.2.** describe, step by step, how the data are interpreted if both the RB probe and LB probe have been used. We refer to **Note 16** for the interpretation of the data, if only the RB probe has been used.

3.2.4.1. SELECTION OF SINGLE-COPY PLANTS

1. For each transformant, determine the number of border fragments obtained with the RB probe and with the LB probe (*see* **Note 17**). The number obtained with both probes is normally equal (*see* **Note 18**) and reveals the number of independently integrated T-DNAs. Indeed, independent T-DNA integration events occur at different places in the genome, so that the size of the border fragments will be different for independently integrated T-DNAs (*see* **Fig. 1**). For a single-copy plant, only one RB fragment and one LB fragment are obtained. Multicopy plants show a number of RB fragments, which normally equals the number of LB fragments.

2. Select the plants for which only one fragment was revealed for both the RB and LB probe.

3. For the selected plants, determine the size of the fragment obtained with the RB probe as follows. Construct a graph using the photograph of the gel (**step 8** of **Subheading 3.2.2.**). For each molecular weight marker, measure its distance from the top of the gel; make a graph by putting this value on the abscissa (linear scale) and its corresponding size in kb on the ordinate (logarithmic scale). Connect the

points; the graph will be linear in the range below 5 kb. Determine for each selected plant the size of the fragment by measuring its distance from the top of the membrane and deducing the corresponding size using the graph (Y value).

4. Calculate the theoretical size of the fragment that would be obtained for two T-DNAs precisely linked in an inverted configuration about the RB (IR_{RB}). In principle, the fragment should have a size twice the distance from the right border to the recognition site of the enzyme site nearest to the right border (*see* **Fig. 1**). This type of repeat structure will normally be revealed very easily if two different probes were used, as two fragments will be revealed with the LB probe. Yet, substantial truncations can be present at the LB of T-DNAs. A truncation at the LB of one of the T-DNAs involved in an IR_{RB} could prevent the detection of this T-DNA by the LB probe. As a consequence, a plant with this kind of IR_{RB} could be taken by mistake for a single-copy plant. Yet, the presence of such a repeat structure can be revealed by comparing the size of the fragment obtained with the RB probe in the selected plants with the theoretical size for an IR_{RB}.

5. For each of the selected plants, check whether the size of the fragment obtained with the RB probe (**step 3**) fits the predicted size for an IR_{RB} (**step 4**). Do this for both restriction digests.

6. Select the plants for which there is no indication for the presence of an IR_{RB}.

3.2.4.2. Selection of Transformants with Physically Unlinked T-DNAs

If no single-copy plants are found, transformants with unlinked T-DNA copies should be selected. These copies will segregate upon sexual propagation if they are also genetically unlinked (*see* **Note 19**), so that progeny plants with only one T-DNA copy can be obtained (*see* **Subheading 3.3.2.**).

1. Select the plants for which only a limited number of bands were revealed by both probes.

2. Select the plants for which the number of fragments revealed by the LB probe equals the number of fragments obtained with the RB probe. This will eliminate the plants with an IR_{RB} or IR_{LB}, because these repeat structures will give an unequal number of bands for both probes. For an IR_{RB}, a single fragment will be obtained for the RB probe, but two bands will be revealed with the LB probe. For an IR_{LB}, a single fragment will be obtained with the LB probe and two bands for the RB probe.

3. For the selected plants, determine the sizes of the fragments obtained with the RB probe for both restriction digests.

4. Check whether these fragments have the size predicted for an IR_{RB}. If such a band is present for both digests, then the presence of an IR_{RB} in which one of the involved T-DNAs has a severe truncation at the LB can be suspected.

5. Select plants for which the data do not indicate the presence of this kind of repeat structure.

6. For the selected plants, determine the sizes of the fragments obtained with the LB probe for both restriction digests.

7. For each restriction digest, separately, compare the sizes of the fragments obtained with the RB probe (**step 3**) with those revealed by the LB probe (**step 6**).
8. Check whether bands of identical size have been detected by the RB and LB probes. An easy way to check this is to superimpose the films obtained with the two different probes. For plants with a DR, the LB and RB probe will detect one fragment of identical size. This will be true for both restriction digests.
9. Select plants for which the data do not indicate the presence of a DR.

3.3. Screening of the Progeny Plants

The progeny of the selected primary transformants (*see* **Subheading 3.2.4.**) are analyzed for the stability of transgene expression. The proposed strategy assumes that the T-DNA contains a selection gene (resistance gene for kanamycin or hygromycin) so that T-DNA-containing seedlings can be selected on a selective medium (*see* **Note 20**).

3.3.1. Analysis of the Progeny of Single-Copy Plants

1. Self-fertilize each of the selected single-copy plants (*see* **Subheading 3.2.4.1.**) and collect the seeds (= R1 seed stock).
2. Preserve the seeds for 1 wk at 4°C (vernalization); this cold treatment will assure a good germination of the seeds.
3. For each transformant, put about 40–50 seeds in a Miracloth filter and make a package by stapling.
4. In a laminar flow, add kanamycin stock solution, to a final concentration of 50 µg/mL (*Arabidopsis*) or 50–100 µg/mL (tobacco), to the germination medium. Alternatively, add hygromycin stock solution, to a final concentration of 20 µg/mL (*Arabidopsis*) or 25 µg/mL (tobacco), to the germination medium. Mix and pour the selective medium in 15-cm Falcon Optilux Petri dishes. About eight selective plates can be poured from 1 L of selective medium. Close the plates after solidification (*see* **Note 21**).
5. In a laminar airflow, put the seed packages in a sterile bottle (at most 20 packages/1-L bottle).
6. Cover the packages with 70% ethanol for 2 min.
7. Pour off the 70% ethanol in a sterile waste container.
8. Cover the packages with bleach solution for 12 min (*Arabidopsis*) or 8 min (tobacco). Shake regularly.
9. Immediately pour off the bleach solution into a sterile waste container.
10. Wash the packages immediately at least four times with sterile H_2O, until the rinse retains its original pH (6–7), pouring off the rinse between the different washes into a sterile waste container.
11. Leave the packages with the seeds in the bottle, covered with sterile H_2O, for 1 h.
12. Open a seed package in a sterile Petri dish and, one by one, put the seeds on the selective medium, using sterile tweezers. Do not put more than 32 seeds on one plate.

13. Seal the plate twice with Urgopore.
14. Incubate the seeds for 4 wk in a culture room (22°C [*Arabidopsis*] or 24°C [tobacco]; 16 h light/8 h dark cycle).
15. Select the kanamycin- or hygromycin-resistant seedlings by looking at the phenotype of the seedlings. Kanamycin-resistant seedlings are plantlets with two pairs of green leaves; kanamycin-sensitive seedlings are bleached, non-developing, and arrested at the cotyledon stage. Hygromycin-resistant seedlings have two pairs of normal expanding leaves, compared to very small, brownish, nonexpanding and rootless, sensitive seedlings.
16. Grow up 10 resistant seedlings and transfer them to the greenhouse by putting them in sterile soil. The tobacco seedlings can also be grown in sterile jars on germination medium.
17. Prepare protein extracts from relevant tissues of each plant. It is advisable to test tissues of a mature plant, because silencing is sometimes only initiated at some point during development.
18. Determine the accumulation level of the heterologous protein in each extract (*see* **Note 22**).
19. Self-fertilize the mature R1 plants (*see* **Note 23**) and collect seeds from each plant.
20. Vernalize the seeds at 4°C for 7 d.
21. Sterilize seeds of each seed stock and sow 32 seeds on selective medium.
22. Incubate the plates for 4 wk in the culture room.
23. For each seed stock, check whether sensitive seedlings are present. For homozygous R1 plants, only resistant seedlings will be obtained. For hemizygous R1 plants, three-fourths of the seedlings will be resistant and one-fourth of the seedlings will be sensitive. On this basis, determine for each seed stock whether the R1 plant was homozygous or hemizygous for the T-DNA. Because the primary transformant contained only one T-DNA copy, one-third of the 10 R1 progeny plants will be homozygous for the T-DNA copy.
24. Compare the heterologous protein levels, determined in **step 18**, for homozygous R1 plants and hemizygous R1 plants. The levels in homozygous R1 plants should be comparable to, or even higher than, the levels in hemizygous R1 plants. If this is not the case, the introduced gene is being silenced in the homozygous state.
25. Continue the transgene expression analysis for some of the subsequent generations of homozygous plants; heterologous protein synthesis sometimes only progressively diminishes in later generations *(31)*.

3.3.2. Analysis of the Progeny of Plants with Physically Unlinked T-DNA Copies

1. Self-fertilize each of the selected primary transformants (*see* **Subheading 3.2.4.2.**) and collect the seeds (= R1 seed stock).
2. Sow 100 seeds on selective medium (*see* **Note 24** and **Subheading 3.3.1.**)
3. Incubate for 4 wk in the culture room.
4. For each primary transformant, determine the number of T-DNA loci by determining the ratio of resistant R1 seedlings to sensitive R1 seedlings. Primary

transformants that contain only one locus will have a ratio of three resistant/one sensitive R1 seedlings; primary transformants with two loci will have a ratio of 15 resistant/1 sensitive R1 seedling.

5. Select the transformants with two loci (*see* **Note 25**). Select among these, preferentially, those for which the number of T-DNA copies revealed by the Southern analysis (*see* **Subheading 3.2.4.2.**) was limited, i.e., ≤3; this assures that at least one of the two loci contains only one T-DNA copy.

6. Grow 20 resistant R1 seedlings of the selected transformants (*see* **Note 26** and **step 16** of **Subheading 3.3.1.**).

7. Prepare genomic DNA (*see* **Subheading 3.2.1.**) from tissue of the R1 plants (*see* **Note 27**).

8. Perform a Southern analysis on the isolated genomic DNA and, if possible, on the genomic DNA isolated from the primary transformant (*see* **Subheading 3.2.1.**). Use one restriction digest (*see* **Subheading 3.2.2.**, **step 1**) in combination with the RB probe.

9. Select plants in which only one T-DNA copy is revealed by Southern analysis. These plants contain only one of the two original loci of the primary transformant because of segregation (*see* **Note 28**), and contain a locus with only one T-DNA copy.

10. Self-fertilize the selected plants and analyze the next generations, as described in **Subheading 3.3.1.**, to identify homozygous lines with stable transgene expression levels.

4. Notes

1. The proposed method starts with the assumption that the primary transformed tissues are already available. Yet, it is important to keep in mind, in the design of the transgene at the beginning of a project, that the choice of the expression signals can have a substantial influence on the likelihood of encountering instability problems. It is advisable to use different expression signals for the gene of interest and the selection marker gene; the presence of homologous promoter sequences can cause gene silencing in a subset of the transformants *(32)*. An overview of the expression signals commonly used for expression of heterologous proteins in plants and their influence on expression efficiency can be found in **ref. 47**.

2. Analysis at the protein level has many advantages compared to the analysis at the RNA level. It usually allows the analysis of many transformants at the same time in a quick, easy, and reproducible manner. It allows screening on the basis of the end product and, as such, gives an indication of the range of heterologous protein levels that can be expected. In this respect, analysis at the RNA level can be misleading, because inefficient translation or instability of the plant-made protein can lead to low heterologous protein levels. Moreover, additional qualitative data on the plant-made protein can be obtained.

3. Different plant tissues can be used for the screening. The choice of a particular tissue will depend on the promoter used to drive the production of the heterolo-

gous protein, the quantity of material available, and the ease of RNA or protein extraction from that tissue. We routinely perform the screening as soon as possible after transformation. For tobacco, it is advisable to do the screening on regenerated shoots and not on callus; there is always a substantial percentage (up to 50%) of tobacco primary calli from which no shoots can be regenerated. Regeneration of shoots from *Arabidopsis* callus poses the same problem. Yet, the screening will have to be done on callus tissue for this species, because shoots of primary *Arabidopsis* transformants tend to have very few leaves.

4. The heterologous protein can be present in very low amounts, or even absent in all primary transformants tested. In this case, a Northern analysis *(45,48)* will help reveal the underlying cause(s). Northern analysis allows us to determine whether any steady-state mRNA can be detected and whether the length of the mRNA is correct. If the mRNA amounts are as expected on the basis of the construct, different reasons could account for the low accumulation level or absence of the heterologous protein. First, the mRNA cannot be efficiently translated because of a bad initiation codon and/or additional AUG codons in the leader. Second, the heterologous protein is not stable in that compartment of the plant cell because of folding problems, degradation, and so on. Third, the heterologous protein is not extracted in the soluble fraction. Fourth, the plant-made protein is not functional, causing negative results in functional assays. If the mRNA levels are low, then the following reasons could account for the low accumulation of the protein. First, the promoter used to drive the production of the heterologous protein has undergone changes, preventing optimal transcription. A deletion in the promoter, for example, can significantly lower the promoter activity *(49)*. This deletion may have occurred during the cloning of the construct in *E. coli* or in *Agrobacterium*. Second, the full-length mRNA cannot be made because of premature transcriptional termination in AT-rich regions. Third, the mRNA is not stable in plants because of posttranscriptional control elements *(50,51)*. Fourth, premature termination of translation because of frame shifts or stop codons, can evoke mRNA destabilization, resulting in low mRNA levels (e.g., *52,53*).

5. In some studies, the effects of different constructs have to be compared. The data obtained for different transgenic populations will have to be interpreted statistically because of the high intertransformant variability. Nap et al. *(54)* described, in a very elegant way, which mistakes are commonly made that lead to erroneous conclusions and how to do statistical analysis to obtain valid conclusions.

6. It is advisable to compare data obtained for plant material harvested at the same time of the day; the amount of mRNA (and protein) may fluctuate diurnally *(20)*. Moreover, transgene expression can be substantially different in different leaves or at different time-points after transformation. Therefore, it is important to work as reproducibly as possible.

7. It is not sufficient to perform a genetic analysis. Genetic data only allow the determination of the number of T-DNA loci and these loci very often contain several linked T-DNAs (e.g., *18*). Therefore, a selection of plants based on genetic data could result in the selection of plants with T-DNA repeat structures, which

should be avoided, because T-DNA repeat structures have been correlated with silencing.

8. An adapted version of the protocol has already been used in our laboratory for the isolation of genomic DNA from 0.2 g of tobacco leaf material, the only adaptation being that the volumes of all solutions were divided by 10.

9. This precipitate will not be visible when only 0.2 g leaf material is used for the isolation of genomic DNA.

10. We routinely use phage λ DNA digested with *Pst*I.

11. We routinely use digested plasmid DNA, for which a fragment of known size is revealed by the probe. The plasmid DNA has to be diluted before loading it on gel. It should be loaded in equimolar amounts as the T-DNA-specific fragment of the digested genomic DNA.

12. In our experience, UV crosslinking of DNA gives much more intense signals than baking upon hybridization. Moreover, the UV crosslinking results in covalent attachment of the DNA to the membrane, so that the filters can be reprobed several times.

13. If only one probe is used, it is advisable to use a probe that detects RB fragments; integrated T-DNA show less truncations at the right T-DNA end than at the left T-DNA end (e.g., *55*).

14. We routinely recover the DNA fragment from an agarose gel, followed by purification using a specifically formulated silica matrix called Glassmilk® (GeneClean kit; BIO 101, La Jolla, CA). We routinely use DNA fragments of 700–1000 bp to prepare probes by the Megaprime™ DNA labeling system (Amersham, Aylesbury, U.K.).

15. We routinely use Bio-Spin columns (Bio-Rad, Hercules, CA).

16. The interpretation of the Southern data will be more difficult if only the RB probe has been used. If only one fragment is observed with the RB probe, then only one type of repeat structure can be present—an IR_{RB} (*see* **Fig. 1**). Therefore, the size of the fragment will have to be determined and compared to the size predicted for an IR_{RB} (*see* **Fig. 1**). In addition, it is necessary to check whether an IR_{RB} is present that gives a fragment different from the one predicted because of the presence of DNA between the linked T-DNAs. This can be done by comparing the data obtained for both restriction digests. When the fragment obtained for both restriction digests deviates in the same way (i.e., 500 bp larger in size) from the size predicted for an IR_{RB}, then the presence of a repeat structure with 500 bp DNA between the linked T-DNAs is very probable. When more than one fragment has been revealed by the RB probe, then the interpretation will be even more difficult. In that case, the sizes of the fragments will have to be compared to the sizes predicted for an IR_{RB} and DR. It is also necessary to check whether an IR_{RB} or DR is present that gives a fragment different from the one predicted because of deletions or the presence of DNA between the linked T-DNAs. Moreover, it will be impossible to screen for the presence of an IR_{LB}; this type of repeat structure gives two junction fragments of unpredictable size with the RB probe (*see* **Fig. 1**). Therefore, plants with two physically unlinked copies cannot

be distinguished from plants with an IR_{LB}, based on the Southern data of the primary transformant.

17. First, check whether the DNA of the untransformed plant gives background signals with the used probe. When this is the case, then these background signals (fragments) should not be taken into account for the determination of the number of border fragments in **step 1**.

18. Sometimes, the number of left and right border fragments are not equal. This is a strong indication for the occurrence of linked T-DNA copies or truncated T-DNAs.

19. It is only possible to deduce from Southern data whether two T-DNAs are physically linked. Two T-DNAs that are not physically linked can still segregate together to the progeny and are then considered to be genetically linked to each other.

20. When the T-DNA does not contain a selection marker, T-DNA-containing seedlings can be selected by PCR analysis, or by using an assay that screens for the presence of the heterologous protein (*see* **Subheading 3.1.**).

21. The plates containing kanamycin can be kept at room temperature for up to 2 wk; the plates containing hygromycin can be kept at room temperature for up to 3 wk.

22. The data will often already indicate which plants are homozygous for T-DNA (the ones with the highest or severely reduced accumulation levels).

23. It can be interesting to perform crosses of the progeny plants with a wild-type plant. If the transgenic progeny plant was homozygous, a homogenous hemizygous seed stock will be obtained in this way.

24. We routinely transfer the primary *Arabidopsis* transformants to the greenhouse in order to obtain seeds. We found that this is necessary to obtain a sufficient amount of seeds, ranging from 50 to more than 100 seeds per primary transformant.

25. Primary transformants that contain only one locus should not be selected, because they contain all the T-DNAs—the number being determined in the Southern analysis (*see* **Subheading 3.2.4.2.**), genetically linked in one locus. These different T-DNAs will not segregate. Moreover, it has been observed that T-DNAs, which are only genetically, and not physically, linked, can also give rise to silencing.

26. If the Southern data gave evidence for only two T-DNA copies, then it will be sufficient to grow 10 plants. However, for plants for which the Southern data indicated the presence of three copies, it is necessary to analyze more plants, because only three out of 15 selected plants will contain the locus with only one T-DNA.

27. For *Arabidopsis* greenhouse plants, we routinely remove the first appearing inflorescence when DNA has to be isolated from the leaves; the plant will then develop more leaf material. After the isolation of the necessary leaf material for the DNA, the inflorescence is no longer removed, so that seeds can be obtained from the plant.

28. The two loci of the primary transformant will have segregated, so that nine out of 15 of the selected R1 plants will contain both loci, three out of 15 of the selected

R1 plants will contain only the first locus, and three out of 15 will contain only the second locus. This will be evident if the pattern of the Southern analysis of each R1 plant is compared to the pattern of the primary transformant. Nine out of 15 plants will have the same pattern as the primary transformant. Three out of 15 plants will reveal the T-DNA(s) of the first locus and three out of 15 will reveal the other T-DNA(s) of the second locus.

Acknowledgments

The authors thank Geert De Jaeger, Willy Dillen, Anni Jacobs, and Matt Sauer for critical reading of the manuscript. This work was supported by grants from the Belgian Programme on Interuniversity Poles of Attraction (Prime Minister's Office, Science Policy Programming, #38) and the Vlaams Actieprogramma Biotechnologie (Emerging Technological Center [ETC] 002). A.-M. B. is indebted to the Vlaams Instituut voor de Bevordering van het Wetenschappelijk-Technologisch Onderzoek in de Industrie for a predoctoral fellowship. H. V. H. is a Research Assistant of the Fund for Scientific Research (Flanders).

References

1. Peach, C. and Velten, J. (1991) Transgene expression variability (position effect) of CAT and GUS reporter genes driven by linked divergent T-DNA promoters. *Plant Mol. Biol.* **17,** 49–60.
2. Songstad, D. D., Somers, D. A., and Griesbach, R. J. (1995) Advances in alternative DNA delivery techniques. *Plant Cell Tissue Organ Cult.* **40,** 1–15.
3. Zupan, J. R. and Zambryski, P. (1995) Transfer of T-DNA from *Agrobacterium* to the plant cell. *Plant Physiol.* **107,** 1041–1047.
4. Gould, J., Devey, M., Hasegawa, O., Ulian, E. C., Peterson, G., and Smith, R. H. (1991) Transformation of *Zea mays* L. using *Agrobacterium tumefaciens* and the shoot apex. *Plant Physiol.* **95,** 426–434.
5. Chan, M.-T., Lee, T.-M., and Chang, H.-H. (1992) Transformation of Indica rice (*Oryza sativa* L.) mediated by *Agrobacterium tumefaciens*. *Plant Cell Physiol.* **33,** 577–583.
6. Hiei, Y., Ohta, S., Komari, T., and Kumashiro, T. (1994) Efficient transformation of rice (*Oryza sativa* L.) mediated by *Agrobacterium* and sequence analysis of the boundaries of the T-DNA. *Plant J.* **6,** 271–282.
7. Ambros, P. F., Matzke, A. J. M., and Matzke, M. A. (1986) Localization of *Agrobacterium rhizogenes* T-DNA in plant chromosomes by *in situ* hybridization. *EMBO J.* **5,** 2073–2077.
8. Robbins, T. P., Gerats, A. G. M., Fiske, H., and Jorgensen, R. A. (1995) Suppression of recombination in wide hybrids of *Petunia hybrida* as revealed by genetic mapping of marker transgenes. *Theor. Appl. Genet.* **90,** 957–968.
9. Finnegan, J. and McElroy, D. (1994) Transgene inactivation: plants fight back! *Bio/Technology* **12,** 883–888.

10. De Loose, M., Danthinne, X., Van Bockstaele, E., Van Montagu, M., and Depicker, A. (1995) Different 5' leader sequences modulate β-glucuronidase accumulation levels in transgenic *Nicotiana tabacum* plants. *Euphytica* **85,** 209–216.
11. van der Hoeven, C., Dietz, A., and Landsmann, J. (1994) Variability of organ specific gene expression in transgenic tobacco plants. *Transgenic Res.* **3,** 159–165.
12. Breyne, P., Gheysen, G., Jacobs, A., Van Montagu, M., and Depicker, A. (1992) Effect of T-DNA configuration on transgene expression. *Mol. Gen. Genet.* **235,** 389–396.
13. Mlynárová, L., Loonen, A., Heldens, J., Jansen, R. C., Keizer, P., Stiekema, W. J., and Nap, J.-P. (1994) Reduced position effect in mature transgenic plants conferred by the chicken lysozyme matrix-associated region. *Plant Cell* **6,** 417–426.
14. Breyne, P., Van Montagu, M., Depicker, A., and Gheysen, G. (1992) Characterization of a plant scaffold attachment region in a DNA fragment that normalizes transgene expression in tobacco. *Plant Cell* **4,** 463–471.
15. Bhattacharyya, M. K., Stermer, B. A., and Dixon, R. A. (1994) Reduced variation in transgene expression from a binary vector with selectable markers at the right and left T-DNA borders. *Plant J.* **6,** 957–968.
16. Depicker, A., Ingelbrecht, I., Van Houdt, H., De Loose, M., and Van Montagu, M. (1996) Posttranscriptional reporter transgene silencing in transgenic tobacco, in *Mechanisms and Applications of Gene Silencing* (Grierson, D., Lycett, G. W., and Tucker, G. A., eds.), Nottingham University Press, Nottingham, UK, pp. 71–84.
17. Gendloff, E. H., Bowen, B., and Buchholz, W. G. (1990) Quantitation of chloramphenicol acetyl transferase in transgenic tobacco plants by ELISA and correlation with gene copy number. *Plant Mol. Biol.* **14,** 575–583.
18. Hobbs, S. L. A., Warkentin, T. D., and DeLong, C. M. O. (1993) Transgene copy number can be positively or negatively associated with transgene expression. *Plant Mol. Biol.* **21,** 17–26.
19. Dean, C., Jones, J., Favreau, M., Dunsmuir, P., and Bedbrook, J. (1988) Influence of flanking sequences on variability in expression levels of an introduced gene in transgenic tobacco plants. *Nucleic Acids Res.* **16,** 9267–9283.
20. Jones, J. D. G., Gilbert, D. E., Grady, K. L., and Jorgensen, R. A. (1987) T-DNA structure and gene expression in petunia plants transformed by *Agrobacterium tumefaciens* C58 derivatives. *Mol. Gen. Genet.* **207,** 478–485.
21. Shirsat, A. H., Wilford, N., and Croy, R. R. D. (1989) Gene copy number and levels of expression in transgenic plants of a seed specific gene. *Plant Sci.* **61,** 75–80.
22. Hobbs, S. L. A., Kpodar, P., and DeLong, C. M. O. (1990) The effect of T-DNA copy number, position and methylation on reporter gene expression in tobacco transformants. *Plant Mol. Biol.* **15,** 851–864.
23. Larkin, P. J., Banks, P. M., Bhati, R., Brettell, R. I. S., Davies, P. A., Ryan, S. A., Scowcroft, W. R., Spindler, L. H., and Tanner, G. J. (1989) From somatic variation to variant plants: mechanisms and applications. *Genome* **31,** 705–711.

24. Angenon, G., Bruyns, A., Jacobs, A., Van Montagu, M., and Depicker, A. (1991) A test system for the molecular analysis of mutants induced by tissue culture or mutagenic treatment in plants. *Med. Fac. Landbouww. Rijksuniv. Gent* **56,** 1393–1401.

25. Heberle-Bors, E., Charvat, B., Thompson, D., Schernthaner, J. P., Barta, A., Matzke, A. J. M., and Matzke, M. A. (1988) Genetic analysis of T-DNA insertions into the tobacco genome. *Plant Cell Rep.* **7,** 571–574.

26. Mittelsen Scheid, O., Paszkowski, J., and Potrykus, I. (1991) Reversible inactivation of a transgene in *Arabidopsis thaliana. Mol. Gen. Genet.* **228,** 104–112.

27. Flavell, R. B. (1994) Inactivation of gene expression in plants as a consequence of specific sequence duplication. *Proc. Natl. Acad. Sci. USA* **91,** 3490–3496.

28. Matzke, M. A. and Matzke, A. J. M. (1995) How and why do plants inactivate homologous (trans)genes? *Plant Physiol.* **107,** 679–685.

29. Meyer, P. (1995) Understanding and controlling transgene expression. *Trends Biotechnol.* **13,** 332–337.

30. Linn, F., Heidmann, I., Saedler, H., and Meyer, P. (1990) Epigenetic changes in the expression of the maize A1 gene in *Petunia hybrida*: role of numbers of integrated gene copies and state of methylation. *Mol. Gen. Genet.* **222,** 329–336.

31. Cherdshewasart, W., Gharti-Chhetri, G. B., Saul, M. W., Jacobs, M., and Negrutiu, I. (1993) Expression instability and genetic disorders in transgenic *Nicotiana plumbaginifolia* L. plants. *Transgenic Res.* **2,** 307–320.

32. Assaad, F. F., Tucker, K. L., and Signer, E. R. (1993) Epigenetic repeat-induced gene silencing (RIGS) in *Arabidopsis. Plant Mol. Biol.* **22,** 1067–1085.

33. Kilby, N. J., Leyser, H. M. O., and Furner, I. J. (1992) Promoter methylation and progressive transgene inactivation in *Arabidopsis. Plant Mol. Biol.* **20,** 103–112.

34. Meyer, P. and Heidmann, I. (1994) Epigenetic variants of a transgenic petunia line show hypermethylation in transgene DNA: an indication for specific recognition of foreign DNA in transgenic plants. *Mol. Gen. Genet.* **243,** 390–399.

35. Neuhuber, F., Park, Y.-D., Matzke, A. J. M., and Matzke, M. A. (1994) Susceptibility of transgene loci to homology-dependent gene silencing. *Mol. Gen. Genet.* **244,** 230–241.

36. de Carvalho Niebel, F., Frendo, P., Van Montagu, M., and Cornelissen, M. (1995) Post-transcriptional cosuppression of β-1,3-glucanase genes does not affect accumulation of transgene nuclear mRNA. *Plant Cell* **7,** 347–358.

37. Jefferson, R. A. (1987) Assaying chimeric genes in plants: the GUS gene fusion system. *Plant Mol. Biol. Rep.* **5,** 387–405.

38. Miner, J. N., Weinrich, S. L., and Hruby, D. E. (1988) Molecular dissection of *cis*-acting regulatory elements from 5'-proximal regions of a vaccinia virus late gene cluster. *J. Virol.* **62,** 297–304.

39. Sleigh, M. J. (1986) A nonchromatographic assay for expression of the chloramphenicol acetyltransferase gene in eucaryotic cells. *Anal. Biochem.* **156,** 251–256.

40. Reiss, B., Sprengel, R., Will, H., and Schaller, H. (1984) A new sensitive method for qualitative and quantitative assay of neomycin phosphotransferase in crude cell extracts. *Gene* **30,** 211–218.

41. D'Halluin, K., De Block, M., Denecke, J., Janssens, J., Leemans, J., Reynaerts, A., and Botterman, J. (1992) The *bar* gene as selectable and screenable marker in plant engineering, in *Recombinant DNA,* part G, vol. 216, *Methods in Enzymology* (Wu, R., ed.), Academic, San Diego, pp. 415–426.
42. Luehrsen, K. R., de Wet, J. R., and Walbot, V. (1992) Transient expression analysis in plants using firefly luciferase reporter gene, in *Recombinant DNA,* part G, vol. 216, *Methods in Enzymology* (Wu, R., ed.), Academic, San Diego, pp. 397–414.
43. Jones, J. D. G., Dunsmuir, P., and Bedbrook, J. (1985) High level expression of introduced chimaeric genes in regenerated transformed plants. *EMBO J.* **4,** 2411–2418.
44. Dellaporta, S. L., Wood, J., and Hicks, J. B. (1983) A plant DNA minipreparation: version II. *Plant Mol. Biol. Rep.* **1,** 19–21.
45. Ausubel, F. M., Brent, R., Kingston, R. E., Moore, D. D., Seidman, J. G., Smith, J. A., and Struhl, K. (1994) *Current Protocols in Molecular Biology,* vol. 1. Current Protocols, New York.
46. McClelland, M., Nelson, M., and Raschke, E. (1994) Effect of site-specific modification on restriction endonucleases and DNA modification methyltransferases. *Nucleic Acids Res.* **22,** 3640–3659.
47. Fütterer, J. (1995) Expression signals and vectors, in *Gene Transfer to Plants* (Springer Lab Manual) (Potrykus, I. and Spangenberg, G., eds.), Springer, Berlin, pp. 311–324.
48. Sambrook, J., Fritsch, E. F., and Maniatis, T. (1989) *Molecular Cloning: A Laboratory Manual,* 2nd ed., Cold Spring Harbor Laboratory, Cold Spring Harbor, NY.
49. Ow, D. W., Jacobs, J. D., and Howell, S. H. (1987) Functional regions of the cauliflower mosaic virus 35S RNA promoter determined by use of the firefly luciferase gene as a reporter of promoter activity. *Proc. Natl. Acad. Sci. USA* **84,** 4870–4874.
50. Green, P. J. (1993) Control of mRNA stability in higher plants. *Plant Physiol.* **102,** 1065–1070.
51. Sullivan, M. L. and Green, P. J. (1993) Post-transcriptional regulation of nuclear-encoded genes in higher plants: the roles of mRNA stability and translation. *Plant Mol. Biol.* **23,** 1091–1104.
52. Jofuku, K. D., Schipper, R. D., and Goldberg, R. B. (1989) A frameshift mutation prevents Kunitz trypsin inhibitor mRNA accumulation in soybean embryos. *Plant Cell* **1,** 427–435.
53. Voelker, T. A., Moreno, J., and Chrispeels, M. J. (1990) Expression analysis of a pseudogene in transgenic tobacco: a frameshift mutation prevents mRNA accumulation. *Plant Cell* **2,** 255–261.
54. Nap, J.-P., Keizer, P., and Jansen, R. (1993) First-generation transgenic plants and statistics. *Plant Mol. Biol. Rep.* **11,** 156–164.
55. Jorgensen, R., Snyder, C., and Jones, J. D. G. (1987) T-DNA is organized predominantly in inverted repeat structures in plants transformed with *Agrobacterium tumefaciens* C58 derivatives. *Mol. Gen. Genet.* **207,** 471–477.

17

Manipulation of Photosynthetic Metabolism

Martin A. J. Parry, Steven P. Colliver, Pippa J. Madgwick, and Matthew J. Paul

1. Introduction

1.1. Possible Targets

Photosynthesis, as the basic process leading to biomass accumulation, is intrinsically limited by the performance of the photosynthetic apparatus under different environmental conditions. Potentially, substantial increases in crop yield and improved efficiency of production could be achieved by increasing leaf photosynthetic rates. An essential prerequisite to improving the efficiency of photosynthesis is an understanding of the individual steps involved, their regulation, and interactions with the external environment. Once potential targets have been identified, techniques enabling stable genetic transformation of important crop plants are available to make many of the specific changes we may require.

One of the major intrinsic limitations in photosynthesis is the low catalytic competence of the enzyme ribulose 1,5-*bis*-phosphate carboxylase/oxygenase (Rubisco), which initiates photosynthesis by the carboxylation of ribulose 1,5-*bis*-phosphate (RuBP). Rubisco has a low specific activity for carboxylation and catalyzes several other reactions. The most important of these is the oxygenation of RuBP, which leads to the production of glycolate, the subsequent metabolism of which results in a massive and energetically wasteful loss of newly assimilated carbon dioxide (CO_2) from the plant, a process called photorespiration. This oxygenase reaction is the most important metabolic constraint on plant productivity, and up to 50% of the carbon fixed by a C_3 plant may be lost in this way *(1)*. In addition, photorespiration generates ammonia within the leaves, some of which is lost to the atmosphere. A large amount of

From: *Methods in Biotechnology, Vol. 3:*
Recombinant Proteins from Plants: Production and Isolation of Clinically Useful Compounds
Edited by: C. Cunningham and A. J. R. Porter © Humana Press Inc., Totowa, NJ

Rubisco is needed to achieve high photosynthetic rates because of its low specific activity for carboxylation; thus, Rubisco dominates the nitrogen (N) requirement of plants (ca. 25% of leaf N). Even with this large investment, the activity of Rubisco still limits CO_2 assimilation under many conditions. Even small increases in the rate of CO_2 fixation through improved Rubisco could have immense benefits, such as improving crop N-use efficiency and water-use efficiency, with larger yields from low inputs. Additionally, enhanced fixation of CO_2 from the atmosphere might counteract global warming.

Clearly, photorespiration and Rubisco are obvious targets for genetic manipulation. One possible strategy is to engineer a more efficient Rubisco by introducing genes into the chloroplast encoding Rubisco with an improved specificity for CO_2. This could be achieved by exploiting the natural variation in the catalytic properties of Rubisco isolated from different species, or by improving the enzyme by site-directed mutagenesis *(2)*. Alternatively, the wasteful consequences of photorespiration could be overcome by introducing a CO_2-concentrating mechanism (this has evolved already in some plants) or by manipulating photorespiratory metabolism.

Since photorespiratory metabolism involves several intracellular compartments, successful manipulation will require recombinant enzymes to be targeted. A number of transit peptide sequences have been identified that permit directed transport to these compartments *(3)*. The introduction into C_3 plants of an effective CO_2-concentrating mechanism similar to that found in C_4 plants would require structural modifications, probably beyond our current capabilities. However, it may be possible to shift the site for the wasteful release of CO_2 and/or ammonia away from the intercellular spaces to improve internal reassimilation. This would require the movement of the glycine decarboxylase multi-enzyme complex from the mitochondria of mesophyll cells to the vicinity of the vascular tissue *(4)*. A more radical alternative might be to introduce genes to abbreviate or modify the photorespiratory pathway to reduce the wasteful consequences of the oxygenase activity. This could be achieved by introducing genes encoding enzymes not normally found; for example, the introduction of glyoxylate carboligase and tartronic semialdehyde reductase, together with the antisense repression of existing photorespiratory enzymes, may prevent the wasteful release of ammonia. A still more speculative approach would be the *ab initio* design of a Rubisco lacking the oxygenase activity.

Models can be of use in deciding targets or in predicting what the result of genetic manipulation might be, but the only way to test whether photosynthetic efficiency can be improved is to directly modify it by genetic manipulation. To demonstrate some of the practical considerations required to produce and analyze plants with modified photosynthetic metabolism, we will consider a test case.

1.2. Manipulation of Rubisco

Models of photosynthesis suggest that, at elevated CO_2 concentrations and at temperatures of 5–25°C, Rubiscos from the photosynthetic bacteria *Chromatium vinosum* and *Anacystis nidulans* should out-perform most higher-plant Rubiscos *(2)*. In higher plants, Rubisco is composed of eight chloroplast-encoded large subunits and eight nuclear-encoded small subunits. Since the catalytic sites of Rubisco are shared between large subunits, the large subunits determine the catalytic properties of the holoenzyme.

We will consider the introduction of *C. vinosum* sequences encoding Rubisco large subunits *(5)* into tobacco. Although higher plant genes of both subunits have been expressed in *Escherichia coli*, together with their respective chaperonins, the expressed proteins have failed to assemble into functional enzymes *(6)*. In contrast, the genes for the bacterial forms have been successfully expressed and assembled in *E. coli*. Furthermore, catalytically competent chimeric enzymes containing the subunit polypeptides from different species have been produced. We can therefore expect the *C. vinosum* large subunits to assemble with higher-plant small subunits and the holoenzyme to retain the catalytic properties of the large subunit.

Because of the large amount of Rubisco present in leaves that will reduce the relative effect of the recombinant Rubisco, we have identified a model system in which the plants are devoid of Rubisco. Tobacco lines lacking functional large subunits have been identified. These lines cannot photosynthesize, but can be maintained in vitro by supplying carbon to the plant in the form of sucrose, and represent ideal material for this study *(7)*. Chloroplast transformation would provide an elegant method for introducing a novel Rubisco large-subunit gene (*rbc*L), but techniques for this are not yet well established and are limited to a single species *(8)*. However, nuclear transformation is routine in many species and novel *rbc*L genes can be integrated into the nuclear genome, provided the polypeptide is then targeted to the chloroplast by an appropriate transit peptide.

1.2.1. Design Strategy

To transform tobacco using *Agrobacterium tumefaciens*-mediated co-transformation, a DNA construct, with the gene of interest under the control of an appropriate promoter, transit peptide (TP), and terminator region, has to be contained within the T-DNA region of a binary plant transformation vector (**Fig. 1**). Numerous plasmids are available that contain some of these elements. We have used the binary plant transformation vectors pROK2 and pROK8, which contain within their T-DNA the constitutive CaMV35S promoter and the inducible, tissue-specific Rubisco small subunit promoter, respectively,

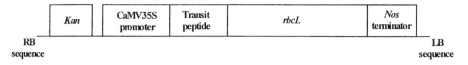

Fig. 1. T-DNA region of the binary plant transformation vector p2TCVL.

together with the Nopaline synthase (*Nos*) terminator region. These plasmids both contain right and left border sequences, derived from the Ti plasmid of *A. tumefaciens*, which mediate the integration of the T-DNA into the nuclear genome. The plasmid pTrans harbors the gene encoding the soybean Rubisco small subunit TP within the multiple-cloning site of pGEM3 (Promega). In this plasmid, a number of restriction sites from the *Sph*I site to *Eco*RI site can be used for subcloning the genes of interest.

The *C. vinosum rbc*L must be fused to a TP-coding sequence to ensure that the *rbc*L is in the same reading frame as the TP, that there is a minimal amount of DNA introduced between the end of the TP and the start of the *rbc*L, that any introduced sequence does not encode a stop codon, and that any residues (e.g., proline) that might be expected to disrupt the normal function of the enzyme are not introduced. In addition, it must be possible to further subclone the TP-*rbc*L fusion. The identification of a restriction enzyme site that would permit an in-frame fusion at the start codon, or just upstream from it, is ideal, but rarely exists. In this case, there are no convenient restriction enzyme sites at the 5' end of the *rbc*L that satisfy these criteria. Therefore, it is necessary to create a unique site by site-directed mutagenesis.

The simplest approach to site-directed mutagenesis is to use a commercially available kit that can be used with many plasmids and relies on the elimination of a unique restriction site. The design of an oligonucleotide primer for mutagenesis must introduce the required changes, bind uniquely and securely to the site to be mutated, and bind to the same strand of the plasmid as the selection primer. **Figure 2** shows an oligonucleotide primer designed to introduce an *Sph*I site just upstream of the start codon of the *rbc*L coding sequence, which will permit it to be subcloned, in-frame, into pTrans downstream of the TP. In this case, mutation of the region around the start codon itself is unwise, because this would modify the N-terminus of the protein by introducing a proline residue at position 2. This may disrupt the polypeptide structure, which may in turn affect import into the chloroplasts and/or catalytic activity.

The binary transformation vectors we have selected are large plasmids with comparatively few cloning sites. Insertion of DNA into these sites is not always straightforward, and in our experience blunt-ended DNA fragments do not

wild-type
sequence   . . . . CTAATCCAGAGGACGATCCAGAATGGCTAAGACGTAC . . . .

mismatches   **XXXXX**

oligonucleotide   CTAATCCAGAGGAC**GCATGC**GAATGGCTAAGACGTAC
                               *Sph*I

Melting temperatures   5' of mismatches, 42°C   3' of mismatches, 50°C

Fig. 2. Oligonucleotide primer design.

readily ligate into the *Sma*I site. Consequently, we recommend using inserts with cohesive ends, even if this involves an extra subcloning step.

2. Materials

All solutions are made up in glass distilled water unless stated otherwise. General laboratory supplies were obtained from Sigma (Poole, Dorset, UK) or Fisons (Loughborough, Leices, UK) unless otherwise noted.

1. Agarose.
2. Alkaline transfer buffer: 250 mM NaOH, 1.5M NaCl.
3. Ampicillin: 100 mg/mL in sterile distilled water (SDW). Store at –20°C.
4. Adenosine triphosphate (ATP): 10 mM in SDW. Store at –20°C.
5. Bacterial strains: *E. coli* DH5α (F⁻, φ80d*lac*ZΔM15, *rec*A1, *end*A1, *gyr*A96, *thi*-1, *hsd*R17, (r$_K$⁻,m$_K$⁺), *sup*E44, *rel*A1, *deo*R, Δ(*lac*ZYA-*arg*F)U169) *(10)*; *A. tumefaciens* LBA4404 (pRAL4404), a derivative of Ach5.
6. Bactoagar (Difco, Detroit, MI).
7. 6-Benzylaminopurine (BAP): Dissolve 25 mg in 0.5 mL 1M NaOH and dilute to 1 mg/mL with SDW. Store at 4°C.
8. 5-Bromo-4-chloro-3-indolyl-β-D-galactopyranoside (X-gal): 50 mg/mL in *N,N*′dimethylformamide. Store at –20°C.
9. (¹⁴C)-CABP.
10. CaCl$_2$6H$_2$O: 50 mM. Store at 4°C.
11. Cefotaxime (Roussel, Dublin, Ireland): 100 mg/mL in SDW. Store at –20 °C.
12. CH$_3$COOK: 5M.
13. CH$_3$COONa: 3M pH to 5.2 with CH$_3$·COOH.
14. Chameleon™ Double-Stranded Site-Directed Mutagenesis Kit (Stratagene, La Jolla, CA).
15. Chloroform.
16. Chloroform solution: chloroform:isoamyl alcohol (24:1[v/v]). Store at 4°C.
17. Denaturing solution: 400 mM NaOH, 1M NaCl.
18. DNA extraction buffer: 15 mL T$_{100}$E$_{10}$, pH 8.0, 1 mL SDS 20% (w/v), 10 μL 2-mercaptoethanol.

19. DNA-equilibrated phenol solution: Molten phenol saturated with 100 mM Tris.base, pH 8.0. Store at 4°C (Gibco, BRL, Paisley, UK).

20. α^{32}P-dCTP solution (ICN, Costa Mesa, CA).

21. dNTP solution: 100 mM. Store at –20°C (Pharmacia, Biotech, St. Albans, UK).

22. Ethanol.

23. Ethidium bromide: 10 mg/mL.

24. Ethylenediaminetetraaceticacid (EDTA): 200 mM.

25. Gene Pulser™ with Pulse Controller Unit (Bio-Rad, Hercules, CA).

26. Gene Pulser™ Cuvette, 0.2 cm (Bio-Rad).

27. Glycerol: 10% (v/v).

28. HCl: 250 mM.

29. H·COOH: 10M.

30. Hybridization oven: Maxi 14 (Life Sciences International, Basingstoke, UK).

31. Isopropanol.

32. Isopropyl β-D-thiogalactopyranoside (IPTG): 100 mM in SDW. Store at 4°C.

33. Kanamycin: 50 mg/mL in SDW. Store at –20°C.

34. LiCl: 8M.

35. Luria-Bertaini (LB) medium: bactotryptone 10 g/L, yeast (Beta Lab, West Mosely, UK) extract 5 g/L, NaCl 5 g/L, pH 7.5, with NaOH.

36. MgCl$_2$: 25 mM.

37. Mineral oil.

38. Miracloth (Calbiochem, Nottingham, UK).

39. MOPS buffer: 20 mM 3-(N-morphlino)propanesulphonic acid (MOPS), 5 mM CH3·COONa, 1 mM EDTA, 6.4% (w/v) formaldehyde.

40. MS plant tissue culture medium (ICN): MS basal plant tissue culture medium 4.7 g/L, sucrose 30 g/L, agargel 5% (w/v), pH 5.8, with KOH.

41. NaCl: 5M.

42. Nappy Booster pads (Boots, own brand).

43. α-Naphthaleneacetic acid (NAA): Dissolve 25 mg in 12 mL absolute EtOH and dilute to 1 mg/mL with SDW, store at 4°C.

44. Nescofilm (Nippon Shoji Kaisha, Japan).

45. Liquid nitrogen.

46. Nuclease-free water: DEPC 0.1% (v/v) treated in autoclaved SDW.

47. Polyethylene glycol 4000 (PEG).

48. Phenol-chloroform extraction solution: Phenol (DNA equilibrated):chloroform: isoamyl alcohol (25:24:1[v/v/v]), store at 4°C.

49. Plant material: Rubisco-minus *Nicotiana tabacum* mutant line Sp25 *(7)*.

50. Plasmid DNA: pCV17 *(5)*, pTrans, pROK2, pROK8, pUC18 *(9)*.

51. Prehybridization solution: 0.5M NaH$_2$PO$_4$, SDS 7% (w/v).

52. Polymerase Chain Reaction (PCR) Machine (Life Sciences, UK).

53. Pronase solution: 2.5 mg/mL in T$_{10}$E$_1$.

54. Prime-It® II Random Primer Labeling Kit (Stratagene).

55. RNA extraction buffer: 1 mL 3M CH$_3$COONa, pH 5.2, 0.3 mL 500 mM EDTA, pH 8.0, and 1.5 mL SDS 10% (w/v) in nuclease-free water.

56. RNA-equilibrated phenol solution: Molten phenol saturated with 12 mM Tris.base, 30 mM NaOH solution. Store at 4°C.
57. RNase A solution: 50 μg/mL in $T_{10}E_1$. Store at –20°C.
58. RNA hybridization solution: 0.12M NaH$_2$PO$_4$, SDS 7% (w/v), 0.25M NaCl, 50% (v/v) formamide.
59. Rubisco extraction buffer: 50 mM Bicine, pH 8.0, 20 mM MgCl$_2$, 1 mM phenylmethlysulphonyl fluoride, 50 mM 2-mercaptoethanol.
60. Rubisco assay mix: 100 mM Bicine, 10 mM MgCl$_2$, 10 mM (^{14}C)-NaHCO$_3$ (7.4 kBq/μmol).
61. Rubisco quantification buffer: 100 mM NaHCO$_3$, 200 mM 2-mercaptoethanol, 200 mM Na$_2$SO$_4$.
62. Rubisco wash solution: PEG 20% (w/v) in 100 mM Bicine, 10 mM NaHCO$_3$, 20 mM MgCl$_2$, pH 8.0.
63. 33 mM Ribulose 1,5-bisphosphate (RuBP).
64. Saranwrap (Dow).
65. Sarkosyl solution: 5% (w/v) in $T_{10}E_1$.
66. Sephaglas™ Band Prep Kit (Pharmacia Biotech).
67. Sodium dodecyl sulphate (SDS): 20% (w/v) in SDW.
68. 20X SSC: 3M NaCl, 300 mM trisodium citrate.
69. Sterile-distilled water.
70. Stratalinker UV crosslinker (Stratagene).
71. $T_{10}E_1$: 10 mM Tris-base, 1 mM EDTA, pH 8.0.
72. $T_{50}E_{20}$: 50 mM Tris-base, 20 mM EDTA, pH 8.0.
73. $T_{100}E_{10}$: 100 mM Tris-base, 10 mM EDTA, pH 8.0.
74. TEA buffer: 40 mM Tris-base, 50 mM EDTA, 0.35% (v/v) glacial acetic acid.
75. Tris(hydroxymethyl)aminomethane (Tris-base).
76. Triton X-100: 1% (v/v).
77. Whatman No. 1 Paper (Whatman, Maidstone, UK).
78. Wizard™ Minipreps DNA Purification System (Promega).
79. Wizard™ Midipreps DNA Purification System (Promega).
80. Wrist action shaker (Stuart Scientific).
81. YEB medium: 1 g/L yeast extract; 5 g/L beef extract, 5 g/L peptone, 8 g/L sucrose, 0.5 g/L MgSO$_4$, pH 7.2.
82. YMB medium: 0.4 g/L yeast extract, 10 g/L mannitol, 0.1 g/L NaCl, 0.2 g/L MgSO$_4$·7H$_2$O, 0.5 g/L K$_2$HPO$_4$·3H$_2$O, pH 7.0.
83. YT medium: 10 g/L yeast extract, 16 g/L tryptone, 5 g/L NaCl.
84. Zeta-Probe® GT blotting membrane (Bio-Rad).

3. Methods

3.1. Growth of E. coli

Grow *E. coli*, harboring plasmids, in LB medium supplemented with the appropriate antibiotic (either ampicillin 100 μg/mL; kanamycin 50 μg/mL), with shaking at 200 rpm, at 37°C, for up to 16 h, as necessary. Flasks should contain no more than 20% of their nominal volume, to ensure adequate aeration. For plate cultures, grow *E. coli* on solidified LB medium containing Bactoagar, 1.5% (w/v), and supplemented with antibiotic, as above, at 37°C for up to 16 h, as necessary.

3.2. Transformation of E. coli

1. Grow *E. coli* in 5 mL of LB medium, with shaking at 200 rpm at 37 °C for 16 h.
2. Transfer 400 μL from this 5 mL culture into a 250 mL flask containing 40 mL LB medium and grow as above for ≈2.5 h to an OD_{550} of ≈0.3.
3. Transfer the cell culture to a sterile 50 mL polypropylene centrifuge tube and pellet cells by centrifugation at 3000*g* for 10 min at 4°C.
4. Discard the supernatant and resuspend the cell pellet in 20 mL 50 m*M* $CaCl_2$ (precooled to 4°C), then place on ice for 30 min.
5. Harvest the cells by centrifugation as above, discard the supernatant, and resuspend the cells in 4 mL 50 m*M* $CaCl_2$ (precooled to 4°C).
6. Place the cells on ice for 1 h prior to using for transformation.
7. To transform competent cells, add plasmid DNA to 60 μL of competent cell suspension in a microcentrifuge tube and place on ice for 40 min.
8. Heat-shock the plasmid DNA/cell suspension mixture at 42°C for 50 s and then place the cells at room temperature for 10 min.
9. Plate cells onto solidified LB medium supplemented with the appropriate antibiotic (*see* **Subheading 3.1.**) to select for recombinant *E.coli*, and incubate at 37°C for 16 h.
10. Purify plasmid DNA from recombinant *E.coli* clones, as described in **Subheading 3.3.**

3.3. Plasmid DNA Preparation

For plasmid DNA preparation from a 50–100 mL *E. coli* culture, for use in further subcloning, we use Wizard midipreps (Promega). For plasmid DNA preparation from a 3 mL *E. coli* culture, for routine analyses and sequencing we use Wizard minipreps (*see* **Note 1**).

3.4. Growth of A. tumefaciens

1. Grow *A. tumefaciens* strain LBA4404 in YT medium at 29°C, with shaking at 200 rpm for 24 h. Flasks should contain no more than 20% of their nominal volume to ensure adequate aeration.
2. Following transformation (*see* **Subheading 3.5.**), grow recombinant *A. tumefaciens* clones on solidified YMB medium containing 1.5% (w/v) bactoagar supplemented with 50 μg/mL kanamycin for 3 d at 29°C.
3. For liquid culture of *A. tumefaciens* clones harboring pROK2-derived binary plant transformation vectors, culture as above in YT medium supplemented with 50 μg/mL kanamycin.

3.5. Transformation of A. tumefaciens

1. Grow *A. tumefaciens* at 29°C, with shaking at 200 rpm for 24–30 h, in YT medium to an OD_{660} of 0.5–0.7. Cool cells on ice and pellet by centrifugation at 16,500g for 2 min.
2. Discard the supernatant and resuspend cells in 1 culture volume of 10% (v/v) glycerol precooled to 4°C.
3. Pellet cells as described above and wash by successively resuspending in 0.5-, 0.2-, and 0.02 culture volume of 10% (v/v) glycerol, as above.
4. Finally, resuspend cells in 0.01 culture volume of 10% (v/v) glycerol to give 10^{11}–10^{12} cells/mL.
5. Transfer a 40-µL aliquot to a 0.2-cm electroporation cuvet that has been precooled to 4°C and add 10–25 ng of plasmid DNA to the cell suspension, mix well, and immediately apply an electric pulse using a Gene Pulser with a Pulse Controller unit. We have successfully transformed cells at a field strength of 12.5 kV/cm, a capacitance of 25 µF, and resistance of 400 mΩ, giving time constants of 6–10 ms (*see* **Note 2**).
6. Add 1 mL of YMB medium, precooled to 4°C, to the cell suspension in the cuvet. Immediately transfer cell suspension to 30 mL bottles and grow at 29°C, with shaking at 125 rpm for 2–3 h.
7. Transfer cell suspension to a microcentrifuge tube and pellet cells by centrifugation at 16,500g, 4°C for 1.5 min. Discard the supernatant and resuspend the cell pellet in 110 µL YMB medium. Plate out aliquots of 10 and 100 µL onto solidified YMB medium, supplemented with 50 µg/mL kanamycin, and culture at 29°C for 3 d.

3.6. Isolation of A. tumefaciens Total DNA

1. Culture selected kanamycin-resistant *A. tumefaciens* clones in 3 mL YEB medium, supplemented with 50 µg/mL kanamycin at 29°C, with shaking at 130 rpm for 24 h. Transfer the cell suspension to two microcentrifuge tubes and harvest cells by centrifugation at 16,500g for 1.5 min at 4°C.
2. Discard the supernatants and resuspend the cell pellets in a total volume of 300 µL $T_{50}E_{20}$. Pool the resuspended cells, add 100 µL of sarkosyl solution and 100 µL of fresh Pronase solution, and incubate at 37°C for a minimum of 1 h, then shear the lysate by passing it rapidly through a 20-gage needle twice.
3. Add an equal volume of DNA-equilibrated phenol to the cleared lysate and vortex the mixture to form an emulsion. Separate the organic and aqueous phases by centrifugation at 16,500g for 5 min at 4°C.
4. Transfer the lysate (aqueous phase) to a sterile microcentrifuge tube and repeat the phenol extraction.
5. Transfer the lysate to a sterile microcentrifuge tube and add an equal volume of chloroform solution, vortex to form an emulsion, and separate the phases by centrifugation, as described in **Subheading 3.6., step 3**.

6. Repeat this chloroform extraction twice.
7. Transfer the cleared lysate to a sterile microcentrifuge tube. Add 0.05 vol 5M NaCl and 2.5 vol absolute ethanol (precooled to –20°C) and, after mixing, place at –20°C for 16 h.
8. Pellet the DNA precipitate by centrifugation at 16,500g for 20 min at 4°C. Discard the supernatant, allow the pellet to air-dry, and resuspend in 50 µL SDW.

3.7. Preparation of Genomic DNA from Plant Tissue

We have successfully used the method described by Robbins and colleagues *(12)* to isolate total plant DNA.

1. Precool a mortar with liquid nitrogen. Fill the precooled mortar with liquid nitrogen and immediately add plant tissue (0.5–1.5 g FW) and grind to a fine powder (*see* **Note 3**).
2. Immediately transfer the frozen powder into a 50 mL Oakridge centrifuge tube containing 16 mL of DNA extraction buffer preheated at 65 °C. After mixing, incubate at 65°C for 10 min.
3. Add 5 mL 5M CH$_3$·COOK and, after mixing, place on ice for 20 min. Clear the lysate by centrifugation at 18,000g for 20 min at 4°C.
4. Filter the supernatant through Miracloth into a sterile tube containing 10 mL of absolute isopropanol (precooled to –20°C). After mixing, place on ice for 20 min.
5. Pellet the DNA precipitate by centrifugation at 18,000g for 20 min at 4°C, discard the supernatant, and resuspend the pellet in 700 µL T$_{10}$E$_1$.
6. Transfer the nucleic acid solution to a sterile microcentrifuge tube. Add 500 µL of phenol-chloroform extraction solution and, after vortexing to form an emulsion, separate the organic and aqueous phases by centrifugation at 16,500g for 5 min at 4°C.
7. Transfer the aqueous phase to a sterile microcentrifuge tube, add 70 µL 3M CH$_3$·COONa and 500 µL isopropanol and, after mixing, place on ice for 20 min.
8. Pellet the nucleic acid precipitate by centrifugation at 16,500g for 20 min at 4°C. Discard the supernatant and wash the pellet in 80% (v/v) ethanol, and centrifuge as above for 5 min.
9. Discard the supernatant and allow to air-dry. Resuspend the pellet in 100 µL RNase A solution.
10. Add 5 µL 5M NaCl and 250 µL ethanol (precooled to –20°C) and, after mixing, leave at –70°C for 30 min.
11. Pellet the DNA precipitate by centrifugation at 16,500g for 20 min at 4 °C. Discard the supernatant, allow the pellet to air-dry, and resuspend in 50 µL T$_{10}$E$_1$.

3.8. Nucleic Acid Manipulation

Restriction and modifying enzymes are used in the buffers provided with the enzymes, according to the manufacturer's instructions.

To concentrate or recover DNA by ethanol precipitation, add 0.05 vol of 5M NaCl and 2.5 vol of absolute ethanol (precooled to –20°C), mix well, place at –20°C for 2 h, then pellet the DNA by centrifugation at 16,500g for 20 min at 4°C. To separate DNA restriction fragments, electrophorese through a 0.8% (w/v) agarose gel in TEA buffer *(11)*. To separate RNA fragments, electrophorese through a 1.5% (w/v) agarose gel in MOPS buffer *(11)*. To permit visualization of DNA by exposure to UV light, add 0.5 µg/mL ethidium bromide to the agarose solution immediately prior to casting the gel. Other routine molecular cloning manipulations were carried out using standard molecular techniques.

3.9. Subcloning of rbcL into pTrans

In this example, to obtain an appropriate fragment for subcloning of the *C. vinosum-rbc*L coding sequence into pTrans, the introduction of an *Sph*I site upstream of the *rbc*L ATG start codon is required. This will permit the subcloning of the *C. vinosum-rbc*L coding sequence in-frame to the soybean small-subunit transit peptide. To introduce this *Sph*I site, we have successfully used the Chameleon double-stranded site-directed mutagenesis kit and the oligonucleotide primer shown in **Fig. 1**.

1. Dilute the oligonucleotide primer in SDW to give a concentration of 100 pmol/µL.
2. To phosphorylate the oligonucleotide primer, add 1 µL T4 polynucleotide kinase (10 U/µL) to 15 µL oligonucleotide primer, 3 µL buffer, 3 µL 10 mM ATP, and 8 µL SDW (final oligonucleotide primer concentration 5 pmol/µL), and incubate at 37°C for 30 min.
3. Following incubation, inactivate the T4 polynucleotide kinase by heating at 65°C for 10 min.
4. Generate a mutagenized pCV17 plasmid in accordance with the instructions provided with the Chameleon double-stranded site-directed mutagenesis kit and isolate mutant plasmid DNA.
5. Confirm the introduction of the required bases by specific restriction (*Sph*I) and/ or DNA sequencing.
6. To isolate the *rbc*L fragment for subcloning, add 2 µL of mutated pCV17 (1 µg/ µL) to 4 µL buffer, 30 µL SDW, and 2 µL each of *Sph*I and *Cla*I. Incubate at 37°C for 120 min. Confirm that restriction is complete by removing a 2 µL aliquot from the reaction and separating restriction fragments through an agarose gel.
7. Electrophorese all of the remaining reaction mix through an agarose gel and excise the ≈1530 bp *Sph*I-*Cla*I *rbc*L fragment. Isolate the DNA from the agarose gel slice by using the Sephaglas Band Prep Kit according to the manufacturer's recommendations.
8. Electrophorese 20% of the ≈1530 bp *Sph*I-*Cla*I *rbc*L fragment solution through a gel to confirm that the fragment isolation was successful.

9. To prepare pTrans vector, add 2 μL pTrans (1 μg/μL) to 2 μL buffer, 14 μL SDW, and 2 μL *Acc*I. Incubate at 37°C for 120 min. Confirm that restriction is complete by taking a 2-μL aliquot and separating restriction fragments by electrophoresis through an agarose gel. Stop the restriction reaction by phenol-chloroform extraction and recover the DNA by ethanol precipitation.

10. Resuspend the *Acc*I linearized pTrans in 16 μL SDW and further restrict the linearized plasmid with 2 μL *Sph*I at 37°C for 2 h.

11. Ligate the isolated *Sph*I-*Cla*I *rbc*L fragment into *Sph*I-*Acc*I restricted pTrans and incubate overnight at 16°C (*see* **Note 4**).

12. Use 5 μL of the ligation mix to transform competent *E. coli*. Add 60 μL cells to 5 μL from the ligation reaction, in a microcentrifuge tube, and mix gently. Place the transformation reaction on ice for 1 h before placing at 42°C for 45 s. After heat-shock, place at room temperature for 10 min, then spread cell suspension onto solidified LB medium containing 100 μg/mL ampicillin, and incubate overnight at 37°C.

Analyze selected putative recombinants by restriction of purified plasmid DNA. Add 2 μL of selected plasmid DNA to 2 μL buffer, 14 μL SDW, and 1 μL each of *Eco*RI and *Hin*dIII. Incubate at 37°C for 120 min. To separate restriction fragments, electrophorese through an agarose gel. Recombinant plasmids should yield an ≈1760 bp TP-*rbc*L fragment on *Eco*RI and *Hin*dIII restriction. Designate correct recombinant plasmids as pTCVL. Confirm that the *rbc*L fragment has been subcloned in-frame to the TP by DNA sequencing.

3.10. Subcloning of Transit Peptide-rbcL Gene Fusion into pUC18

1. To obtain a TP-*rbc*L fragment for subcloning, add 2 μL pTCVL (1 μg/μL) to 2 μL buffer, 14 μL SDW, and 1 μL each of *Hin*dIII and *Bam*HI. Incubate at 37°C for 120 min.

2. After restriction, in-fill the protruding ends by adding 1 μL of dNTP solution (2 m*M*) and 1 μL Klenow fragment of DNA polymerase I (5 u/μL) to the reaction mix, and incubate at 25°C for 30 min. Stop the reaction by adding 2.2 μL of 200 m*M* EDTA. Add an equal volume of phenol-chloroform extractionsolution, vortex to form an emulsion, and separate phases by centrifugation. Transfer the aqueous phase to a fresh tube and recover the DNA by ethanol precipitation.

3. Resuspend DNA in 20 μL SDW and separate in-filled restriction fragments by electrophoresis through an agarose gel. Excise the in-filled ≈1760 bp *Hin*dIII-*Bam*HI TP-*rbc*L fragment and isolate the DNA from the agarose slice by Sephaglas Band Prep extraction.

4. To prepare the pUC18 vector, add 2 μL pUC18 DNA (1 μg/μL) to 2 μL of buffer, 14 μL SDW, and 2 μL of *Sma*I, and incubate at 37°C for 120 min. Confirm that restriction is complete by removing a 2 μL aliquot from the reaction and separating the restriction fragments by electrophoresis through an agarose gel.

5. To dephosphorylate, add 20 µL of the *Sma*I restricted pUC18 vector to 5 µL buffer, 24 µL SDW, and 1 µL of calf intestinal phosphatase (0.1 u/µL), and incubate at 37°C for 15 min. Stop the reaction by adding EDTA to a final concentration of 5 m*M* and heat to 70°C for 10 min. Purify the DNA by extraction with phenol-chloroform solution and recover the DNA by ethanol precipitation as in **Subheading 3.8.**

6. Ligate the in-filled ≈1760 bp *Hin*dIII-*Bam*HI TP-*rbc*L fragment into *Sma*I restricted and dephosphorylated pUC18 in a reaction volume of 10 µL, and incubate overnight at 4°C.

7. Use 5 µL of the ligation mix to transform competent *E. coli*, as described in **Subheading 3.2.** Plate out cells onto solidified LB medium, supplemented with 100 µg/mL ampicillin, X-gal, and IPTG, which permits blue/white color selection of recombinant clones.

8. Analyze selected white putative recombinant clones by restriction of purified plasmid. Add 2 µL of plasmid DNA to 2 µL buffer, 14 µL SDW, and 1 µL *Acc*I, and incubate at 37°C for 120 min. To separate restriction fragments, electrophorese through an agarose gel. Recombinant plasmids in the correct orientation for directional subcloning into pROK2 should yield an ≈800 bp fragment on *Acc*I restriction. Designate correct recombinant plasmids as p18TCVL.

3.11. Sub-Cloning of Transit Peptide-rbcL Gene Fusion into pROK2

1. Add 2 µL p18TCVL (1 µg/µL) to 2 µL buffer, 14 µL SDW, and 1 µL *Kpn*I, and incubate at 37°C for 120 min, then add 2 µL of 1*M* NaCl and 1 µL *Bam*HI and incubate at 37°C for a further 120 min.

2. Separate restriction fragments by electrophoresis through an agarose gel and excise the ≈1760 bp *Bam*HI-*Kpn*I TP-*rbc*L fragment. Isolate the DNA from the agarose slice by Sephaglas Band Prep extraction.

3. To prepare pROK2, add 2 µL pROK2 (1 µg/µL) to 2 µL buffer, 14 µL SDW, and 1 µL *Kpn*I, and incubate at 37°C for 120 min. Then add 2 µL 1*M* NaCl and 1 µL *Bam*HI, and incubate at 37°C for a further 120 min. Confirm that restriction is complete by removing a 2 µL aliquot and separating the restriction fragments by agarose gel electrophoresis.

4. To dephosphorylate, add 20 µL of the *Bam*HI-*Kpn*I restricted pROK2 to 5 µL buffer, 24 µL SDW, and 1 µL calf intestinal phosphatase (0.1 U/µL) and incubate at 37°C for 15 min (*see* **Subheading 3.10., step 5**).

5. Ligate the *Bam*HI-*Kpn*I TP-*rbc*L fragment into the *Bam*HI-*Kpn*I restricted and dephosphorylated pROK2 in a reaction volume of 10 µL and incubate overnight at 16°C.

6. Use 5 µL of the ligation mix to transform competent *E. coli*, as above. Plate out cells onto solidified LB medium supplemented with 50 µg/µL kanamycin and incubate at 37°C for 16 h.

7. Analyze selected putative recombinant clones by restriction of purified plasmid DNA. Add 2 μL of selected plasmid DNA to 2 μL buffer, 14 μL SDW, and 1 μL each of *Eco*RI and *Hin*dIII, and incubate at 37°C for 120 min. To separate restriction fragments, electrophorese through an agarose gel. Recombinant plasmids should yield an ≈3000 bp CaMV35S/TP-*rbc*L/*Nos* fragment. Designate the correct recombinant plasmid as p2TCVL.

3.12. Transformation of A. tumefaciens

1. Use high-voltage electroporation to introduce p2TCVL into *A. tumefaciens*, as described in **Subheading 3.5.**
2. Select a number of kanamycin-resistant clones for molecular characterization.

3.13. Characterization of A. tumefaciens Harboring p2TCVL

Characterization of the DNA isolated from the putative recombinant *A. tumefaciens* clones by restriction and Southern blot hybridization should yield characteristic fragments that permit the integrity of the promoter-insert-terminator region in the binary plant transformation vector to be confirmed (**Fig. 3**). Southern blot hybridization is required to distinguish p2TCVL restriction fragments from *A. tumefaciens* fragments, because the number of restriction fragments generated from *A. tumefaciens* DNA is large. Restrict the *A. tumefaciens* harboring p2TCVL total DNA and purified plasmid DNA with *Hin*dIII-*Eco*RI (CaMV35S promoter/TP-*rbc*L/*Nos* terminator), *Hin*dIII-*Kpn*I (CaMV35S promoter/TP-*rbc*L), *Bam*HI-*Kpn*I (TP-*rbc*L), and *Bam*HI-*Eco*RI (TP-*rbc*L/*Nos* terminator).

1. Restrict 4 μg of *A. tumefaciens* harboring p2TCVL total DNA and 4 ng of p2TCVL as a control with *Hin*dIII-*Eco*RI, *Hin*dIII-*Kpn*I, *Bam*HI-*Kpn*I, and *Bam*HI-*Eco*RI in reaction volumes of 25 μL at 37°C for 4 h.
2. Separate restriction fragments by electrophoresis through an agarose gel and then depurinate the DNA by immersing the gel in 250 m*M* HCl for 20 min.
3. Rinse the gel in SDW and then denature the DNA in the gel by placing the gel in denaturing solution for 30 min at room temperature before transferring the gel into alkaline-transfer buffer for 15 min, prior to constructing the blotting apparatus.
4. Construct a blotting apparatus composed of a glass dish containing approx 1 L of alkaline transfer buffer, with an inverted gel-forming tray in the center. Overlay the gel-forming tray with three layers of Whatman No. 1 paper soaked in alkaline transfer buffer to form a wick.
5. Remove any air bubbles from the wick by rolling out to the edges with a glass rod. Place Nescofilm strips along the edges of the gel-forming tray to prevent buffer flowing directly from the reservoir to the blotting stack.
6. Place the gel onto the wick, so that no air bubbles form between the gel and the wick, and so that the gel is not distorted. Cut the Zeta-Probe® GT blotting

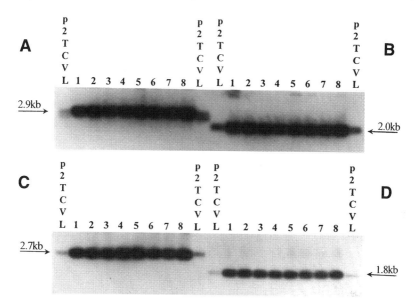

Fig. 3. Characterization of the DNA isolated from the putative recombinant *A. tumefaciens* clones. Southern blot hybrization of restricted recombinant *A. tumefaciens* DNA and control p2TCVL DNA. The restriction combinations shown: **(A)** *Hin*dIII-*Eco*RI; **(B)** *Bam*HI-*Eco*RI; **(C)** *Hin*dIII-*Kpn*I; **(D)** *Bam*HI-*Kpn*I give rise to TP-*rbc*L cross-hybridizing fragments which show that the integrity of the CaMV35S promoter/TP-*rbc*L/*Nos* terminator region in the recombinant *A. tumefaciens* clones is maintained.

    membrane to the exact size of the gel and soak it in SDW immediately prior to placing it on the gel (*see* **Note 5**).

7. Place Whatman No. 1 paper (cut to the size of the gel) soaked in SDW on the membrane. Remove any air bubbles by rolling out with a glass rod.

8. Place a further two layers of dry Whatman No. 1 paper and two layers of absorbent nappy booster pad (Boots) cut to the size of the gel, on the blotting stack. Finally, place a perspex plate and a 0.5 kg weight on the blotting stack.

9. Leave for 16 h, to allow complete transfer of the DNA fragments from the gel to the membrane.

10. Dismantle the blot apparatus and wash the membrane in 2X SSC. Air-dry the membrane, wrap in Saranwrap, and crosslink the transferred DNA to the membrane. We have successfully used a Stratalinker UV crosslinker.

11. Prehybridize the membrane, rolled within a fine mesh, in a hybridization bottle containing 25 mL of prehybridization solution, by rotating at 6 rpm for 4 h at 65°C.

12. Prepare the probe fragment by radiolabeling with α-^{32}P dCTP. We have successfully used the Stratagene Prime It II random-primed labeling kit to generate a radiolabeled *Bam*HI-*Kpn*I TP-*rbc*L fragment of high specific activity.

13. Decant the prehybridization solution and replace with hybridization solution (as per prehybridization solution with the addition of radiolabeled probe). Hybridize the membrane by rotating at 6 rpm for 16 h at 65°C.
14. Decant the hybridization solution and wash the membrane in 2X SSC, 0.1% (w/v) SDS for 30 min at 65°C in the hybridization bottle.
15. Remove nonspecific background cross-hybridization from the membrane by washing with increasing stringency in: 1X SSC, 0.1% (w/v) SDS, 0.5X SSC, 0.1% (w/v) SDS, and 0.1X SSC, 0.1% (w/v) SDS. All washes should be carried out at 65°C.
16. After washing, wrap the membranes in Saranwrap to keep moist (*see* **Note 6**).
17. Autoradiograph the membrane at –70°C for 16–48 h, then develop the film.

3.14. A. tumefaciens-*Mediated Co-Transformation of Tobacco Leaf Explants*

We have successfully used the method described in **ref.** *13* for transformation of wild-type and Rubisco-minus mutant tobacco lines.

1. In a laminar flow bench, excise a leaf from the recipient wild-type or Rubisco-minus mutant line, Sp25. Immerse the leaf in recombinant *A. tumefaciens* cell suspension (O.D$_{600}$ ≈ 1). Cut the leaf into 1-cm-square sections using a sterile scalpel and leave immersed for 1 min (*see* **Note 7**).
2. Recover the inoculated leaf explants with sterile forceps and remove excess *A. tumefaciens* suspension by blotting onto sterile filter paper.
3. Place the leaf explants, upper leaf surface uppermost, onto MS medium supplemented with BAP (1 mg/L). Ensure a good contact between each leaf disk and the culture medium. Seal the Petri dish with Nescofilm and incubate at 25°C, 16 h daylight, for 2 d.
4. Transfer the leaf explants to MS medium, supplemented with 1 mg/L BAP, 1 mg/L NAA, and 200 mg/L cefotaxime, to kill off *A. tumefaciens*, and 100 mg/L kanamycin to select for co-transformed plants. Incubate as in **Subheading 3.14., step 3**, until shoots regenerate.
5. After 1 or 2 wk, the leaf explants expand and some of the explant may lift away from the agar. At this stage, cut the explant into four and press the pieces into MS medium supplemented as in **Subheading 3.14., step 4**, ensuring good contact.
6. Shoots should form after 3–6 wk. When the shoots are at least 5 mm long, excise them at the base and place them individually in tubes containing 10 mL MS medium supplemented as in **Subheading 3.14., step 4**, At this stage, a large number (approx 50) of putative co-transformants for each construct, which form roots in the presence of kanamycin, should be maintained until molecular characterization has confirmed their co-transformation status.

3.15. Molecular Characterization of Co-Transformed Plants

Before analysis of gene expression of transferred genes in transgenic plants is carried out, we determine the structure and copy number of the introduced transgene by PCR and Southern blot hybridization analyses. Initial molecular characterization of the selected co-transformed plants should demonstrate their co-transformation status. Molecular screening of co-transformants by PCR, using construct-specific primers, is convenient. This demonstrates the presence of the transgene in the plants. Further characterization by Southern blot hybridization to a transgene-specific probe should be undertaken to confirm the presence and to determine the number of copies of the integrated transgene.

3.15.1. Amplification from Plant Genomic DNA to Demonstrate the Presence of the Transgene

PCR analysis for the presence of introduced transgenes may be carried out employing CaMV35S and *Nos* construct-specific oligonucleotide primers (*see* **Note 8**).

1. Use 300 ng (30 ng/μL) of genomic DNA as a target for amplification, using a CaMV35S-specific primer (TGGCTCCTACAAATGCCATCATTGC) and a *Nos*-specific primer (TAATCATCGCAAGACCGGCAACAGG).
2. To target DNA in 10 μL, add 40 μL of PCR reaction mix (0.1 m*M* dNTPs; 0.75 m*M* MgCl$_2$; 1 μ*M* CaMV35S primer; 1 μ*M Nos* primer; 3 U Amplitaq-polymerase [Perkin-Elmer]; 4 μL buffer; SDW to 40 μL).
3. Overlay each reaction with 50 μL sterile mineral oil before thermo-cycling (94°C for 3 min, followed by 94°C, 1 min; 55°C, 1 min; and 72°C, 1.5 min for 10 cycles, then 94°C, 1 min; 50°C, 1 min and 72 °C, 2.5 min for 25 cycles).
4. Following thermocycling, separate amplified products by electrophoresis of a 20-μL aliquot from the amplification mixture through an agarose gel. Identify definitive co-transformants by the presence of a specific ≈1760 bp amplification product.

3.15.2. Southern Blot Hybridization Analysis to Determine the Structure and Copy Number of the Transgene

The structure and copy number of the introduced transgene can be determined by restriction of genomic DNA isolated from selected co-transformed plants with an enzyme that is unique to the binary plant transformation construct multiple-cloning site.

1. Restrict 11 μg genomic DNA isolated from each selected co-transformed line with *Bam*HI. Separate restriction fragments by electrophoresis through an agarose gel.
2. Transfer DNA fragments to Zeta-Probe® GT blotting membrane (*see* **Subheading 3.13.**).
3. Determine the integrity and copy number of the transgene by cross-hybridization to a radiolabeled transgene-specific *rbc*L probe (*see* **Subheading 3.13.**).
4. The number of cross-hybridizing bands approximately correlates with the transgene copy number, because *Bam*HI restricts uniquely in the multiple cloning site of p2TCVL.

3.15.3. Determination of the Level of Transgene Expression

The level of transgene expression in selected lines can be determined using Northern blot hybridization by cross-hybridization to transgene-specific DNA or strand-specific riboprobe. We have successfully used a modification of the method described in **ref. *14*** to isolate total plant RNA.

1. Precool a mortar with liquid nitrogen. Fill the precooled mortar with liquid nitrogen and immediately add plant tissue (0.5–1.5 g FW) and grind to a fine powder (*see* **Note 3**).
2. Immediately transfer the frozen powder into a 50 mL Oakridge centrifuge tube containing 4 mL of RNA extraction buffer and 4 mL of RNA-equilibrated phenol per gram of tissue, mix, and incubate at 65°C for 30 min.
3. Following incubation, shake the samples vigorously, using a wrist-action shaker, for 30 min at room temperature. Add 4 mL of chloroform per gram of tissue and continue shaking for a further 10 min.
4. Separate the aqueous and organic phases by centrifugation at 36,000*g* for 30 min at 20°C. Discard the lower organic phase, leaving the interface intact. Add 16 mL of chloroform and shake on a wrist-action shaker for 10 min.
5. Separate the phases by centrifugation at 36,000*g* for 30 min at 20°C. Transfer the aqueous phase and repeat this chloroform extraction.
6. Transfer the aqueous phase to a sterile centrifuge tube and add 8*M* LiCl to a final concentration of 2*M*. Mix well and leave for 16 h at 4°C.
7. Pellet the RNA precipitate by centrifugation at 36,000*g* for 30 min at 4°C. Discard the supernatant. Add 2 mL of 2*M* LiCl and agitate to wash the pellet, then centrifuge as just mentioned at 4°C and discard the supernatant. Repeat this wash once with 2*M* LiCl and twice with 80% (v/v) ethanol.
8. Finally, lyophilize the pellet and resuspend in 200 μL nuclease-free water.
9. Separate RNA by electrophoresis through an agarose gel containing formaldehyde and visualize the RNA on a UV transilluminator.
10. Transfer RNA fragments from the gel to Zeta-Probe® GT blotting membrane by (capillary transfer (*see* **Subheading 3.13.**) except the transfer buffer is 10X SSC and the Whatman No. 1 paper is soaked in 10X SSC.
11. Prehybridize the membrane, rolled within a fine mesh, in a hybridization bottle by rotating at 6 rpm for 16 h at 42°C in 25 mL RNA hybridization solution.
12. Decant the RNA hybridization solution and replace with fresh solution, with the radiolabeled probe added. Hybridize membranes for 16 h, as described in **Subheading 3.15.3., step 11**.
13. Following incubation, decant the RNA hybridization solution and remove nonspecific background cross-hybridization from the membrane by washing with increasing stringency in: 2X SSC, 0.1% (w/v) SDS; 1X SSC, 0.1% (w/v) SDS; 0.5X SSC, 0.1% (w/v) SDS; and 0.1X SSC, 0.1% (w/v) SDS. All washes should be carried out at 42°C. Keep washed membranes moist by wrapping in Saranwrap.
14. Autoradiograph and develop (*see* **Subheading 3.13.**).

3.16. Biochemical and Physiological Characterization of Co-Transformed Plants

After molecular characterization, it is important to determine whether the recombinant Rubisco assembles into active enzyme, by measuring the amount and activity of the enzyme. Changes at other levels (e.g., amounts of mRNA) do not always result in differences in amounts of protein. Plants without Rubisco activity will be unable to survive out of tissue culture.

3.16.1. Extraction of Rubisco

1. Punch a leaf disk (2.5 cm^2) directly into liquid nitrogen using a liquid nitrogen cooled clamp.
2. Grind the disks in 1 mL of rubisco extraction buffer.
3. Clarify the extracts by centrifugation at 10,000g for 3 min at 4°C.

3.16.2. Determination of Rubisco Total Activity

1. Incubate 25 µL of the supernatant for 3 min in 475 µL of Rubisco assay mix.
2. Start the assay of Rubisco activity by adding RuBP to a final concentration of 0.33 mM.
3. After 30 s, add 200 µL H·COOH to stop the reaction.
4. Evaporate to dryness and determine ^{14}C incorporation into 3-phosphoglyceric acid by liquid scintillation spectrometry.

3.16.3. Determination of Amount of Rubisco

Rubisco active sites can be quantified using the radiolabeled transition state analog, 2-carboxyarabinitol 1,5-bisphosphate, which binds almost irreversibly to Rubisco *(15)*.

1. Add 200 µL of clarified extract to 200 µL of Rubisco quantification buffer and an approximate 10-fold molar excess of (^{14}C)-CABP, and incubate for 15 min at 4°C.
2. To precipitate the Rubisco, add 288 µL PEG, 60% (w/v), and incubate for 30 min at 4°C.
3. Pellet the Rubisco precipitate by centrifugation at 10,000g for 10 min at 4 °C and discard the supernatant.
4. Wash the pellet twice with Rubisco wash solution to remove any nonspecifically bound (^{14}C)-CABP.
5. Finally, resuspend the pellet in 1% (v/v) Triton X-100 and determine the amount of (^{14}C)-CABP bound by liquid scintillation spectrometry.

3.17. Further Analyses

Once the functional activity of the recombinant Rubisco has been established in the transformed lines, rooted plantlets may be removed from culture medium and placed in earth or vermiculite, to be grown in greenhouse conditions under optimal conditions (e.g., elevated CO_2 and cool temperature). Simply observe the plants for any phenotype. Once sufficient growth has occurred, the important physiological questions to address in these plants include: How is Rubisco activity and photosynthetic metabolism affected in vivo; how do changes in Rubisco activity and photosynthetic metabolism affect whole-plant assimilation rates, N- and water-use efficiency, and growth, allocation, and yield; and how are these parameters affected by growth environment.

4. Notes

1. We have often found it necessary to further purify the DNA prior to sequencing. We have successfully used phenol-chloroform extraction (*see* **Subheading 3.6.**) and ethanol precipitation (*see* **Subheading 3.8.**).
2. To ensure a high transformation efficiency, with a time constant of 6–10 ms, it is important that salts are removed from both the DNA and cell suspension. To prepare DNA for transformation, ethanol-precipitate (*see* **Subheading 3.8.**) and resuspend in SDW.
3. Do not allow the tissue to warm up. Keep the sample under liquid nitrogen.
4. To maximize the likelihood of successful ligation, prepare a series of ligation reactions containing molar ratios of vector:insert of 1:1, 1:3, and 3:1, respectively. It is important to include the following control reactions: no DNA (to confirm that the buffers are not contaminated with plasmid DNA); vector DNA alone (to determine the background amount of unrestricted vector); insert DNA alone (to confirm that the insert is not contaminated with plasmid DNA); and vector DNA plus ligase (to determine the amount of self-ligation possible following phosphatase treatment).
5. Extreme care must be taken not to touch the Zeta-probe GT blotting membrane with bare hands.
6. It is a good idea to keep membranes moist, because they can then be stripped and reprobed, should the initial hybridization be unsuccessful.
7. When excising the leaf explants, avoid the midrib and any damaged areas of the leaf. Allow 50–100 explants per construct to ensure sufficient primary co-transformed shoots. Only one shoot should be excised from each explant to ensure isolation of independent co-transformants.
8. The thermocycling conditions should be optimized for your specific target DNA and primers. We have successfully used the conditions described.

Acknowledgments

I.A.C.R. receives grant-aided support from the Biotechnology and Biological Research Council of the United Kingdom. We are grateful to: A. Gatenby, H. Kobayashi, and T.A. Dyer for supplying plasmids; H. Dulieu for supplying the SP25 tobacco line; and B.G. Forde, C. Foyer, A.J. Keys, and A. Yokota for helpful discussion and advice.

References

1. Keys, A. J. (1983) Prospects for increasing photosynthesis by control of photorespiration. *Pesticide Sci.* **19**, 313–316.
2. Bainbridge, G., Madgwick, P. J., Parmar, S., Mitchell, R., Paul, M. J., Pitts, J., Keys, A. J., and Parry, M. A. J. (1995) Engineering Rubisco to change its catalytic properties. *J. Exper. Bot.* **46**, 1269–1276.
3. Douwe de Boer, A., and Weisbeek P. J. (1991) Chloroplast protein topogenesis: import, sorting and assembly. *Biochem. Biophys. Acta* **1071**, 221–253.
4. Bauwe, H. (1984) Photosynthetic enzyme activities in C3 and C3-C4 intermediate species of Moricania and Panicum milioides. *Photosynthetica* **18**, 201–209.
5. Viale A. M., Kobayashi H., and Akazawa T. (1990) Distinct properties of *Escherichia coli* products of plant-type Ribulose-1,5-bisphosphate carboxylase/oxygenase directed by two sets of genes from the photosynthetic bacterium *Chromatium vinosum. J. Biol. Chem.* **265**, 18,386–18,392.
6. Gutteridge, S. and Gatenby, A. A. (1995) Rubisco synthesis, assembly, mechanism and regulation. *Plant Cell* **7**, 809–819.
7. Brangeon, J., Nato, A., and Forchioni, A. (1989) Ultrastructural detection of Rubisco and target mRNAs in wild type and holoenzyme-deficient *Nicotiana* using immunogold and in-situ hybridisation. *Planta* **177**, 151–159.
8. Svab, Z. and Maliga, P. (1993) High frequency plastid transformation in tobacco by selection for a chimaeric *aadA* gene. *Proc. Nat. Acad. Sci. USA* **90**, 913–917.
9. Yanisch-Perron, C., Vieira, J., and Messing, J. (1985) Improved M13 phage cloning vectors and host strains- nucleotide-sequences of the M13mp18 and pUC19 vectors. *Gene* **33**, 103–119.
10. Hanahan, D. (1983) Studies on transformation of *Escherichia coli* with plasmids. *J. Mol. Biol.* **166**, 557–580.
11. Sambrook, J., Fritsch, E. F., and Maniatis T. (1989) *Molecular Cloning—A Laboratory Manual*, 2nd ed. Cold Spring Harbour Laboratory, Cold Spring Harbour, NY.
12. Robbins, M. P., Evans, T. E., Morris, P., and Carron, T. R. (1991) Some notes on the extraction of genomic DNA from transgenic *Lotus. Lotus Newsletter* **22**, 18–21.
13. Gallois, P. and Marinho, P. (1995) Leaf disk transformation using Agrobacterium tumefaciens-expression of heterologous genes in tobacco, in *Methods in Molecular Biology*, vol. 49: *Plant Gene Transfer and Expression Protocols* (Jones, H., ed.), Humana, Totowa, NJ, pp. 39–48.
14. Ougham, H. J. and Davies, T. G. E. (1990) Leaf development in *Lolium temulentum:* gradients of RNA complement plastid and non-plastid transcripts. *Physiol. Plantarum* **79**, 331–338.
15. Yokota, A. and Canvin, D. T. (1985) Ribulose bisphosphate carboxylase/oxygenase content determined with (^{14}C) Carboxypentitol bisphosphate in plants and algae. *Plant Physiol.* **77**, 735–739.

18

Quantification of Heterologous Protein Levels in Transgenic Plants by ELISA

Anne-Marie Bruyns, Myriam De Neve, Geert De Jaeger,
Chris De Wilde, Pierre Rouzé, and Ann Depicker

1. Introduction

An IgG-type antibody is a Y-shaped protein whose arms form two identical antigen-binding sites that are highly variable between different molecules. The stem of the Y is part of the constant region of an antibody and has very limited diversity, which can be used to detect and quantitate antibodies. The interaction of an antibody with its antigen involves multiple noncovalent bonds, such as hydrogen bonds, electrostatic, Van der Waals, and hydrophobic attractive forces (1). The high specificity of this interaction enables an antibody to recognize its antigen even in the presence of a huge amount of contaminating antigens. This makes antibodies indispensable tools in a wide range of disciplines, including molecular biology.

Immunoassays are immunological methods that make use of the binding specificity of an antibody to measure either the antigen or the antibody. To quantitate the reaction, either the antigen or the antibody is labeled. In an enzyme-linked immunosorbent assay (ELISA), the label is an enzyme with high turnover number, such as horseradish peroxidase, alkaline phosphatase, or β-galactosidase. In most applications of ELISA, one of the components is adsorbed passively to a solid phase, most often plastic in the form of 96-well microtiter plates, in a first step. Subsequent reagents are added and, after a period of incubation, unreacted material is simply washed away. The enzyme-labeled molecule is added and in a final step a relevant enzyme substrate is added. The substrate is converted by the bound enzyme conjugate to a colored or fluorescent product, which can be quantified by a spectrophotometer or a

From: *Methods in Biotechnology, Vol. 3:*
Recombinant Proteins from Plants: Production and Isolation of Clinically Useful Compounds
Edited by: C. Cunningham and A. J. R. Porter © Humana Press Inc., Totowa, NJ

fluorometer. The amount of product generated in a microtiter well is proportional to the amount of bound analysate.

The stepwise attachment of reagents to the solid phase allows great versatility for ELISA, since the various components of assays can be used in different combinations and in different phases. ELISAs can be classified under four headings: direct, indirect, sandwich, and competition. In a direct ELISA, the antigen attached to the solid phase is reacted directly with an enzyme-labeled antiserum (directly labeled antibody ELISA), or, in reverse, enzyme-labeled antigen react with plastic-bound antibodies (directly labeled antigen ELISA). In an indirect ELISA, the antigen coated on the solid phase is reacted with antibodies against this antigen, and finally the bound antibodies are detected with antispecies antibody conjugated to an enzyme. The number of steps can be further increased (amplified assays) to enhance the sensitivity of the assay or to gear the assay to available reagents. For this purpose, other intermolecular interactions (e.g., avidin-biotin) can be included. To quantify an antigen, both sandwich and competition ELISAs may be used. In a sandwich ELISA, the antigen to be measured is captured by a first antibody bound to the solid phase and quantified by a second enzyme-labeled antibody. In competition assays, the antigen to be measured is added as the second step of a direct ELISA, together with the labeled reagent, inhibiting the formation of the labeled antigen–antibody complex. In both cases, the reaction can be turned into an indirect (or amplified) assay by performing the labeled antibody step in two (or more) steps, using species-specific anti-Ig-labeled antibodies. An excellent overview of the various ELISAs can be found in **ref. 2**.

Antibodies can be produced against almost any substance, including proteins, carbohydrates, nucleic acids, and even smaller molecules such as peptides, hormones, and haptens *(1)*. The repeated injection of a purified immunogen into an animal results in the production of highly specific polyclonal antibodies. Immunization requires sufficient amounts of purified antigen; at least 0.2–1 mg of antigen is needed. When the cDNA for the antigen is available, overexpression in bacteria provides an excellent source of antigen. Good results are obtained by T7 promoter-driven expression in commercially available plasmids *(3)*. Methods for the isolation and purification of the recombinant protein from the bacteria can be found in **ref. 4**. A simple method, giving sufficiently pure antigen for immunization, is electroelution of the protein out of an sodium dodecyl sulfate (SDS)-polyacrylamide gel.

Rabbits are the most commonly used animals for the production of polyclonal antibodies, because of the ease of handling and care and the relatively large yield of serum. At least two animals should be injected with the same immunogen, since the intensity of the immune response and the cross-reaction with plant proteins can vary markedly from animal to animal. Proteins are commonly injected subcutaneously at different sites in 50–500-μg doses per injected animal. The use of an adjuvant is essential to induce a strong immune response against soluble antigens. In addition, small antigens, such as peptides or hormones, should be polymerized or coupled to an immunogenic protein carrier to enhance the immune response *(5,6)*. Technical information and time/injection schedules can be found in **ref. 7**.

Besides polyclonal antibodies, monoclonal antibodies can also be used in immunoassays. The decision to use monoclonal or polyclonal antibodies should be guided by the evaluation of the advantages and disadvantages of both *(4,8)*. When both a mouse monoclonal and a rabbit polyclonal antibody against the antigen are available, an indirect sandwich ELISA can be set up without the need to conjugate one of the antibodies, since the detection can be performed with commercially available species-specific antibody conjugates. Monoclonal antibodies can be produced by making hybridoma lines *(7)* or by phage display of antibodies *(9)*.

One of the applications of antibodies in molecular biology is their use for the characterization and quantification of heterologous proteins produced in transgenic organisms. Western blotting can be used for the biochemical characterization of the heterologous protein. By the use of SDS gels and isoelectric-focusing gels, it is possible to investigate whether the protein has undergone changes in molecular weight and isoelectric point, respectively, as a result of processing within the cell. However, Western blotting is not a reliable quantitative method. Immunoblotting is capable of giving an estimate of the amount of antigen present in a sample, but for an accurate and valid result, sandwich ELISA is the method of choice *(7,8)*. Yet, sandwich ELISA is rarely used for the quantification of heterologous proteins in transgenic plants, as shown by the few publications describing its use *(10–14)*. Instead, researchers are tempted to use Western blotting or dot blots for quantitative measurements *(15–19)*. Competition ELISAs *(7,8)* are valid alternatives for sandwich ELISA, but sandwich ELISAs are more precise and reproducible.

Sandwich ELISA is an easy, safe, and economical method, and allows the quantification of a particular protein in plant extracts in a reproducible, sensitive, and specific manner. The following protocol has been used routinely in our laboratory for the quantification of different antigens in tobacco and *Arabidopsis* extracts.

2. Materials
2.1. Special Laboratory Tools and Materials

1. 96-Well microtiter plates or 8-well microtiter modules (Maxisorp Nunc; Gibco BRL, Gaithersburg, MD).
2. Titertek plate sealers (ICN Biomedicals, Costa Mesa, CA).

3. Multichannel pipet and/or multi-pipet.
4. Reagent reservoirs for multichannel pipets (Costar, Cambridge, MA).
5. Microtiter plate spectrophotometer with 405 nm filter.
6. Sephadex G-25 column (Pharmacia, Uppsala, Sweden).

2.2. Reagents and Column Materials

1. Secondary antispecies antibody conjugated to alkaline phosphatase (*see* **Subheading 3.1.1.**).
2. Sodium azide (S2002; Sigma, St. Louis, MO).
3. *p*-Nitrophenyl phosphate (pNPP) tablets of 5 mg (N9389; Sigma).
4. Protein A-Sepharose CL-4B (Pharmacia).
5. Biotin Labeling Kit (Boehringer Mannheim, Mannheim, Germany).
6. Antigen (purified or partially purified).
7. Streptavidin-alkaline phosphatase conjugate (Boehringer Mannheim).

2.3. Buffers

1. Phosphate-buffered saline (PBS): 140 mM NaCl, 2.7 mM KCl, 4 mM Na$_2$HPO$_4$, 1.8 mM KH$_2$PO$_4$, pH 7.2.
2. Blocking buffer: 3% (w/v) bovine serum albumin (BSA) (A9647; Sigma) in PBS.
3. Substrate buffer: 1M diethanolamine, 0.5 mM MgCl$_2$, pH 9.8, stored at 4°C.

3. Methods

3.1. Titration, Purification, and Labeling of Antibodies

3.1.1. Titration of the Sera

After immunization, the animal with the best immunological response can be selected by determining the antigen-specific antibody titer in the serum of each animal in an indirect ELISA. A constant amount of antigen is adsorbed to wells and antisera are titrated against the coated antigen as a dilution series. Bound antibodies are detected by addition of a constant amount of antispecies conjugate.

1. Dilute the antigen to a final concentration of 10–100 µg/mL in PBS (*see* **Note 1**).
2. Dilute a protein extract of an untransformed plant in PBS to a final concentration of 20 µg/mL (*see* **Notes 2, 3,** and **4**).
3. For each serum, coat six wells with the antigen by pipeting (*see* **Note 5**) 100 µL of the antigen dilution (**step 1**) per well, and six wells with soluble proteins of an untransformed plant by pipeting 100 µL of the dilution (**step 2**) per well.
4. Cover the plate with an adhesive plastic and incubate for 2 h at room temperature.
5. Remove the coating solution and wash the plate twice by filling the wells to the top with PBS (*see* **Note 6**). The solutions or washes are removed by flicking the plate over a sink. The remaining drops are removed by patting the plate on a paper towel.

6. Fill six uncoated wells with 300 µL blocking buffer for each serum. Fill all coated wells (**step 3**) with 300 µL blocking buffer (*see* **Note 7**). Cover the plate with an adhesive plastic and incubate for at least 2 h at room temperature or, if more convenient, overnight at 4°C. In this step, the remaining protein-binding sites on the plate are saturated.
7. Make 10-fold dilutions (1:10, 1:100, 1:1,000, 1:10,000, 1:100,000, and 1:1,000,000) of the sera in blocking buffer (*see* **Note 8**).
8. Wash the plate twice with PBS.
9. Apply 100 µL of each dilution to an uncoated blocked well, 100 µL to an antigen-coated well, and 100 µL to a well coated with proteins of an untransformed plant.
10. Cover the plate with an adhesive plastic and incubate for 2 h at room temperature.
11. Wash the plate four times with PBS.
12. To all wells, add 100 µL of the secondary antispecies antibody conjugated to alkaline phosphatase, diluted at the optimal concentration (according to the manufacturer) in blocking buffer immediately before use (*see* **Note 9**). These conjugated antibodies recognize antibodies of the animal species in which the immune response was elicited.
13. Cover the plate with an adhesive plastic and incubate for 2 h at room temperature.
14. Wash the plate four times with PBS.
15. Dissolve the pNPP tablets at a concentration of 1 mg/mL in substrate buffer (*see* **Note 10**). Add 100 µL of the substrate solution per well with a multichannel pipet or a multipipet.
16. Measure the absorbance at 405 nm, using a microtiter plate spectrophotometer. Perform an end-point measurement after 1 h (*see* **Note 11**).
17. Calculate the titer of the sera. The titer can be defined as the dilution of serum giving an optical density (OD) of 0.2 above the background of the ELISA after a 1-h reaction. Check the signals obtained for the uncoated blocked wells. If background signals occur for all dilutions, then try other blocking agents (*2,8*). Check also the signals obtained for the wells coated with the proteins of an untransformed plant. These wells were included in the test to check for cross-reaction of the serum with the plant extract. For the development of the sandwich ELISA, select the serum for which the signals for the plant protein-coated wells are lowest (*see* **Note 12**), and which gives the highest signal:noise ratio.

3.1.2. IgG Purification from the Serum Using Protein A-Sepharose

The polyclonal antibodies need to be purified from contaminating proteins in the serum for their use in immunoassays and other immunological methods (*8*). The purification of IgG antibodies on protein A beads is an effective and easy method. This protocol was adapted from the low-salt protocol of **ref. 8**.

1. Weigh out the required amount of protein A-Sepharose and suspend it in 100 mM Tris-HCl, pH 8.0.
2. Pour the slurry in the column. Try to avoid the introduction of air bubbles.

3. Fill the remainder of the column with 100 m*M* Tris-HCl, pH 8.0, and connect the column to the pump.
4. Open the bottom outlet of the column and set the pump to run at the desired flow rate, according to the manufacturer of the beads.
5. Maintain the packing flow rate until a constant bed height is reached.
6. Adjust the pH of the crude antibody preparation, selected in **Subheading 3.1.1.**, to 8.0 by adding one-tenth vol of 1.0*M* Tris-HCl, pH 8.0.
7. Pass the antibody solution through the protein A bead column. This column binds approx 10–20 mg of antibody per mL of wet beads. Serum contains approx 10 mg/mL of total IgG.
8. Wash the beads with 10 column volumes of 100 m*M* Tris-HCl, pH 8.0.
9. Wash the beads with 10 column volumes of 10 m*M* Tris-HCl, pH 8.0.
10. Elute the column with 100 m*M* glycine, pH 3.0. Add this buffer stepwise, approx 500 µL per sample. Collect the eluate in Eppendorf tubes containing 50 µL of 1*M* Tris-HCl, pH 8.0. Mix each Eppendorf tube gently to bring the pH back to neutral.
11. Determine the concentration of IgG in the preparation by measuring the absorbance at 280 nm (1 OD = approx 0.8 mg/mL).
12. Store the purified antibodies as back-up stocks at –20°C; they can be stored as such for years. Repeated freezing and thawing should be avoided. For regular use, the antibodies can be stored at 4°C, upon addition of sodium azide, to a final concentration of 0.02%.

3.1.3. Labeling Antibodies with Biotin

Antibodies can be labeled with different moieties for their use as secondary reagents. In our hands, labeling with biotin has been shown to be easy and very effective (*see* **Note 13**). We routinely use the Biotin Labeling Kit, commercially available from Boehringer Mannheim (*see* **Note 14**). All equipment is delivered with the kit. The principle of this labeling is the reaction of free amino groups of the antibodies with D-biotinoyl-ε-aminocaproic acid-N-hydroxysuccinimide ester (biotin-7-NHS), resulting in the formation of a stable amide bond. Nonreacted biotin-7-NHS is removed by separation on a Sephadex® G-25 column.

1. The antibodies, purified by protein A-Sepharose (*see* **Subheading 3.1.2.**), are dialyzed extensively against PBS. The dialysis is performed once overnight and twice 4 h in a PBS volume corresponding to 100 times the volume of the antibody sample.
2. Take a dialyzed antibody sample containing 1 mg antibody and add PBS to a final volume of 1 mL (*see* **Note 15**).
3. Dissolve biotin-7-NHS in dimethyl sulphoxide (DMSO) at 20 µg/mL.
4. Add 7.6 µL of biotin-7-NHS solution in steps to the antibody solution, shaking between each addition. Incubate end-over-end, rotating at room temperature for 2 h.

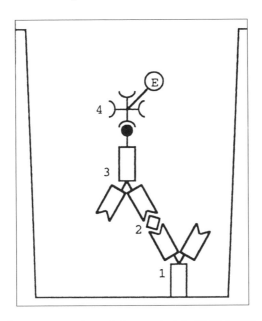

Fig. 1. Schematic representation of the sandwich ELISA. (1) Primary antibody, polyclonal rabbit antigen-specific antibody; (2) antigen; (3) secondary antibody, biotinylated polyclonal rabbit antigen-specific antibody; the biotin molecule is symbolized by a solid circle; (4) streptavidin-alkaline phosphatase conjugate; the alkaline phosphatase is symbolized by E.

5. Set up a Sephadex G-25 column and block it by applying 5 mL blocking solution to the column.
6. Equilibrate the column with 30 mL PBS.
7. Apply the biotinylated antibody sample on the column, followed by 1.5 mL PBS.
8. Elute the antibody by applying 3.5 mL PBS on the column. Collect 0.5 mL samples in Eppendorf tubes.
9. Determine which fractions contain antibody by measuring the absorbance at 280 nm and pool them. It is not possible to calculate correct antibody concentrations, because biotin absorbs at 260–280 nm. Therefore, calculate a theoretical concentration by determining the total volume of the collected antibody-containing fractions, and by assuming that all the antibodies (1 mg/labeling reaction) are eluted from the column. Concentrations of the labeled antibody (secondary antibody) in the next sections will refer to this theoretical concentration (*see* **Note 16**).

3.2. Development of the Sandwich ELISA

The sandwich ELISA that we used routinely for the quantification of an antigen in transgenic plants is schematically represented in **Fig. 1** and is based on the protocol for the two-antibody sandwich assay described in **ref. 8**. The assay consists of five stages separated by washings. First, the wells of a microtiter plate are coated with protein A-purified polyclonal antibodies (primary antibody). Second, a serial dilution of reference antigen or plant extract is supplied. Third, bound antigen is detected with biotinylated secondary antibodies. These antibodies are from the same batch as the primary antibodies and are labeled with biotin, as described in **Subheading 3.1.3.** Fourth, the biotinylated secondary antibodies are detected with a streptavidin-alkaline phosphatase conjugate (Boehringer Mannheim). Finally, the substrate solution for alkaline phosphatase is added.

Three different pilot ELISAs will have to be done to develop the sandwich ELISA for the antibodies and antigen used. These pilot ELISAs are essential to minimize nonspecific binding and to check for the interference or background problems caused by plant components. It also allows the establishment of the conditions for which the highest sensitivity is obtained, if this is needed for the intended purpose.

3.2.1. Determination of the Optimal Concentration of the Primary and Secondary Antibody

This first pilot ELISA determines the optimal set of primary and secondary antibody concentrations. The plate set-up is clarified in **Fig. 2**. In order to test four different primary antibody and four different secondary antibody concentrations, two plates will have to be made using the same plate set-up. Three types of blanks are included as controls for each set of primary and secondary antibody dilution (*see* **Fig. 2**).

1. Dilute the primary antibody at 12.5 μg/mL, 2.5 μg/mL, 0.5 μg/mL, and 0.1 μg/mL in PBS (*see* **Note 17**).
2. Fill the wells for blank 3 with 100 μL PBS each.
3. For each primary antibody dilution, coat 36 wells by pipeting 100 μL of the dilution per well (*see* **Fig. 2**).
4. Cover the plates with an adhesive plastic and incubate for 2 h at room temperature.
5. Wash the plates twice with PBS.
6. Fill the wells with 300 μL blocking buffer, cover the plates with an adhesive plastic, and incubate for at least 2 h at room temperature or, if more convenient, overnight at 4°C.
7. Make fourfold serial dilutions of the antigen in blocking buffer, covering a wide range of antigen concentrations. As an example, we used dilutions from 512 ng/mL through 0.125 ng/mL for a protein with a molecular mass of 48 kDa.
8. Wash the plates twice with PBS.
9. Fill the wells for blanks 1 and 2 with 100 μL blocking buffer each (*see* **Fig. 2**).
10. Add 100 μL of the highest antigen concentration to the wells for blank 3.

Antibody
conc. (μg/mL)

p	s		1	2	3	4	5	6	7	8	9	10
0.1	0.125	A	blank 1	blank 2	blank 3	512	128	32	8	2	0.5	0.125
0.1	0.5	B	blank 1	blank 2	blank 3	512	128	32	8	2	0.5	0.125
0.1	2.0	C	blank 1	blank 2	blank 3	512	128	32	8	2	0.5	0.125
0.1	8.0	D	blank 1	blank 2	blank 3	512	128	32	8	2	0.5	0.125
0.5	0.125	E	blank 1	blank 2	blank 3	512	128	32	8	2	0.5	0.125
0.5	0.5	F	blank 1	blank 2	blank 3	512	128	32	8	2	0.5	0.125
0.5	2.0	G	blank 1	blank 2	blank 3	512	128	32	8	2	0.5	0.125
0.5	8.0	H	blank 1	blank 2	blank 3	512	128	32	8	2	0.5	0.125

Fig. 2. Plate set-up of a sandwich ELISA for the determination of the optimal set of primary and secondary antibody concentrations. Numbers within the plate represent antigen concentrations in ng/mL applied to each well. Blank 1: primary antibody, no antigen, secondary antibody, streptavidin-alkaline phosphatase conjugate. Blank 2: primary antibody, no antigen, no secondary antibody, streptavidin-alkaline phosphatase conjugate. Blank 3: no primary antibody, antigen (512 ng/mL), secondary antibody, streptavidin-alkaline phosphatase conjugate. Abbreviations: p, primary antibody; s, secondary antibody.

11. Add the complete antigen dilution series to the remaining wells of each row of the microtiter plates, pipeting 100 μL per well (*see* **Fig. 2**).
12. Cover the plates with an adhesive plastic and incubate for 2 h at room temperature.
13. Wash the plates four times with PBS.
14. Dilute the biotinylated secondary antibody in blocking buffer to a concentration of 0.125 μg/mL, 0.5 μg/mL, 2 μg/mL, and 8 μg/mL (*see* **Note 17** and **Subheading 3.1.3., step 9**).
15. Add 100 μL blocking buffer to the wells for blank 2.
16. Add the biotinylated secondary antibody dilutions as shown in **Fig. 2**, so that each secondary antibody concentration is tested in combination with each primary antibody concentration. Apply 100 μL per well.
17. Cover the plates with an adhesive plastic and incubate for 2 h at room temperature.
18. Wash the plates four times with PBS.
19. To each well, add 100 μL of the streptavidin-alkaline phosphatase conjugate, diluted in blocking buffer at the recommended concentration for ELISA just prior to use (*see* **Note 9**).
20. Cover the plates with an adhesive plastic and incubate for 1 h at room temperature.
21. Wash the plates four times with PBS.
22. Dissolve the pNPP tablets at a concentration of 1 mg/mL in substrate buffer. Add 100 μL of the substrate solution per well with a multichannel pipet or a multipipet.
23. Incubate the plates at 37°C. Perform end-point measurements at 405 nm, using a microtiter plate spectrophotometer, making sure that the different plates are mea-

sured at the same time-points (i.e., 30 min and 1 h) after the addition of the substrate (*see* **Note 18**).
24. Choose the optimal set of primary and secondary antibody dilutions. Normally this choice will be a compromise between performance (OD value for a particular antigen concentration) and the cost (and/or availability) of the antibodies. If it is important to make the ELISA as sensitive as possible (*see* **Note 19**), one should choose the combination that distinguishes easily between the lowest detectable and the fourfold higher antigen concentration, and that gives a high signal-to-noise ratio for the lowest detectable antigen concentration. Make sure that the amount of secondary antibody is in excess when compared to the amount of antigen, to assure that the assay is quantitative. **Figure 3** shows standard curves obtained when the secondary antibodies are limited (curve B) and in excess (curve A). Check also whether the blanks of the chosen set of primary and secondary antibody dilution give background signals (*see* **Note 20**). If the same background problems occur for different sets, then adaptations to the basic protocol will have to be made (*see* **Note 20**).

3.2.2. Determination of the Linear Range of the Standard Curve

The second pilot ELISA determines accurately the linear range of the standard curve, using the conditions established in **Subheading 3.2.1.** The described protocol assumes that no adaptations had to be made to cope with background problems.

1. Dilute the primary antibody in PBS to the optimal concentration (*see* **Subheading 3.2.1.**).
2. Coat wells of a microtiter plate with the primary antibody by pipeting 100 μL of the dilution (**step 1**) in each well.
3. Cover the plate with an adhesive plastic and incubate for 2 h at room temperature.
4. Wash the plate twice with PBS.
5. Fill the wells with 300 μL blocking buffer, cover the plate with an adhesive plastic, and incubate for at least 2 h at room temperature or, if more convenient, overnight at 4°C.
6. Make for this pilot ELISA more dilutions, i.e., twofold serial dilutions of the antigen in blocking buffer, covering the range tested in **Subheading 3.2.1.**
7. Wash the plate twice with PBS.
8. Add blocking buffer to two coated wells (100 μL to each well), which will serve as blanks. Add the complete antigen dilution series to the remaining wells coated with the primary antibody by pipeting 100 μL per well. Run each dilution in duplicate.
9. Cover the plate with an adhesive plastic and incubate for 2 h at room temperature.
10. Wash the plate four times with PBS.
11. To each well, apply 100 μL of the secondary antibody diluted in blocking buffer to the optimal concentration (*see* **Subheading 3.2.1.**).

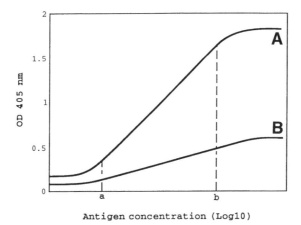

Antigen concentration (Log10)

Fig. 3. Representative standard curves for a sandwich ELISA with the antigen concentration on the abscissa and the absorbance at 405 nm on the ordinate. Curve A represents a standard curve obtained with the secondary antibody in excess; in curve B, the secondary antibody was present in limiting amounts compared to the amount of antigen. The region between the antigen concentrations (a) and (b) represents the linear range of standard curve A. The region above antigen concentration (b) reaches a plateau (constant high OD), meaning that the primary antibody on the plate is saturated with antigen. The plateau region below antigen concentration (a) represents the plate background and is the result of the change in substrate color independent of any enzymatic activity. When the secondary antibody is limited compared to the amount of antigen (curve B), the slope of the linear range is smaller and the plateau OD at high antigen concentrations is lower.

12. Cover the plate with an adhesive plastic and incubate for 2 h at room temperature.
13. Wash the plate four times with PBS.
14. Add 100 μL of the streptavidin-alkaline phosphatase conjugate to each well, diluted in blocking buffer as in **Subheading 3.2.1.**
15. Cover the plate with an adhesive plastic and incubate for 1 h at room temperature.
16. Wash the plate four times with PBS.
17. Dissolve the pNPP tablets at a concentration of 1 mg/mL in substrate buffer. Add 100 μL of the substrate solution per well with a multichannel pipet or a multipipet.
18. Measure the absorbance at 405 nm, using a microtiter plate spectrophotometer. Perform an end-point or a kinetic measurement (*see* **Note 21**) at 37°C.
19. Determine the linear range of the standard curve. ELISA gives a sigmoid standard curve of absorbance vs antigen concentration (*see* **Fig. 3**). The curve shows a plateau at low antigen concentrations, a linear range, and a plateau at high antigen concentrations. Only the linear range of the standard curve can be used for quantitative determinations of the antigen concentration in unknown samples.

3.2.3. Interference and Background Problems Caused by Plant Components

This third pilot ELISA is performed to check whether plant components interfere with the binding of the antigen to the primary antibody and, if interference occurs, at which concentration of total protein. Moreover, this ELISA determines which concentrations of total plant protein give background signals in the ELISA. The protocol below assumes that no changes had to be made to the basic protocol to cope with background problems.

1. Dilute the primary antibody in PBS to the optimal concentration (*see* **Subheading 3.2.1.**).
2. Coat 27 wells with the primary antibody by pipeting 100 µL of the dilution in each well.
3. Cover the plate with an adhesive plastic and incubate for 2 h at room temperature.
4. Wash the plate twice with PBS.
5. Fill the wells with 300 µL blocking buffer and incubate for at least 2 h at room temperature or, if more convenient, overnight at 4°C.
6. Dilute the antigen in blocking buffer to the highest, midpoint, and lowest concentration of the linear range of the standard curve (*see* **Subheading 3.2.2.**).
7. Make a dilution series in blocking buffer so that the dilutions contain a fixed amount of antigen but different amounts of total protein of an untransformed plant (see **Notes 2**, **3**, and **4**), e.g., 20, 50, 100, 200, 400, or 600 µg/mL. Three dilution series are made, containing either the highest, midpoint, or lowest antigen concentration of the linear range of the standard curve.
8. Make an analogous dilution series of the extract of the untransformed plant in blocking buffer, as described in **step 7**, but without the addition of antigen.
9. Wash the plate twice with PBS.
10. Add 100 µL of each dilution to a coated well.
11. Cover the plate with an adhesive plastic and incubate for 2 h at room temperature.
12. Wash the plates four times with PBS.
13. To each well, apply 100 µL of the secondary antibody diluted in blocking buffer to the optimal concentration (*see* **Subheading 3.2.1.**).
14. Cover the plate with an adhesive plastic and incubate for 2 h at room temperature.
15. Wash the plate four times with PBS.
16. Add 100 µL of the streptavidin-alkaline phosphatase conjugate to each well, diluted in blocking buffer as for the previous pilot ELISAs.
17. Cover the plate with an adhesive plastic and incubate for 1 h at room temperature.
18. Wash the plate four times with PBS.
19. Dissolve the pNPP tablets at a concentration of 1 mg/mL in substrate buffer. Add 100 µL of the substrate solution per well with a multichannel pipet or a multipipet.
20. Measure the absorbance at 405 nm, using a microtiter plate spectrophotometer. Perform an end-point or a kinetic measurement at 37°C.

21. For each antigen concentration, compare the signal obtained for the antigen alone (**step 6**) to the signal obtained for the wells that contained 2, 5, 10, 20, 40, or 60 μg of total protein of an untransformed plant in addition to the antigen (**step 7**). If the signal is significantly lower in the presence of a certain amount of plant proteins, then this amount of protein interferes with the binding of the antigen to the primary antibody. Determine from which concentration of total plant protein interference occurs.

22. Check if plant proteins give background in the ELISA (**step 8**) and, if so, from which concentration.

3.3. Determination of the Antigen Concentration in Plant Extracts

Once the ELISA has been calibrated, it is possible to analyze many plant extracts in a quick, reproducible, and quantitative way. If the antigen concentration is expected to be substantially different for the different plant extracts, then it is advisable to first do a screening ELISA and subsequently a more accurate ELISA. The protocol below is the basic protocol, assuming that no adaptations had to be made in the pilot ELISAs.

1. Dilute the primary antibody in PBS to the optimal concentration (*see* **Subheading 3.2.1.**).
2. Coat the wells of a microtiter plate with the primary antibody by pipeting 100 μL of the dilution to each well.
3. Cover the plate with an adhesive plastic and incubate for 2 h at room temperature.
4. Wash the plate twice with PBS.
5. Fill the wells with 300 μL blocking buffer and incubate for at least 2 h at room temperature or, if more convenient, overnight at 4°C.
6. Make a standard series by diluting the antigen in blocking buffer. Make twofold serial dilutions of the antigen in blocking buffer, covering the linear range, and one or two concentrations higher and lower (*see* **Subheading 3.2.2.**).
7. Make four different dilutions of the plant extracts (*see* **Notes 2, 3**, and **22**) in blocking buffer, including an extract of an untransformed plant. As a rule of thumb, half the magnitude of the linear range of the standard curve is taken as the spacing between two subsequent dilutions. For example, if the linear range is 10–1 ng, appropriate dilutions would be 1/2, 1/10, 1/50, and 1/250.
8. Wash the plate twice with PBS.
9. Fill one coated well with 100 μL blocking buffer, which serves as the blank. Add 100 μL of each standard dilution and each plant extract dilution to a well.
10. Cover the plate with an adhesive plastic and incubate for 2 h at room temperature.
11. Wash the plates four times with PBS.
12. To each well, apply 100 μL of the secondary antibody diluted in blocking buffer to the optimal concentration (*see* **Subheading 3.2.1.**).
13. Cover the plate with an adhesive plastic and incubate for 2 h at room temperature.

14. Wash the plate four times with PBS.
15. Add to each well 100 µL of the streptavidin-alkaline phosphatase conjugate, diluted in blocking buffer as for the pilot ELISAs, prior to use.
16. Cover the plate with an adhesive plastic and incubate for 1 h at room temperature.
17. Wash the plate four times with PBS.
18. Dissolve the pNPP tablets at a concentration of 1 mg/mL in substrate buffer. Add 100 µL of the substrate solution per well with a multichannel pipet or a multipipet.
19. Measure the absorbance at 405 nm, using a microtiter plate spectrophotometer. Perform an end-point or a kinetic measurement at 37°C.
20. Estimate the concentration of the antigen for each extract, using the standard curve (known concentrations of the antigen vs signals obtained). Be aware that only the dilutions within the linear range give accurate results.
21. Perform a second, more accurate ELISA, using the same protocol as stated above, but applying twofold dilutions of the plant extract in blocking buffer, which cover the linear range of the standard curve. Run standard and sample dilutions in duplicate. Make sure that the total protein content in the dilutions is lower than the protein concentration that interfered with the binding of the antigen to the primary antibody (*see* **Subheading 3.2.3., step 21**). If the total protein concentration is higher than the one giving background signals in the ELISA (*see* **Subheading 3.2.3., step 22**), add total protein of an untransformed plant to the standard and sample dilutions, so that the total protein amount is the same in all wells. In this way, the background will be equal in all wells.

Determine the concentration of the antigen in the samples by comparing the absorbance of the serially diluted sample to the linear range of the standard curve of absorbance vs antigen concentration. Take only values within the linear range into account for calculations. Calculate the quantity of antigen expressed per total soluble protein amount (*see* **Notes 4** and **23**).

4. Notes

1. For some antigens, better coating efficiencies are obtained upon dilution in 50 mM NaHCO$_3$, pH 9.6. Other coating buffers can be found in **ref. 2**.
2. Use a protein extract of an untransformed plant of the same species as the one used for the transformation with the heterologous gene. Make a protein extract of the plant tissue for which you want to determine the antigen accumulation in the transformed plants.
3. Most commonly used procedures for extraction of proteins from plant material are compatible with ELISA. Components in extraction buffers that are possibly interfering with ELISA are: extreme pH values, high salt concentrations, and high detergent concentrations. Many plant species contain polyphenols and alkaloids in the vacuoles; these compounds may inhibit enzyme activity and can be removed from the plant extract by complexing with insoluble poly (vinylpyrrolidone) or by gel filtration (*4*).

4. The total protein amount in the plant extracts is determined with a Bio-Rad Protein Assay *(20)* or a Bio-Rad DC Protein Assay *(21)*.
5. To avoid bubble and foam formation, which interfere with the accuracy and reproducibility of the ELISA, pipeting with a Pipetman or multichannel pipet should always be done as follows: Push the button a bit further than the first stop and take up the liquid; touch the edge of the well(s) with the top of the tip(s), and dispense the liquid to the well(s) by pushing until the first stop; agitate the plate gently to obtain a similar meniscus in all the wells.
6. The washing buffer should be at room temperature. A multichannel pipet, a multipipet, a 500-mL squirt bottle, or a commercially available hand washing device (Nunc) are convenient for washing. Do not let the plate dry at any moment, because this will give high background values and unreliable results. Other washing buffers can be found in **ref.** *2*.
7. Different blocking buffers can be found in **refs.** *2* and *8*.
8. Preimmune serum of each animal can be used as a control.
9. The recommended concentration for ELISA can be found on the product sheet. If a concentration range is advised, a good starting point is to use the midpoint concentration. If the resulting sensitivity is not satisfactory, a higher concentration can be tried.
10. The temperature of the substrate buffer should be the same for every test. This can be achieved by preincubating the substrate buffer in a water bath at a constant temperature before dissolving the tablets.
11. For an end-point measurement, the microtiter plate is incubated for a certain time at a certain temperature and the reaction is stopped by the addition of $3N$ NaOH, to a final concentration of $1N$. Commonly used conditions are 30 min or 1 h at room temperature. To measure samples that contain very different concentrations, it is advisable to omit the addition of the stop solution, so that the plate can be measured after different incubation times. The advantage of an end-point measurement is the possibility to develop different plates simultaneously.
12. If all sera seem to give background signals in the wells coated with plant proteins, there are still ways to circumvent this problem later in the ELISA; one can either adsorb the purified antibodies with an acetone powder protein preparation of an untransformed plant before their use as secondary antibody *(8)*, or add total protein of a negative extract, being an extract without antigen, so that all wells (standards and samples) contain the same amount of total protein. In this way, the background will be equal in all wells.
13. Alternatively, the antibodies can be directly labeled with alkaline phosphatase or peroxidase *(2,7,8)*. Alkaline phosphatase is recommended because of its rapid catalytic rate, excellent intrinsic stability, and the linearity of its reaction within time *(7)*. Peroxidase easily gives background problems with plant extracts, because of the presence of plant peroxidases. Furthermore, the peroxidase reaction rate tends to slow down relatively quickly and the results are less easy to reproduce.

14. The reagents can also be purchased separately and biotinylation protocols, described in **refs.** *7* and *8*, can be followed.

15. When planning the analysis of many plant extracts, perform this labeling reaction at least in duplicate. After elution of the biotinylated antibodies, the antibody-containing fractions of all the labeling reactions are pooled and this pooled stock is used for the development of the ELISA.

16. The biotinylated antibodies are stored at 4°C, with the addition of 10% sodium azide, to a final concentration of 0.02%.

17. The mentioned concentrations were used for polyclonal antibodies which were protein A-purified from rabbit antiserum with an antigen-specific antibody titer of 10^{-5}. If the antigen-specific antibody titer of the serum used is higher or lower, the concentrations of the antibody solution might need to be lower or higher, respectively.

18. Be aware that very low signals can be caused by a bad labeling of the secondary antibody with biotin.

19. The sensitivity of the ELISA, if not satisfactory, can be increased by using longer incubation times and/or higher incubation temperatures, although this might also increase nonspecific binding; by the use of another coating buffer *(2)*; by the use of the fluorogenic substrate, 4-methylumbelliferyl phosphate, instead of the chromogenic substrate pNPP for alkaline phosphatase *(22,23)*; or by the use of other antibodies with a higher specificity or affinity.

20. The use of different blanks (*see* **Fig. 2**) allows the pointing out of the compound(s) and/or condition(s) giving high backgrounds. If only blank 1 gives background, then a compound in the secondary antibody solution is binding to the primary antibody. Since the primary and secondary antibodies are originally the same, they cannot recognize each other. Probably, some free biotin-7-NHS is present in the secondary antibody solution, which is reacting with the primary antibody on the plate. This background problem can be solved by removing the biotin-7-NHS from the secondary antibody solution by a second run of the antibody solution on a Sephadex G-25 column, or by extensive dialysis of the antibody solution against PBS + 0.02% sodium azide, or by the addition of free amino groups (tris[hydroxymethyl]aminomethane or glycine, up to 1*M*), which will react with the biotin-7-NHS. If only blank 3 gives background, then the antigen is reacting with the blocking agent or the plate. First try other blocking agents *(2,8)* and then other plates *(8)*. If only blanks 1 and 3 give background, then the secondary antibody is binding nonspecifically to the blocking agent or the plate. This background problem is rather likely to occur at high concentrations of secondary antibody. Yet, the background problems should disappear for lower secondary antibody concentrations. If this is not the case, then other blocking agents *(2,8)*, other plates *(8)*, or another conjugate should be tried. If all blanks give background, this may be explained by inefficient blocking of the wells, so that all the binding sites in the well are not occupied by the blocking agent, and/or nonspecific binding of the streptavidin-alkaline phosphatase to the blocking agent or the plate. Different blocking agents *(2,8)* and then other plates should be tested.

If background problems occur, it is advisable to include more blanks in which one or more compounds are systematically left out. This will allow the discrimination between different possibilities. In general, the following processes can be tested to reduce nonspecific binding in the step(s) in which the problem occurs: Wash and/or incubate with the addition of up to 0.2% Tween-20; wash and/or incubate with the addition of NaCl or $MgCl_2$ up to 500 mM; use shorter incubation times and/or lower incubation temperatures. All these methods might also reduce the signal strength of the ELISA and the best conditions are those that give an increase in signal:noise ratio. If background problems persist, other antibodies will have to be used.

21. We routinely perform two kinetic measurements at 37°C: A short measurement (every minute for 10 min) and a long measurement (every 5 min for 50 min), to determine the concentration in samples containing high and low amounts, respectively. If desired, the measurement can go on for another 50 min, and so on. An advantage of a kinetic measurement is that samples containing very different amounts can be measured with the same standardized protocol. Another advantage is that the fault, caused by the fact that the substrate solution is not added at exactly the same time point to the different wells, is minimized by comparing the maximal reaction rate, instead of the absorbance as such.

22. The reproducibility of the extraction procedure and the compatibility of the extraction buffer with the ELISA should be checked specifically for each protein. The reproducibility of the extraction procedure should be checked carefully, since the largest errors in ELISA result from unreproducible extractions. The effect of the extraction buffer on the ELISA can be examined by comparing a standard curve in PBS and a standard curve in extraction buffer. Nonsoluble proteins (e.g., membrane-bound) pose peculiar problems for extraction and measurement.

23. We successfully determined concentrations of soluble antigens in extracts of leaves, roots, seeds, calli, and whole plants. Results can only be compared for samples obtained from the same type of plant tissue, and care should be taken when comparing different tissues. The fault on the determination of the total protein amount might differ significantly between extracts from different tissues. Furthermore, the efficacy of antigen extraction from different tissues might vary.

Acknowledgments

This work was supported by grants from the Belgian Programme on Interuniversity Poles of Attraction (Prime Minister's Office, Science Policy Programming, #38) and the Vlaams Actieprogramma Biotechnologie (ETC 002). A.-M. B., G. D. J., and C. D. W. are indebted to the Vlaams Instituut voor de Bevordering van het Wetenschappelijk-Technologisch Onderzoek in de Industrie for a predoctoral fellowship. M. D. N. was a Research Assistant of the National Fund for Scientific Research (Belgium). P. R. is a Research Director of the Institut National de la Recherche Agronomique (France).

References

1. Janeway, C. A. and Travers, P. (1994) *Immunobiology: The Immune System in Health and Disease.* Current Biology, London, and Garland, New York.
2. Crowther, J. R. (1995) in *Methods in Molecular Biology, ELISA: Theory and Practice,* vol. 42. Humana, Totowa, NJ.
3. Studier, F. W., Rosenberg, A. H., Dunn, J. J., and Dubendorff, J. W. (1990) Use of T7 RNA polymerase to direct expression of cloned genes, in *Gene Expression Technology: Methods in Enzymology,* vol. 185 (Goeddel, D. V., ed.), Academic, San Diego, pp. 60–89.
4. Deutscher, M. P. (1990) *Guide to Protein Purification: Methods in Enzymology,* **182,** Academic, San Diego.
5. Erlanger, B. F. (1980) The preparation of antigenic hapten-carrier conjugates: a survey, in *Immunochemical Techniques, Part A: Methods in Enzymology,* **70,** (Van Vunakis, H. and Langone, J. J., eds.), Academic, New York, pp. 85–104.
6. Reichlin, M. (1980) Use of glutaraldehyde as a coupling agent for proteins and peptides, in *Immunochemical Techniques, Part A: Methods in Enzymology,* **70,** (Van Vunakis, H. and Langone, J. J., eds.), Academic, New York, pp. 159–165.
7. Coligan, J. E., Kruisbeek, A. M., Margulies, D. H., Shevach, E. M., and Strober, W. (1994) *Current Protocols in Immunology,* vol. 1, Current Protocols, New York.
8. Harlow, E. and Lane, D. (1988) *Antibodies: A Laboratory Manual.* Cold Spring Harbor Laboratory, Cold Spring Harbor, NY.
9. Hoogenboom, H. R., Marks, J. D., Griffiths, A. D., and Winter, G. (1992) Building antibodies from their genes. *Immunol. Rev.* **130,** 41–68.
10. De Neve, M., De Loose, M., Jacobs, A., Van Houdt, H., Kaluza, B., Weidle, U., Van Montagu, M., and Depicker, A. (1993) Assembly of an antibody and its derived antibody fragment in *Nicotiana* and *Arabidopsis. Transgenic Res.* **2,** 227–237.
11. da S. Conceição, A., Van Vliet, A., and Krebbers, E. (1994) Unexpectedly higher expression levels of a chimeric 2S albumin seed protein transgene from a tandem array construct. *Plant Mol. Biol.* **26,** 1001–1005.
12. Matzke, M. A., Moscone, E. A., Park, Y.-D., Papp, I., Oberkofler, H., Neuhuber, F., and Matzke, A. J. M. (1994) Inheritance and expression of a transgene insert in an aneuploid tobacco line. *Mol. Gen. Genet.* **245,** 471–485.
13. Jordan, E. T., Hatfield, P. M., Hondred, D., Talon, M., Zeevaart, J. A. D., and Vierstra, R. D. (1995) Phytochrome A overexpression in transgenic tobacco. *Plant Physiol.* **107,** 797–805.
14. Wahl, M. F., An, G., and Lee, J. M. (1995) Effects of dimethyl sulfoxide on heavy chain monoclonal antibody production from plant cell culture. *Biotechnol. Lett.* **17,** 463–468.
15. Kohno-Murase, J., Murase, M., Ichikawa, H., and Imamura, J. (1994) Effects of an antisense napin gene on seed storage compounds in transgenic *Brassica napus* seeds. *Plant Mol. Biol.* **26,** 1115–1124.
16. Florack, D. E. A., Dirkse, W. G., Visser, B., Heidekamp, F., and Stiekema, W. J. (1994) Expression of biologically active hordothionins in tobacco. Effects of

pre- and pro-sequences at the amino and carboxyl termini of the hordothionin precursor on mature protein expression and sorting. *Plant Mol. Biol.* **24,** 83–96.

17. Shewmaker, C. K., Boyer, C. D., Wiesenborn, D. P., Thompson, D. B., Boersig, M. R., Oakes, J. V., and Stalker, D. M. (1994) Expression of *Escherichia coli* glycogen synthase in the tubers of transgenic potatoes (*Solanum tuberosum*) results in highly branched starch. *Plant Physiol.* **104,** 1159–1166.

18. Firek, S., Draper, J., Owen, M. R. L., Gandecha, A., Cockburn, B., and Whitelam, G. C. (1993) Secretion of a functional single-chain Fv protein in transgenic tobacco plants and cell suspension cultures. *Plant Mol. Biol.* **23,** 861–870.

19. Bucher, M., Brändle, R., and Kuhlemeier, C. (1994) Ethanolic fermentation in transgenic tobacco expressing *Zymomonas mobilis* pyruvate decarboxylase. *EMBO J.* **13,** 2755–2763.

20. Bradford, M. M. (1976) A rapid and sensitive method for the quantitation of microgram quantities of protein utilizing the principle of protein-dye binding. *Anal. Biochem.* **72,** 248–254.

21. Lowry, O. H., Rosebrough, N. J., Farr, A. L., and Randall, R. J. (1951) Protein measurement with the folin phenol reagent. *J. Biol. Chem.* **193,** 265–275.

22. Ward, C. M., Chan, H. W.-S., and Morgan, M. R. A. (1988) The potential of fluorescence detection in ELISA, in *Immunoassays for Veterinary and Food Analysis*, vol. 1 (Morris, B. A., Clifford, M. N., and Jackman, R., eds.), Elsevier, Amsterdam, p. 275.

23. Roberts, I. M., Stephen, L. J., Premier, R. R., and Cox, J. C. (1991) A comparison of the sensitivity and specificity of enzyme immunoassays and time-resolved fluoroimmunoassay. *J. Immunol. Meth.* **143,** 49–56.

19

Analysis of *N*- and *O*-Glycosylation of Plant Proteins

Anne-Catherine Fitchette-Lainé, Lise-Anne Denmat, Patrice Lerouge, and Loïc Faye

1. Introduction

In eukaryotic cells, secreted proteins can be subjected to numerous post-translational modifications. One of these, glycosylation, consists of the addition of sugar residues to the protein backbone, while the protein enters or travels within the secretory pathway. Glycosylation of secreted proteins can be of two types: *N*- or *O*-glycosylation, depending on the linkage involved between the oligosaccharide moiety and the protein backbone.

N-glycosylation is the most studied of the glycosylation events. It occurs as soon as the nascent secreted protein enters the endoplasmic reticulum (ER), by transfer of an oligosaccharide precursor from a lipid carrier onto a specific Asn residue *(1)*. This Asn residue belongs to a *N*-glycosylation consensus sequence: Asn-X-Ser/Thr, in which X can be any amino acid except Pro or Asp. The structure of the oligosaccharide precursor is $Glc_3Man_9GlcNAc_2$ (*see* **Note 1** and **Fig. 1**). Once *N*-linked to the protein backbone, this oligosaccharide is progressively matured into a high mannose-type and eventually further modified into a complex-type glycan (**Fig. 1**). These maturation steps occur while the secreted protein travels through the different compartments constitutive to the secretory pathway. Plant complex-type *N*-glycans differ from mammalian ones, an example of which is presented in **Fig. 1**, by the presence of a β 1–>2 xylose on the β mannose of the core, the presence of an α 1–>3 fucose (rather than α 1–>6) on the proximal *N*-acetylglucosamine residue of the core, and the absence of neuraminic acid (**Fig. 1**).

From: *Methods in Biotechnology, Vol. 3:*
Recombinant Proteins from Plants: Production and Isolation of Clinically Useful Compounds
Edited by: C. Cunningham and A. J. R. Porter © Humana Press Inc., Totowa, NJ

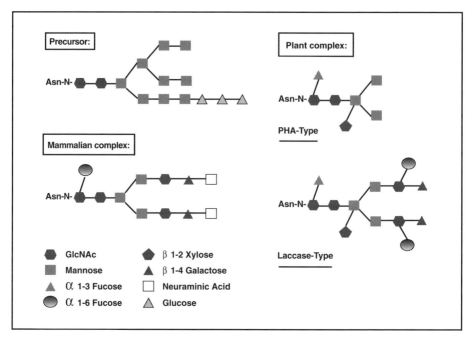

Fig. 1. Structure of various *N*-linked oligosaccharides. The precursor oligosaccharide is composed of the (GlcNAc$_2$Man$_3$) core, six additional mannose residues and three terminal glucose residues. Plant complex glycans are of two types: PHA- and laccase- types. They differ from their mammalian counterparts by the presence of a β 1→2 xylose on the β mannose of the core, the presence of an α 1→3 fucose on the proximal *N*-acetylglucosamine residue of the core and the absence of neuraminic acid.

In plant cells, as in mammalian cells, *N*-glycan maturation starts in the ER with the removal of the three terminal glucose residues from the oligosaccharide precursor. This cleavage is performed by two ER enzymes: glucosidase I and glucosidase II (**Fig. 2**) *(2,3)*. A potential reglucosylation can occur at this level, in plant and in mammalian cells *(4)*, under the control of a ER glucosyltransferase, leading to the formation of monoglucosylated *N*-linked glycans. These glycans seem to be involved in mammalian glycoprotein retention within the ER lumen, via an interaction with the transmembrane protein, calnexin. This retention could be an element of a protein quality control and thus would help the secreted glycoproteins to fold properly *(5)*. Calnexin does exist in plants *(6)*, but its involvement as a glycoprotein chaperone has not yet been demonstrated in plant cells. Before leaving the ER compartment, glycoproteins can be further matured by the possible trimming of one terminal mannose residue (**Fig. 2**).

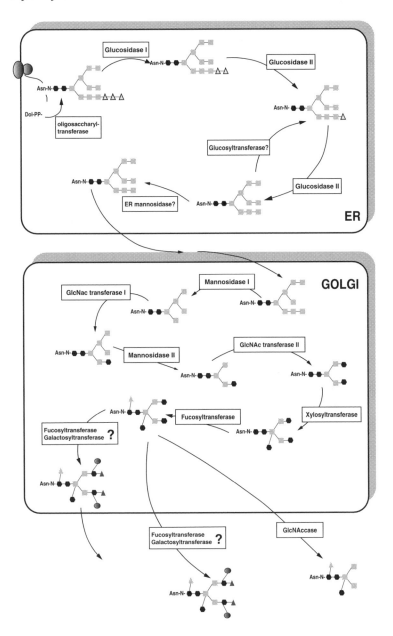

Fig. 2. *N*-glycan maturation events in microsomal membranes from plant cell.
(adapted from **refs.** *1,3,4,9,33,* and *34*) (symbols for sugar residues, *see* **Fig. 1**).

Upon glycoprotein arrival in the Golgi apparatus, other mannose residues can be cleaved, yielding to high mannose-type N-glycans, from $Man_8GlcNAc_2$ (M8) to $Man_5GlcNAc_2$ (M5). The M5 oligosaccharide is the precursor of complex-type glycans in plants, as well as in mammals. In both organisms, M5 is futher modified within the Golgi apparatus by three enzymes that act successively: N-acetylglucosamine transferase I, α-mannosidase II, and N-acetylglucosamine transferase II. This leads to the formation of the N-glycan structure $GlcNAc_2Man_3GlcNAc_2$ (**Fig. 2**). Beyond this point, the maturation steps responsible for the complex glycan formation differ in plant and in mammalian cells.

In plant cells, N-glycans are modified by addition of β 1–>2 xylose on the β-mannose and α 1–>3 fucose on the proximal N-acetylglucosamine of the core (**Fig. 2**). These two events have been localized within the Golgi apparatus of sycamore cells. Xylosylation has been shown to be mainly a medial Golgi event, but α 1–>3 fucose is added to complex N-glycans mostly in the trans-Golgi compartment *(7–9)*. The maturation of plant complex N-glycans can then be achieved by two different pathways. Terminal N-acetylglucosamine residues can be removed in a post-Golgi compartment, allowing the formation of $Man_3XylFucGlcNAc_2$, as demonstrated for bean PHA *(10)*. N-glycans can also be matured by addition of terminal α 1–>6 fucose and β 1–>4 galactose residues yielding to biantennary laccase-type glycan structures *(11)*.

O-glycans are present on glycoproteins depicted as hydroxyprolin-rich glycoproteins (HRGPs). These HRGPs represent three groups of proteins: extensins, ArabinoGalactan Proteins (AGPs), and solanaceous lectins. The extensins and the solanaceous lectins contain repeating sequences consisting of Ser-(Hyp)$_4$ in which the serine can be O-galactosylated and hydroxyproline can be substituted by arabinosyl chains of 1–4 residues *(12)*. In extensins, arabinosylation occurs in a stepwise manner at the Golgi apparatus level *(13)*. The AGP family is composed of extracellular and plasma-membrane proteins. The protein moiety of AGPs is rich in Hyp, Ser, Ala, Thr and Gly, and contains Ala-Hyp repeats. Many of the Hyp residues of the AGPs are substituted with polysaccharide chains, with β1–>3 linked galactose residues branched with β1–>6 galactose side chains, which are in turn branched with arabinose or other less abundant monosaccharides (*see* **ref.** *12* for review).

The fact that a protein bears N- or O-glycans implies that it has passed through the compartments of the ER and the Golgi apparatus. Consequently, the presence of glycans, N- or O-linked to a protein, is considered as a marker of its transport through the secretory pathway, although it should be kept in mind that not all the secretory proteins are glycosylated (for example, barley α-amylase, pea vicilin, legumin, and aquaporin are not glycoproteins). The aim of this chapter is to introduce several methods for determining whether a protein is glycosylated or not, and the type of linkage involved between the glycan and the protein backbone.

2. Materials
2.1. Special Reagents

1. ECL™ Glycoprotein Detection System (Amersham, Little Chalfont, UK).
2. DIG Glycan Detection Kit (Boehringer Mannheim, Mannheim, Germany).
3. GlycoTrack™ (Oxford GlycoSystem, Oxford, UK).
4. Trifluoromethanesulfunic acid (Sigma, St. Louis, MO).
5. Anhydrous anisole (Sigma).
6. Diethyl ether (Aldrich, Milwaukee, WI).
7. Pyridine (Sigma).
8. 4-Chloro-1-naphtol (HRP-color development reagent; Bio-Rad, Hercules, CA).
9. α-Methyl mannoside (Sigma).
10. Sodium periodate (Sigma).
11. Sodium borohydride (Sigma).
12. Endoglycosidase H (Boehringer Mannheim).
13. PNGase F (Boehringer Mannheim).

2.2. Proteins

1. Concanavalin A (Sigma).
2. Horseradish peroxidase (Sigma).
3. Ovalbumin (Serva, Heidelberg, Germany).
4. Biotinylated *Galanthus nivealis* agglutinin (Oxford GlycoSystem).
5. Conjugate streptavidin/peroxidase (BRL, Bethesda, MD).
6. Ribonuclease B (Sigma).
7. Soybean agglutinin (Sigma).
8. Phytohemagglutinin L (Sigma).
9. Gelatin (Bio-Rad).
10. Phosholipase A_2 from honeybee venom (Sigma).
11. Snail hemocyanin (Sigma).
12. Goat anti-rabbit IgG/horseradish peroxidase conjugate (Bio-Rad).

2.3. Immunsera

1. Serum directed against *Phaseolus vulgaris* lectin (PHA) (Sigma).
2. Serum directed against horseradish peroxidase (Sigma).
3. Serum directed against honeybee venom (Sigma).

2.4. Buffers and Other Solutions

1. TBS buffer: 10 mM Tris-HCl, pH 7.4, and 0.5M NaCl.
2. TTBS buffer: 10 mM Tris-HCl, pH 7.4, 0.5M NaCl, and 0.1% (v/v) Tween-20.

3. Con A buffer: 10 mM Tris-HCl, pH 7.4, 0.5M NaCl, 0.1% (v/v) Tween-20, 1 mM CaCl$_2$, and 1 mM MnCl$_2$.
4. PBS: 50 mM phosphate buffer, pH 7.5, and 0.15M NaCl.
5. 100 mM sodium acetate buffer, pH 4.5.
6. 150 mM sodium acetate buffer, pH 5.7.
7. 2X Electrophoresis sample buffer: 125 mM Tris-HCl, pH 6.8, 2% (w/v) sodium dodecyl sulfate, 20% (v/v) glycerol, and 10% (v/v) 2-mercaptoethanol.
8. PNGase buffer: 50 mM Tris-acetate, pH 8.4, 5 mM EDTA, and 20 mM 2-mercaptoethanol, or 50 mM phosphate buffer, pH 8.4, 5 mM EDTA, and 20 mM 2-mercaptoethanol.

2.5. Special Laboratory Tools and Materials

Reaction vial with teflon-faced screw cap (Sigma).

3. Methods
3.1. Is My Protein Glycosylated?

Two approaches can be used to answer this general question. First, N- or O-glycans linked to a protein can be detected directly with general detection kits after electrophoretic separation and blotting of the proteins onto a PVDF membrane. In a second approach, the presence of a N- or an O-glycan can be shown from the reduction in molecular weight of a glycoprotein, when this latter has been chemically deglycosylated. This molecular weight modification is then visualized by an increase in the protein mobility in SDS-PAGE.

3.1.1. General Detection Kits

After SDS-PAGE and transfer onto PVDF membrane, glycoproteins can be detected and quantified on electroblots after oxidation of their glycans using sodium periodate. This reagent specifically oxidizes vicinal hydroxyl groups of sugar residues constitutive of glycans. Mild periodate oxidation produces aldehyde functions on the oligosaccharide moiety, which can react in a second step with biotin-hydrazide to give covalently linked biotin groups via a hydrazone bond. Glycoproteins are then revealed on the blot, using either streptavidin-peroxidase coupled to a chemoluminescent detection system (Amersham), as illustrated on **Fig. 3**, or streptavidin-alkaline phosphatase (Oxford GlycoSystems). In another detection kit, the aldehyde groups generated on the oligosaccharide moiety of glycoproteins by mild periodate oxidation react with digoxigenin-hydrazide, which is immunodetected in the last step with antibodies specific for digoxigenin conjugated with alkaline phosphatase (Boehringer).

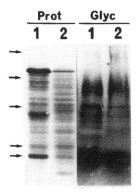

Fig. 3. Detection of glycoproteins from *Arabidopsis thaliana* with the ECL™ detection system from Amersham. Protein extracts from *Arabidopsis thaliana* var. Landsberg erecta, grown under light (lanes 1) or dark conditions (lanes 2), were separated by SDS-PAGE and stained with Coomassie blue (panel Prot) or transfered onto a PVDF membrane (panel Glyc). After electroblotting, the transferred glycoproteins were detected with the Amersham ECL detection kit. Arrows on left indicate the positions of the molecular weight standards, respectively, from top to bottom: 68, 43, 29, 18.4, and 14.3 kDa.

3.1.2. Chemical Deglycosylation

N- or *O*-linked glycans can be cleaved from a plant glycoprotein by a very efficient deglycosylation procedure using trifluoromethanesulfonic acid (TFMS), according to the following method modified from **ref. *14*.**

1. When TFMS arrives from the manufacturer, aliquot it into dry ampules (0.5 mL/ ampule) flushed with dry nitrogen. Seal the ampules and store at –20°C (it is easier to buy it directly in small volumes from the company).
2. Use a new ampule for each experiment.
3. Transfer 2 mg (or less) of protein into a reaction vial with teflon-faced screw cap by weighing the freeze-dried protein in it, or precipitating the protein in the vial with acetone (do not hesitate to spin the vial, or precipitating the protein in the vial with 10% TCA and washing the pellet with acetone.
4. Dry the protein in the vial under vacuum, then flush the vial with dry nitrogen (water traces should be very carefully removed before treatment with TFMS).
5. Mix TFMS with dry anisole in a volume ratio of 2:1 under nitrogen. The solution becomes slightly yellow. Add 200 µL of this mixture to the protein, flush with nitrogen, and close with a cap (it does not hurt to use an excess of this reagent mixture relative to protein, i.e., 500 µL mix to 2 mg of protein).
6. Put the reaction vial in ice for 3 h. Mix with a vortex every 15 min. The reaction mixture turns slight purple. A dark purple color would indicate that one of the

reagents is wet. In this case, the protein will be degraded and it is better to start over, using a new batch of glycoprotein.

7. Stop the reaction by adding 600 µL of diethyl ether kept at –40°C. The solution becomes colorless.

8. Extract the protein by adding 600 µL of an aqueous solution of pyridine (1/1, v/v). Mix with vortex and spin for 10 s for phase separation. Transfer the lower phase into a new vial with a Pasteur pipet and extract again with 1 mL diethyl ether.

9. Dialyse the lower phase against water and freeze-dry the deglycosylated protein.

10. This deglycosylation can be visualized by migration of the protein and the TFMS-treated sample on SDS-PAGE, followed by Coomassie blue staining (*see* **Fig. 4**) or by migration on SDS-PAGE, transfer onto nitrocellulose membrane and detection of the protein with specific antibodies *(15)*.

3.2. What Is the Glycan Linkage?

Many procedures are available to detect the presence of glycans N-linked to plant glycoproteins. *N*-glycans can be easily detected on blots using lectins (affinodetection) or glycan-specific antibodies (immunodetection). *N*-glycans can also be removed by specific enzyme treatments. This removal will be detected on SDS-PAGE by the observation of a reduced molecular weight for the deglycosylated protein. Moreover, the detection of *N*-glycans on a glyco-protein can also be performed using in vivo glycan-specific labeling proce-dures with radiolabeled monosaccharide precursors. Unfortunately, no procedure is currently available to specifically detect the presence of an *O*-glycan on a plant glycoprotein.

3.2.1. N-Glycan Specific Detections on Blot

3.2.1.1. AFFINODETECTION OF HIGH-MANNOSE TYPE *N*-LINKED GLYCANS

As most lectin specificities have been characterized on the basis of their affinity for mammalian glycoproteins, their use as plant glycoprotein detection reagents is still very limited. We will recommend only two of them as unam-biguous probes to study plant high mannose *N*-glycans: concanavalin A (Con A) and *Galanthus nivealis* agglutinin (GNA).

Con A is specific for terminal mannose and glucose residues and widely recognizes all high-mannose type *N*-glycans. Affinodetection with Con A is an inexpensive procedure, thanks to the ability of this lectin to bind horseradish peroxidase.

GNA is highly specific for terminal $\alpha 1 \to 3$ mannose residues *(16,17)*. This lectin binds to ribonuclease B (M5 and M6; *see* **Fig. 5A**), but not to soybean agglutinin (M9; *see* **Fig. 5B**). In M9, all $\alpha \to 3$ mannose residues are substituted by terminal mannose which prevent GNA binding. Complex-type *N*-glycans

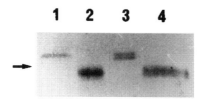

Fig. 4. Deglycosylation by TFMS shows that recombinant PHA L expressed in tobacco cells is a glycoprotein. PHA L is a subunit of phytohemagglutinin from bean seeds. It has been expressed as a recombinant protein in transgenic tobacco cells *(15)*. Bean PHA L (lanes 1 and 2) and recombinant PHA L produced in tobacco (lanes 3 and 4) have been purified on a thyroglobulin-Sepharose column. An aliquot of each PHA has been deglycosylated by TFMS. Native bean and recombinant PHA Ls (lanes 1 and 3, respectively) and TFMS-treated bean and recombinant PHA Ls (lanes 2 and 4, respectively) were then analyzed by SDS-PAGE and Coomassie blue staining. The arrow on left shows the position of the 29 kDa molecular weight standard.

attached to honeybee venom phospholipase A_2 are also detected by GNA (*see* **Fig. 5** for structure). In contrast, xylose-containing complex-type *N*-glycan from horseradish peroxidase is not recognized (*see* **Fig. 5** for structure), which suggests that the presence of xylose on the core inhibits all recognition of the proximal $\alpha 1->3$ Man by GNA.

We developed an affinodetection procedure allowing the specific detection of M5. GNA binds to M5 with a very high affinity and, to a much lesser extent, to other high mannose-type *N*-glycans. The recognition of glycans from M6 to M8 can be inhibited using increasing concentration of α-methyl mannoside, as shown in **Fig. 6**. GNA is commercially available in a biotinylated form.

The Con A/peroxidase procedure (modified from **ref. 18**) (specificity: all high-mannose type *N*-glycans):

1. Separate the proteins by SDS-PAGE and transfer onto a nitrocellulose membrane.
2. Saturate the blot for 1 h in a TTBS buffer at room temperature (*see* **Note 2**).
3. Incubate the membrane in Con A buffer containing Con A (25 µg/mL) for 2 h at room temperature.
4. Wash the membrane four times for 15 min each in TTBS.
5. Incubate the membrane in TTBS containing horseradish peroxidase (50 µg/mL) for 1 h at room temperature.
6. Rinse the blot four times for 15 min in TTBS and once in TBS for 15 min.
7. Prepare extemporaneously the peroxidase development reaction mixture by dissolving, in one beaker, 30 mg of 4-chloro-1-naphtol in 10 mL methanol, and mixing in another beaker 30 µL H_2O_2 with 50 mL TBS. Mix the two solutions just prior to pouring onto the membrane, and shake gently to optimize the development reaction.

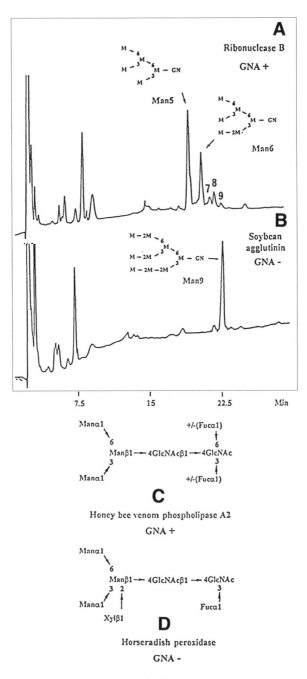

Fig. 5.

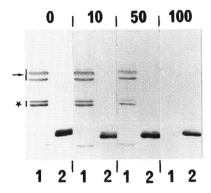

Fig. 6. Specific detection of M5 glycans with *Galanthus nivealis* agglutinin (GNA) in the presence of α-methyl mannoside. Proteins from bean seeds (lanes 1) and bovine ribonuclease B (lanes 2) have been subjected to SDS-PAGE and transferred onto a nitrocellulose membrane. The blots have been affinodetected with GNA in the presence of increasing amounts of α-methyl mannoside (0–100 m*M*).

When α-methyl mannoside is added to a final concentration of 100 m*M*, GNA does not detect bean phytohemagglutinin (star) or phaseolin (arrow) anymore, as both contain M6–M8 high-mannose type *N*-glycans. The ribonuclease B, which is mostly *N*-glycosylated by a M5 glycan (*see* Fig. 5A), is still detected by GNA in these conditions.

8. Stop the reaction by discarding the development mixture and rinsing the blot several times with distilled water. Dry the membrane and store between two sheets of filter paper. Glycoprotein affinodetection using this procedure is illustrated on **Fig. 7B**.
9. Controls for specificity:
    a. Ovalbumin can be used as a positive control glycoprotein.
    b. Lectin binding to a protein is not necessarily glycan-specific. In this procedure, you should run a control by adding a Con A competitive inhibitor (i.e., 0.3*M* α-methyl mannoside) in the solutions used in **step 3** and first rinsing of **step 4**. If the binding is caused by glycans, no color should develop in the control.

Fig. 5. *(previous page)* Binding specificity of *Galanthus nivealis* agglutinin (GNA). **(A)** and **(B)**: High pH anion exchange chromatography with pulsed-amperometric detection (HPAEC-PAD) chromatograms and structures of high-mannose type *N*-glycans released from (A) ribonuclease B and (B) soybean agglutinin. Both samples were analyzed according to **ref. 35**. Glycoproteins were purified by SDS-PAGE, electroblotted onto PVDF membranes, and stained. The stained bands were excised and treated with Endo H. The released *N*-glycans were then analyzed by HPAEC-PAD. **(C)** *N*-glycans from honeybee venom phospholipase A$_2$. **(D)** *N*-glycan from horseradish peroxidase.

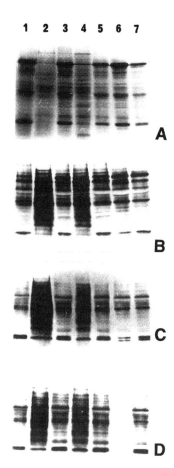

Fig. 7. Analysis of glycoprotein patterns in various elongation mutants of *Arabidopsis thaliana* grown in light or dark conditions. Protein extracts from *A. thaliana* var. Columbia (wild type, lanes 1–2; mutant *prc 1-1*, lanes 3–4; and mutant *bot 1-1* (Höfte, unpublished results), lane 5) or from *A. thaliana* var. Landsberg erecta (wild type, lane 7, and mutant deficient in L-fucose *[36]*, lane 6) were separated in SDS-PAGE. Plants were grown in light (lanes 1, 3, 5, 6, and 7) or dark conditions (lanes 2 and 4). After the electrophoretic separation, the proteins were stained in the gel with Coomassie blue **(A)** or transferred onto a nitrocellulose membrane. The blots were then treated for glycan detection by the Con A/peroxidase method **(B)**, or by immunodetection with anti-β1–>2 xylose antibodies **(C)**, or anti-α1–>3 fucose antibodies **(D)**. All the analyzed extracts contain glycoproteins detected by Con A and anti-xylose antibodies. Anti-fucose antibodies detect glycoproteins in all the *A. thaliana* extracts, except in the fucose-deficient (Fuc⁻) mutant (lane 6, D). The Fuc⁻ mutant is not deficient in *N*-glycosylation, as Con A detects most of its glycoproteins (compare lanes 6 and 7, B). *N*-glycans are also matured in this Fuc⁻ mutant as antibodies directed against β1–>2 xylose are able to detect glycoproteins just as well as in the wild type extract (compare lanes 6 and 7, C). This Fuc⁻ mutant is only devoid of L-fucose and is not able to fucosylate glycoproteins or cell wall components *(37)*.

The GNA-biotin/Streptavidin-peroxidase procedure (specificity: plant M5 high-mannose type *N*-glycans):

1. Separate the proteins to be studied by SDS-PAGE and transfer onto nitrocellulose membrane.
2. Saturate the blot with TTBS buffer.
3. Incubate the blot in TTBS containing GNA-biotin (0.1 mg/20 mL) and 100 m*M* α-methyl mannoside for 2 h at room temperature, for specific detection of glycoproteins containing M5 *N*-glycans. Avoid the α-methyl mannoside for detection of Man-5 to M8 *N*-glycans.
4. For M5 *N*-glycan detection, rinse the membrane two times with TTBS containing 100 m*M* α-methyl mannoside and two times with TTBS alone, for 15 min each. Wash four times in TTBS for a general high-mannose *N*-glycan detection.
5. Incubate the blot with a conjugate streptavidin-peroxidase diluted 1:1000 in TTBS for 1 h at room temperature.
6. Wash the blot in TTBS (4 × 15 min) and once in TBS prior to development.
7. Develop the peroxidase reaction as described in the Con A/peroxidase procedure.
8. Controls for specificity:
   a. It might be advisable to use Ribonuclease B (mostly M5 glycans; *see* **ref. 19** and **Fig. 5**) as a positive control protein, soybean agglutinin (mostly M9 glycans; *see* **Fig. 5**) as a general negative control and bean phytohemagglutinin (mostly M6, M7, and M8 glycans; **ref. 15**) as a negative control in presence of 100 m*M* α-methyl mannoside.
   b. Lectin binding specificity should also be checked by running the affinodetection with GNA in presence of 0.3*M* α-methyl mannoside, as described above for the Con A/peroxidase procedure.

3.2.1.2. IMMUNODETECTION OF PLANT COMPLEX N-LINKED GLYCANS

It has been shown that the β1–>2 xylose and the α1–>3 fucose epitopes of N-linked complex glycans are highly immunogenic in rabbit *(20)*. As a consequence, sera prepared against plant glycoproteins presenting complex glycans generally contain antibodies directed against the β1–>2 xylose and/or the α1–>3 fucose.

We have prepared several of these glycan-specific sera in our laboratory *(20)* and their use is illustrated on **Fig. 7**. Our recent analysis of some commercial immunsera raised against plant or insect glycoproteins showed that these

---

Fig. 7. *(continued)* The *N*-glycosylation of proteins in the *A. thaliana* elongation mutants, *prc 1-1* and *bot 1-1*, is not different from what is observed in wild type *A. thaliana* (compare lanes 3 and 5 to lane 1 on B, C, and D). One can observe that more glycoproteins are detected on blots when wild type or mutant plants of *A. thaliana* are grown in darkness (compare lane 2 to lane 1 and lane 4 to lane 3).

sera have the same characteristics as our home-made probes. For instance, commercially available immunsera directed against PHA or horseradish peroxidase can be used as probes specific for plant complex glycans with α1–>3 fucose and β1–>2 xylose residues. More specific is the immunserum raised against honeybee venom proteins, which can be used as a specific probe for α1–>3 fucose-containing plant complex glycans. These differences in specificities can be easily explained through our recent results *(20)*.

1. Proteins are separated by SDS-PAGE and transferred onto a nitrocellulose membrane.
2. Saturate the blot with 3% gelatin prepared in a TBS buffer for 1 h at room temperature.
3. Incubate the blot in TBS buffer containing 1% gelatin and the immunserum at a convenient dilution for 2 h at room temperature. We use commercial immunsera described above at the following dilutions:
   a. Anti-bean phytohemagglutinin, 1:300.
   b. Anti-horseradish peroxidase, 1:300.
   c. Anti-honeybee venom, 1:200.
4. Wash the blot with TTBS buffer four times for 15 min each.
5. Incubate the blot in TBS buffer containing 1% gelatin and a goat anti-rabbit IgG conjugate coupled to horseradish peroxidase at a suitable dilution for 1 h at room temperature. We generally use the GAR-HRP conjugate from Bio-Rad at a dilution of 1:2000.
6. Wash the blot four times with TTBS buffer, for 15 min each, and once in TBS prior to development.
7. Develop the peroxidase reaction as described for the Con A/peroxidase procedure.
8. Control for *N*-glycan specificity of the immunodetection: It might be wise to check for the specificity of sera toward *N*-glycans attached to the studied protein. This can be performed by doing a mild periodate oxidation on blot prior to immunodetection (*see* **Note 3**). Mild periodate treatment oxidizes glycans and abolishes any recognition of the glycoproteins by the antiglycan antibodies (*see* **Fig. 8**). Any remaining signal will be the consequence of a protein backbone/antibody recognition *(21)*.
9. Controls for fucose or xylose specificity: Some proteins can be used as positive controls for the *N*-glycan immunodetection. Snail hemocyanin is a glycoprotein bearing a β1–>2 xylose, and is deficient in α1–>3 fucose *(22)*. Phospholipase A$_2$ from honeybee venom contains the α1–>3 fucose residue, and is devoided of β1–>2 xylose *(23)*. PHA L is a glycoprotein containing both β1–>2 xylose and α1–>3 fucose.

3.2.3. N-*Glycan Specific Removal*

Another alternative to show that a plant protein is *N*-glycosylated consists of the removal of the potential *N*-glycans by specific enzymes: endoglycosidase

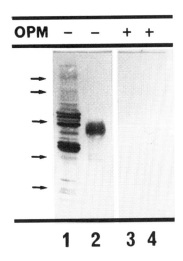

Fig. 8. Mild periodate oxidation performed on blot abolishes any recognition of glycans by antibodies. Proteins from bean cotyledon extract (lanes 1 and 3) and horseradish peroxidase (lanes 2 and 4) were subjected to SDS-PAGE and transferred onto a nitrocellulose membrane. The antiserum used for immunodetection was raised against horseradish peroxidase and contains mostly anti-glycan antibodies *(20)*. Mild periodate oxidation, performed on the blot prior to the immunodetection with this antiserum, totally abolishes glycan-specific cross-reaction with bean glycoproteins (compare lanes 1 and 3) and greatly reduces peroxidase detection (compare lanes 2 and 4). Arrows on the left indicate the positions of the molecular weight standards, respectively, from top to bottom: 97.4, 68, 43, 29, and 18.4 kDa.

H (endo H) or peptide *N*-glycosidases (PNGases). Endo H is able to cleave a *N*-linked glycan from its protein carrier by disrupting the linkage between the two proximal *N*-acetylglucosaminyl residues of the core. This enzyme requires a minimal tetrasaccharide Manα1–>3Manα1–>6Manβ1–>4GlcNAc structure as glycan specificity *(24,25)*. This means that only high-mannose-type glycans will be removed by endo H from a plant glycoprotein.

Peptide *N*-glycosidases (PNGases) cleave *N*-glycans between the proximal *N*-acetylglucosamine and the Asn residue of the protein carrier. PNGase A from almond is capable of releasing any type of *N*-glycans, but intact proteins are poor substrates. Glycoproteins must be digested into glycopeptides before PNGase A treatment. Therefore, this enzyme cannot be used for electrophoretical analysis of glycoproteins. In contrast, PNGase F from *Flavobacterium meningosepticum* is able to directly release *N*-glycans from glycoproteins, but it is inactive on α1–>3 fucose containing complex *N*-glycans, which are usually found in plant glycoproteins *(26)*. Therefore, PNGase F can only be applied for deglycosylation of nonfucosylated plant complex *N*-glycans *(27,28)*.

Deglycosylation by endo H or PNGase F will be visualized on SDS-PAGE by the increase in the electrophoretic mobility of the protein caused by the decrease in its apparent mol wt.

Digestion with endoglycosidase H (endo H):

1. Prior to enzymic deglycosylation with endo H, the protein sample (sample volume 10–100 µL) is denatured by heating 5 min at 100°C in the presence of 1% (w/v) SDS.
2. Following this pretreatment, the sample is diluted fivefold with a 150 m*M* sodium acetate buffer (pH 5.7), 167 pkat (10 milliunits) of endo H are added, and the mixture is incubated at 37°C.
3. After a 6 h incubation, an equal volume of 2X concentrated electrophoresis sample buffer is added.
4. The mixture is heated in boiling water for 3 min and an aliquot is analyzed by SDS-PAGE.

Digestion with peptide-glycosidase F (PNGase F):

1. Prior to enzymic deglycosylation with PNGase F, the protein sample is denatured for 5 min at 100°C in 100 µL of PNGase buffer containing 0.1% SDS (the SDS amount has to be as low as possible, as the PNGase F is very sensitive to this detergent).
2. Add 100 µL of PNGase buffer containing 0.5% Nonidet P40.
3. Hydrolysis by PNGase F (1 mU, 16 pkat) is performed at 37°C.
4. Sample is analyzed as described for the endo H digestion.

3.2.4. N-Glycan Specific In Vivo Labeling

Glycosylation of a plant protein can also be identified using in vivo metabolic labeling. Proteins to be studied are labeled in parallel with [^{3}H] *N*-glucosamine, [^{3}H] mannose, or [^{3}H] fucose, and a radioactive amino acid such as [^{35}S]methionine. After in vivo labeling, the protein of interest is immunoprecipitated by specific antibodies, and incorporation of both labeled substrates in the immunopurified protein can then be visualized by SDS-PAGE and fluorography. So far, GlcNAc has only been found in *N*-linked glycans; therefore, a protein that incorporates [^{3}H]GlcNAc is likely to be a *N*-linked glycoprotein. This labeling technique can also be performed in presence of drugs perturbing the *N*-glycosylation or some *N*-glycan maturation events. The presence of *N*-glycans on the protein of interest will be determined by observing a modification in the electrophoretic mobility between the treated and the nontreated samples. The drugs used and their effects on glycoprotein processing are:

1. *Tunicamycin:* Tunicamycin inhibits *N*-glycosylation *(29)* by preventing the biosynthesis of the precursor oligosaccharide onto the dolichol carrier.

2. *Castanospermine, deoxymannojirimycin*, or *swainsonin:* these drugs inhibit early steps of the *N*-glycan maturation and lead to the formation of unmature *N*-linked glycans without affecting the secretion of plant glycoproteins *(30,31)*.

Several requirements are needed before performing such analysis. First, it should be possible to label in vivo the protein of interest. Second, antibodies directed against the studied protein must be available and their specificity has to be suitable to realize immunofishing experiments.

4. Future Perspectives and Conclusions

Using the methods outlined in this chapter, preliminary information about the presence, the linkage, and the composition of glycans constitutive of a plant glycoprotein can be obtained. The approaches discussed here will certainly continue to be useful, but it is necessary to develop new probes that are particularly needed to identify and characterize glycans *O*-linked to plant proteins. It is also necessary to improve and further adapt enzymatic methods for the cleavage and the sequencing of plant complex *N*-linked glycans.

5. Notes

1. Abbreviations used: Fuc: fucose; Glc: glucose; GlcNAc: *N*-acetylglucosamine; Man: mannose; Xyl: xylose.
2. The solution used for blocking remaining binding sites on the nitrocellulose needs to be devoid of glycoproteins. In this respect, we have observed differences from batch to batch, using gelatin as a blocking agent as described in *(18)*. This is the reason why we recommend using Tween-20 to coat the nitrocellulose *(32)* in this modified procedure.
3. Mild periodate oxidation performed on blot *(21):*
   a. After saturation with gelatin, incubate the blot in a 100 m*M* sodium acetate buffer, pH 4.5, containing 100 m*M* sodium metaperiodate for 1 h in darkness at room temperature, changing the incubating solution after 30 min.
   b. Incubate the blot in PBS containing 50 m*M* sodium borohydride for 30 min at room temperature.
   c. Rinse the blot with TBS buffer, saturate for 15 min with TBS buffer containing 1% gelatin, and perform the immunodetection as described in **Subheading 3.2.1.1.**

Acknowledgments

We thank Herman Höfte and Wolf-Dieter Reiter for providing *A. thaliana* mutant plants. This work was in part financed by Institut National de la Recherche Agronomique Action Incitative Programmée "Biologie du développement."

References

1. Abeijon, A. and Hirschberg, C. B. (1992) Topography of glycosylation reactions in the endoplasmic reticulum. *Trends Biochem. Sci.* **17,** 32–36.
2. Szumilo, T. Kaushal, G. P., and Elbein, A. D. (1986) Purification and properties of a glucosidase I from mung bean seedlings. *Arch. Biochem. Biophys.* **247,** 261–271.
3. Sturm, A., Johnson, K. D., Szumilo, T., Elbein, A. D., and Chrispeels, M. J. (1987) Subcellular localization of glycosidases and glycosyltransferases involved in the processing of N-linked oligosaccharides. *Plant Physiol.* **85,** 741–745.
4. Trombetta, S. E., Bosch, M., and Parodi, A. J. (1991) Glucosylation of glycoproteins by mammalian, plant, fungal and trypanosomatid protozoa microsomal membranes. *Biochemistry* **28,** 8108–8116.
5. Hammond, C., Braakman, I., and Helenius, A. (1994) Role of N-linked oligosaccharide recognition, glucose trimming, and calnexin in glycoprotein folding and quality control. *Proc. Natl. Acad. Sci. USA* **91,** 913–917.
6. Huang, L., Franklin, A. E., and Hoffman, N. E. (1993) Primary structure and characterization of an *Arabidopsis thaliana* calnexin-like protein. *J. Biol. Chem.* **268,** 6560–6566.
7. Lainé, A.-C., Gomord, V., and Faye, L. (1991) Xylose-specific antibodies as markers of subcompartmentation of terminal glycosylation in the Golgi apparatus of sycamore cells. *FEBS Lett.* **295,** 179–184.
8. Zhang, G. F. and Staehelin, L. A. (1992) Functional compartmentation of the Golgi apparatus of plant cells. *Plant Physiol.* **99,** 1070–1083.
9. Fitchette-Lainé, A.-C., Gomord, V., Chekkafi, A., and Faye, L. (1994) Distribution of xylosylation and fucosylation in the plant Golgi apparatus. *Plant J.* **5,** 673–682.
10. Vitale, A. and Chrispeels, M. J. (1984) Transient N-acetylglucosamine in the biosynthesis of phytohemagglutinin: attachment in the Golgi apparatus and removal in the protein bodies. *J. Cell Biol.* **99,** 133–140.
11. Takahashi, N., Hotta, T., Ishihara, H., Mori, M., Tejima, S., Bligny, R., Akazawa, T., Endo, S., and Arata, Y. (1986) Xylose-containing common structural unit in N-linked oligosaccharides of laccase from sycamore cells. *Biochemistry* **25,** 388–395.
12. Schowalter, A. M. (1993) Structure and function of plant cell wall proteins. *Plant Cell* **5,** 9–23.
13. Moore, P. J., Swords, K. M. M., Lynch, M. A., and Staehelin, L. A. (1991) Spatial organization of the assembly pathways of glycoproteins and complex polysaccharides in the Golgi apparatus of plants. *J. Cell Biol.* **112,** 467–479.
14. Edge, A. S. B., Faltyneck, C. R., Hof, L., Le Reichet, J. R., and Weber, P. (1981). Deglycosylation of glycoproteins by trifluoromethane sulfonic acid. *Anal. Biochem.* **118,** 131–137.
15. Rayon, C., Gomord, V., Faye, L., and Lerouge, P. (1996) N-glycosylation of bean and recombinant phytohemagglutinin. *Plant Physiol. Biochem.* **34,** 273–281.

16. Shibuya, N., Goldstein, I. J., van Damme, E. J. M., and Peumans, W. J. (1988) Binding properties of a mannose-specific lectin from the snowdrop (*Galanthus nivealis*) bulb. *J. Biol. Chem.* **263,** 728–734.
17. Animashaun, T., Mahmood, N., Hay, A. J., and Hughes, R. C. (1993) Inhibitory effect of novel mannose-binding lectins on HIV-infectivity and syncitium formation. *Antiv. Chem. Chemother.* **4,** 145–153.
18. Faye, L. and Chrispeels, M. J. (1985) Characterization of N-linked oligosaccharides by affinoblotting with concanavalin A-peroxidase and treatment of the blots with glycosidases. *Anal. Biochem.* **149,** 218–224.
19. Hsieh, P., Rosner, M. R., and Robbins, P. W. (1983) Selective clivage by endo β-N-acetylglucosaminidase H at individual glycosylation sites of Sindbis virion envelope glycoproteins. *J. Biol. Chem.* **258,** 2555–2561.
20. Faye, L., Gomord, V., Fitchette-Lainé, A.-C., and Chrispeels, M. J. (1993) Affinity purification of antibodies specific for Asn-linked glycans containing α–>3 fucose or β1–>2 xylose. *Anal. Biochem.* **209,** 104–108.
21. Lainé, A.-C. and Faye, L. (1988) Significant immunological cross-reactivity of plant glycoproteins. *Electrophoresis* **9,** 841–844.
22. van Kuik, J. A., van Halbeek, H., Kamerling, J. P., and Vliegenthart, J. F. G. (1985) Primary structure of the low-molecular-weight carbohydrate chains of *Helix pomatia* α-hemocyanin. *J. Biol. Chem.* **260,** 13,984–13,988.
23. Kubelka, V., Altmann, F., Staudacher, E., Tretter, V., März, L., Hard, K., Kamerling, J. P., and Vliegenthart, J. F. G. (1993) Primary structures of the N-linked carbohydrate chains from honeybee venom phospholipase A₂. *Eur. J. Biochem.* **213,** 1193–1204.
24. Kobata, A. (1979) Use of endo- and exoglycosidases for structural studies of glycoconjugates. *Anal. Biochem.* **100,** 1–14.
25. Claradas, M. H. and Kaplan, A. (1984) Maturation of α-mannosidase in *Dictyostelium discoideum*: acquisition of endoglycosidase H resistance and sulfate. *J. Biol. Chem.* **259,** 14,165–14,169.
26. Tretter, V., Altman, F., and März, L. (1991) Peptide-N4-(N-acetyl-β-glucosaminyl) asparagine amidase F cannot release glycans with fucose attached α 1–3 to the asparagine-linked N-acetylglucosamine residue. *Eur. J. Biochem.* **141,** 97–104.
27. Woodward, J. R., Basic, A., Jahnen, W., and Clarke, A. E. (1989) N-linked glycan chains on S-allele-associated glycoproteins from *Nicotiana alata*. *Plant Cell* **1,** 511–514.
28. D'Andrea, G., Salucci, M. L., Pitari, G., and Avigliano, L. (1993) Exhaustive removal of N-glycans from ascorbate oxidase: effect on the enzymatic activity and immunoreactivity. *Glycobiology* **3,** 563–565.
29. Hori, H. and Elbein, A. D. (1981) Tunicamycin inhibits protein glycosylation in suspension cultured soybean cells. *Plant Physiol.* **67,** 882–886.
30. Elbein, A. D. (1987) Inhibitors of the biosynthesis and processing of N-linked oligosaccharide chains. *Annu. Rev. Biochem.* **56,** 497–537.

31. Driouich, A., Gonnet, P., Makkie, M., Lainé, A.-C., and Faye, L. (1989) The role of high-mannose and complex asparagine-linked glycans in the secretion and stability of glycoproteins. *Planta* **180,** 96–104.

32. Bird, C. R., Gearing, A. J. H., and Thorpe, R. (1988) The use of Tween-20 alone as a blocking agent for the immunoblotting can cause artefactual results. *J. Immunol. Meth.* **106,** 175–179.

33. Johnson, K. D. and Chrispeels, M. J. (1987) Substrate specificities of N-acetylglucosaminyl-, fucosyl-, and xylosyltransferases that modify glycoproteins in the Golgi apparatus of bean cotyledons. *Plant Physiol.* **84,** 1301–1308.

34. Tezuka, K., Hayashi, M., Ishihara, H., Akazawa, T., and Takahashi, N. (1992) Studies on synthetic pathway of xylose-containing N-linked oligosaccharides deduced from substrate specificities of the processing enzymes in sycamore cells (*Acer pseudoplatanus* L.). *Eur. J. Biochem.* **203,** 401–413.

35. Weitzhandler, M., KadWeitzhandler, M., Kadlecek, D., Avdalovic, N., Forte, J. G., Chow, D., and Townsend, R. R. (1993) Monosaccharide and oligosaccharide analysis of proteins transferred to polyvinylidene fluoride membranes after sodium dodecyl sulfate-polyacrylamide gel electrophoresis. *J. Biol. Chem.* **268,** 5121–5130.

36. Reiter, W.-D., Chapple, C. C. S., and Sommerville, C. R. (1993) Altered growth and cell walls in a fucose-deficient mutant of *Arabidopsis. Science* **261,** 1032–1035.

20

Two-Dimensional Electrophoresis

*A Tool for Protein Separation and Processing
for Microsequencing*

N. P. Eswara Reddy and M. Jacobs

1. Introduction

Electrophoresis is defined as the migration of a charged particle in an electric field. Electrophoretic techniques have been used to analyze, separate, and characterize proteins, peptides, and nucleic acids. A number of different one-dimensional electrophoresis techniques have been developed using various kinds of gel, including starch, agarose, or cellulose acetate. A major breakthrough in electrophoresis methodology occurred with the development of polyacrylamide gels *(1)*. Polyacrylamide gels are nonionic polymers, which are chemically inert and stable over a wide pH range, temperature, and ionic strength. Moreover, they allow gels to be prepared with a wide range of pore sizes, facilitating the separation of proteins of different mol wt. A polyacrylamide gel is formed by the polymerization of monomers of acrylamide with monomers of a suitable bifunctional crosslinking agent, usually *N,N′*-methylene-*bis*-acrylamide *(Bis)*. The concentration of acrylamide and *bis*-acrylamide will determine the physical properties of the gel, especially pore size. Polyacrylamide gels are transparent, allowing excellent visualization of the separated proteins after electrophoresis. Polyacrylamide gel electrophoresis (PAGE) can thus be consider as a major protein purification technique, allowing the separation of proteins according to their mol wt, charge, and geometric configuration.

The following one-dimensional PAGE techniques are commonly used: native PAGE (nondenaturing electrophoresis)—separation according to charge

From: *Methods in Biotechnology, Vol. 3:*
Recombinant Proteins from Plants: Production and Isolation of Clinically Useful Compounds
Edited by: C. Cunningham and A. J. R. Porter © Humana Press Inc., Totowa, NJ

and geometric configuration; sodium dodecyl sulfate polyacrylamide gel electrophoresis (SDS-PAGE)—separation of proteins according to their mol wt; isoelectric focusing (IEF)—separation of proteins based on the electric charge in the form of a defined pH gradient in the gel. Under these conditions, proteins migrate according to their charge properties until they reach the pH values at which they have their isoelectric points (pI). The proteins will, therefore, attain a steady state of zero migration and will be concentrated or focused into narrow zones. The different pH gradients for IEF are generated by incorporation of low mol wt amphoteric compounds, known as synthetic carrier ampholytes, into a polyacrylamide gel matrix. In the absence of an electric field, the ampholytes are randomly distributed throughout the gel. When an electric field is applied, the ampholyte molecules will start to migrate to one or other of the electrodes, depending on their net charge, and create a continuous pH gradient. The carrier ampholyte preparations are available in the range of pH 2.5–11.0. A wide range of pH gradient (pH 3.0–10.0) gels are generally used for complex mixture of proteins having a wide range of pI values (*see* **Note 1**).

A major disadvantage of one-dimensional electrophoretic separation is that its resolution is restricted because many proteins share a common isoelectric point or mol wt. This limitation was overcome by the introduction of two-dimensional electrophoresis, in 1975 *(2)*. Two-dimensional electrophoresis is a combination of two unidimensional gel separation techniques, in particular, IEF, followed by SDS-PAGE. Since its introduction, this technique has become very popular, being used extensively for comparing a spectrum of proteins from different tissues under various conditions.

Recently, reverse genetics strategies have utilized information about protein sequence to clone a range of genes. Purification is a required step for the characterization of proteins. The common methods used in protein purification are ammonium sulphate precipitation, electrophoresis, gel filtration, and chromatography, used either individually or in combination. The purified protein can be used for different purposes, such as sequencing, amino acid analysis, and for the production of antibodies. In the past, electrophoresis followed by electroelution *(3,4)* was used to purify the proteins for sequence determination and for crystallographic studies. Electroelution presents several drawbacks, including low recovery, artificial modification of proteins, and contamination with other proteins. These problems have been minimized with the introduction of protein-electroblotting procedures, which allow the transfer of proteins onto chemically resistant membranes *(5)*. In addition, the transfer of proteins from polyacrylamide gels onto immobilizing supports *(6)* has been extensively used for identifying proteins with antibodies *(7,8)*, and for protein microsequencing.

Table 1
Different Classes of Seed Storage Proteins in *S. bicolor*
Based on Their Solubility

Solvent	Class
Water	Albumins
1% NaCl	Globulins
60% t-butanol	Kafirins
60% t-butanol + 1% ME	Alcohol soluble reduced glutelins (ASRG)
Borate (pH 10) + 0.6% ME + 0.5% SDS	Alcohol insoluble reduced glutelins (AIRG)

ME: β-mercaptoethanol.

Two-dimensional electrophoresis is one of the most powerful techniques available to separate a protein from a complex mixture. The major problem when using two-dimensional electrophoresis, however, is identifying the protein of interest. This can be overcome by co-migration of partially purified and already characterized protein in the protein mixture *(9)*, the use of differentially migrating isozymes, or by the use of specific antibodies *(10)*. The separated proteins can be transferred onto Immobilon-P (Millipore, Bedford, MA) membrane and can be readily subjected to Edman degradation for sequencing or for amino acid analysis *(11,12)*. In some cases, the NH_2-terminal of the protein is not accessible for sequencing because of in vivo modifications. The problem of NH_2-terminal blockage can be circumvented by digesting the protein and analyzing the generated internal peptides *(13)*. Microsequencing of these fragments generally provides information on 10–30 amino acid residues.

We are applying two-dimensional gel electrophoresis to characterize seed storage proteins from lysine-rich cultivars of *Sorghum bicolor*. The seed storage proteins in *S. bicolor* have been grouped into different classes, based on their solubility (**Table 1**). The prolamine fraction of *S. bicolor*, known as kafirins, has a poor lysine content and includes 50–60% of total endosperm storage proteins. Different classes of these proteins were characterized, both biochemically and immunologically, from lysine-rich cultivars and low-lysine cultivars, respectively. Our results show that, in defined cultivars, the reduced synthesis of alcohol-soluble fractions (kafirins and ASRG), and a relative increase of nonkafirin fractions (albumins, globulins, and AIRG), are associated with an absolute and relative increase of lysine. Furthermore, these results indicate the presence of two kafirin proteins (25.3 and 25.9 kDa; **Fig. 1**).

Our ultimate aim is to isolate the gene(s) responsible for the synthesis of the 25.3 kDa kafirin protein and 25 kDa albumin protein for further characterization.

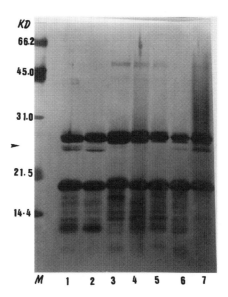

Fig. 1. One-dimensional SDS-PAGE of total kafirins from different cultivars of *S. bicolor*. M: low mol wt marker, lane 1 P 721o (high lysine mutant); lane 2 White Martin (low lysine cultivar); and lanes 3–7 are lysine-rich cultivars: IS 21702, CVS 365, G 1058, G 205, CVS 549.

Initially, different techniques, including electroelution after one-dimensional electrophoresis and chromatography, were used in an attempt to purify these proteins, but were not successful. Finally, we succeeded in sequencing both proteins by combining two-dimensional gel electrophoresis and electroblotting.

We describe here the protocol for the two-dimensional separation (first dimension as IEF and second dimension as SDS-PAGE) and electroblotting of seed storage proteins from *S. bicolor*. The albumins and kafirins (prolamines) are separated (**Fig. 2**) from lysine-rich cultivars of *S. bicolor*, as well as from the low-lysine cultivar for comparison. After electroblotting, proteins were directly sequenced. We limit ourselves here to the procedures up to and including electroblotting, and show that purified material can be readily used for sequencing following a chemical procedure described elsewhere *(14–16)*. The 2D-electrophoresis (IEF and SDS-PAGE) can also be used for the analysis of gene expression at the protein level and to evaluate the influence of environmental conditions on the protein composition of plant cells *(17,18)*.

2. Materials
2.1. Seed Material

1. Low lysine *S. bicolor* cultivar: White Martin.
2. High-lysine *S. bicolor* cultivars: IS 21702, CVS 365, G 1058, G 205, and CVS 549.

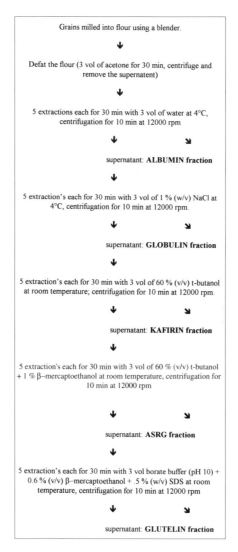

Fig. 2. Flowchart showing sequential extraction of *S. bicolor* proteins.

S. bicolor seeds should be harvested after maturation and air-dried. The flour was obtained by grinding the whole seeds in a Waring blender and passing the product through a 40-mesh screen.

2.2. Apparatus and Materials

1. Waring blender, (Vel, Belgium).
2. Mesh (40-mesh screen).
3. Dialysis tube (type depends on the mol wt of the protein, Polyab, Belgium).

4. Filter papers (Whatman No. 91).
5. Conical flasks (Duran, Germany).
6. Shaker (New Brunswick Scientific Edition, Edison, NJ).
7. Microcentrifuge (Heraeus).
8. Centrifuge (Sorvall).

2.3. Protein Electrophoresis

2.3.1. First Dimension: Isoelectric Focusing (IEF)

1. Stock solutions for IEF gels:
   a. 30% (w/v) acrylamide + *bis*-acrylamide (28.38% [w/v] acrylamide and 1.62% [w/v] *bis*-acrylamide).
   b. 10% (v/v) Nonidet-P40 (NP40) (Boehringer, Mannheim, Germany). Store at room temperature.
   c. Ampholines, pH 6.0–8.0. Store at 4°C (Pharmacia, Sweden).
   d. Ampholines, pH 3.5–10.0. Store at 4°C (Pharmacia).
   e. *N,N,N',N'*- tetramethylethylenediamine (TEMED). Store at 4°C.
   f. 10% (w/v) Ammonium persulphate (APS). Fresh solution is preferable. It can be stored for a maximum of 2 wk at 4°C.
2. Lysis buffer (or) sample buffer: $9.5M$ urea, 2% (w/v) NP40, 2% (w/v) ampholines (1.6%, pH 5.0–7.0 or pH 6.0–8.0, and 0.4%, pH 3.5–10.0), and 100 mM dithiothreitol (DTT) or 5% (v/v) β-mercaptoethanol. This solution should be stored as small aliquots at –20°C.
3. Sample overlay buffer: $6M$ urea, 1% ampholines (0.8% pH 5.0–7.0 or pH 6.0–8.0 and 0.2% pH 3.5–10.0), 0.1% (w/v) bromophenol blue. This solution should be stored as small aliquots at –20°C.
4. Electrode buffers:
   a. Anode: 10 mM H_3PO_4: Prepared by diluting 0.63 mL of H_3PO_4 (sp gr 1.75) in water to make a final volume of one liter.
   b. Cathode: 20 mM NaOH.

2.3.2. Second Dimensions (SDS-PAGE)

1. Stock solutions:
   a. 30% acrylamide (Acros, NJ).
   b. 0.8% *bis*-acrylamide (Sigma).
   c. $1.5M$ Tris-HCl, pH 8.8, with 0.4% (w/v) SDS.
   d. $0.5M$ Tris-HCl, pH 6.8, with 0.4% (w/v) SDS.
   e. 10% (w/v) ammonium persulphate (Bio-Rad).
   f. TEMED (ICN, Cleveland OH).
   g. 1% (w/v) low melting point agarose.
2. 10X Electrode buffer: $0.25M$ Tris-glycine buffer, pH 8.6, with 1% (w/v) SDS.
3. Equilibration buffer: 62.5 mM Tris-HCl, 1% (w/v) SDS, 10% (v/v) glycerol, and 5% (v/v) β-mercaptoethanol or DTT.

2.3.3. Protein-Electroblotting on Immobilon-P PVDF Membranes

1. Pre-equilibration buffer: 50 m*M* boric acid and 0.1% (w/v) SDS, pH 8.0, adjusted with NaOH.
2. Transfer buffer: 50 m*M* boric acid and 50 m*M* Tris.
3. Amido black (Sigma): 0.1% (w/v) Amido black, 10% (v/v) methanol, and 2% (v/v) acetic acid.
4. Coomassie brilliant blue: 0.1% (w/v) Coomassie brilliant blue R-250 (Sigma), 45% (v/v) methanol, and 10 % (v/v) acetic acid.
5. Destaining solutions: 45% methanol and 7% acetic acid.

3. Methods
3.1. Protein Extraction

1. Extract different fractions sequentially from the flour as per the protocol shown in Fig. 2. The extracted albumins and kafirins are used for two-dimensional separation and subsequently blotted onto the Immobilon-P PVDF membrane (*see* **Note 2**).
2. Dialyse the kafirins from each cultivar for 48 h at 4°C, with several changes of water. Dissolve in 60% (v/v) t-butanol, after lyophilization, in order to obtain the appropriate protein concentration (*see* **Note 3**). The protein concentration can be measured by any standard method, including that of Bradford *(19)*. Compare the protein profiles on SDS-PAGE (**Fig. 1**).

3.2. First Dimension Gel (IEF)
3.2.1. Preparation of Glass Tubes

1. Immerse the tubes in 1*M* NaOH for 1 h.
2. Rinse with water, followed by immersion in 1*M* acetic acid for 1 h.
3. Rinse with water and air-dry. The bottom of the tube should be sealed with parafilm and a small rubber ring, to avoid leakage of the gel.
4. Using a permanent marker, mark the tubes 2 cm from the top.

3.2.2. Preparation of First-Dimension Gel Solution

1. Add 5.5 g urea, 1.33 mL 30% acrylamide + *bis*-acrylamide, 2.0 mL 10% (v/v) NP40, 1.97 mL water, 0.4 mL ampholines, pH 6–8, and 0.1 mL ampholines, pH 3.5–10, to a 100-mL side-arm flask, and shake till the urea is completely dissolved (*see* **Note 4**). This volume is sufficient for approx 10 tubes. Degas the solution after the urea is completely dissolved.
2. Add 20 µL of 10% (w/v) APS and 12 µL of TEMED. Using a long narrow-gage needle, pour the gel solution into the tubes up to the mark, avoiding the formation of air bubbles.
3. Overlay each gel with 20 µL of water and allow to polymerize.

3.2.3. Preparation of Protein Samples for First Dimension

Add 57 mg of urea to each 100 μL of protein sample (*see* **Subheading 3.1.**) in order to obtain a final urea concentration of 9.5M. Add an equal volume of lysis buffer and allow the urea to dissolve.

3.2.4. Running the Gel

1. Remove the parafilm present at the bottom of the gel tube and fix the tubes (carefully) in the holes of the electrophoresis chamber. Stopper the remaining holes.
2. Fill the lower electrode chamber with 10 mM phosphoric acid and ensure that the bottom of the tube makes contact with the lower buffer.
3. Carefully remove the air bubbles trapped at the bottom of the gel with a bent syringe needle.
4. Remove the water layer from the top of the gel and add 20 μL of lysis buffer; fill the remaining volume with 20 mM NaOH buffer.
5. Completely fill the upper chamber with 20 mM NaOH buffer and connect to the cathode electrode.
6. Connect the lower chamber, containing 10 mM phosphoric acid, to the anode electrode.
7. Prerun the gel using the following parameters: 15 min at 200 V; 30 min at 300 V; and 30 min at 400 V.
8. After the prerun, remove both the buffer in the upper chamber and the buffer above the gel in the gel tubes.
9. Apply 10–100 μL of protein samples to the top of the gel and cover with 20 μL of overlay buffer.
10. Refill the upper chamber with fresh 20 mM NaOH buffer.
11. Connect the apparatus to the electrodes, as described above, and run the gel at 400 V for 15–18 h, i.e., overnight.
12. Increase the voltage to 800 V for 1 hr to obtain sharp bands. The second-dimension gels can be prepared while the first-dimension gels are running.

3.3. Preparation of Second-Dimension Gels (SDS-PAGE)

1. Wash the glass plates thoroughly with detergent and ethanol to make them grease-free. Fix 1.5-mm thick spacers between two glass plates with the help of Vaseline, and hold the plates together with clamps.
2. Prepare the required quantity of separation gel solution in a 100-mL side-arm flask by mixing the components as outlined in **Table 2**. Mix all the components, except APS and TEMED, and degas.
3. Add 100 μL of APS and 20 μL of TEMED and pour the solution to 2–3 cm from the top of the glass plates.
4. Overlay the gel carefully, either with isopropanol or water, and allow to polymerize. A thin line can be seen between the water and gel solution, indicating polymerization has taken place.

Table 2
Composition of the Separation Gel (for 20 mL)

Acrylamide Concentration	7.5%	10%	12.5%	15%	18%
Water	10	8.33	6.77	5.00	3.00
Acrylamide + *bis*-acrylamide	5.00	6.67	8.33	10.00	13.33
Lower gel	5	5	5	5	5
Ammonium persulphate	0.1	0.1	0.1	0.1	0.1
TEMED	0.02	0.02	0.02	0.02	0.02

All volumes are in milliliters.

Table 3
Composition of the Stacking Gel (for 5 mL)

Acrylamide Concentration	3%	3.5%	4%	4.5%	5.0%
Water	3.25	3.17	3.08	3.00	2.92
Acrylamide + *bis*-acrylamide	0.5	0.58	0.67	0.75	0.83
Upper gel	1.25	1.25	1.25	1.25	1.25
Ammonium persulphate	0.015	0.015	0.015	0.015	0.015
TEMED	0.005	0.005	0.005	0.005	0.005

All volumes are in milliliters.

5. Prepare the required quantity of stacking gel by mixing the different components as outlined in **Table 3**. Decant the isopropanol or water after polymerization and pour stacking gel onto the separation gel, up to 0.5 cm from the top of the glass plates. Overlay the gel with isopropanol or water and allow to polymerize.

3.3.1. Running the Second-Dimension Gel (SDS-PAGE)

1. After polymerization, remove the overlaying isopropanol or water, rinse once with water, and mount the glass plates containing the gel to the electrophoresis apparatus.
2. Remove the tubes from the IEF-apparatus and extrude the gel from the tube. Use a 5-mL syringe containing water to push the gels out of the glass tubes onto parafilm.
3. Equilibrate the first-dimension gel for 10 min in second-dimension equilibration buffer (*see* **Note 5**).
4. Place the equilibrated gel on top of the second-dimension gel and press gently to assure firm contact between both gels.
5. Place 1 mL of equilibration buffer on top of the gel and start electrophoresis at 100 V for 6–8 h.
6. Stop the electrophoresis when the bromophenol blue completely migrates out of the gel.

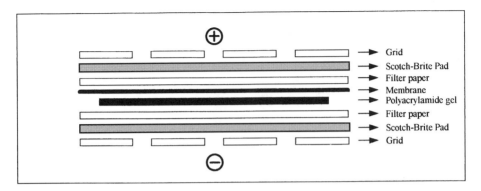

Fig. 3. Electroblot sandwich layout.

3.3.2. Staining

1. Remove the gels from the electrophoresis apparatus. Gels can either be stained or electroblotted. Do not stain the gels used for electroblotting.
2. Stain gels with Coomassie blue staining solution for 1–2 h.
3. Using the destain solution, destain the gels until the protein bands are clearly visible.

3.4. Protein Electroblotting onto Immobilon-P Membranes

1. Pre-equilibrate the second-dimension gels for 1 h in pre-equilibration buffer.
2. Cut one Immobilon-P membrane (PVDF) and four Whatman 3MM papers to fit the size of the gel. Soak the membrane for 5–10 min in 100% methanol and wash with water for about 30 min.
3. Prepare the blotting sandwich as shown in **Fig. 3**, with the material presoaked in transfer buffer.
4. Avoid air bubbles while preparing the gel sandwich. Place the hydrophilic side of the membrane on the gel in order to avoid air bubbles, and insert into the gel holder of the blotting apparatus (Bio-Rad Transblot Apparatus). Remember to connect the gel slide to the cathode electrode.
5. Fill the apparatus with transfer buffer and connect to the power supply. Perform the transfer for at least 8 h, but preferably overnight (*see* **Note 6**).
6. Disassemble the blotting sandwich and rinse the membrane once with distilled water to remove the adhering gel pieces.
7. Stain by immersing the membrane in staining solution (either with Coomassie blue or Amido black) for 5–10 min.
8. Destain with destaining solution (Coomassie blue staining) or wash with distilled water (amido black staining) until the protein spots are clearly visible on a white background.
9. Dry the stained blots and locate the protein spot to be sequenced. The number of blots required depends on the size of the protein spot after the Amido black or Coomassie blue staining. Alternatively, the dried blots can be stored for several months at room temperature.

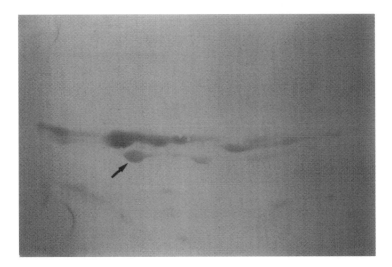

Fig. 4. Two-dimensional electrophoretic (IEF and SDS-PAGE) pattern of kafirins from cultivar White Martin (lane 2 from **Fig. 2**). The protein used for microsequence determination is indicated by an arrow.

In our studies, kafirins and albumins were electroblotted by following the above procedure (**Figs. 4** and **5**). For the NH_2-terminal sequence, 3–4 and 8–10 blots were used for the protein spots indicated in **Figs. 4** and **5**, respectively. The proteins immobilized on these membranes were used directly for the NH_2-terminal sequence.

4. Notes

1. The pH gradient in the first-dimension electrophoresis (IEF) can be optimized by mixing a range of ampholytes. This gives better protein separation, but values vary. In our case, the pH ranges of 6.0–8.0 for kafirins and 5.0–7.0 for albumins gave best results.
2. The composition of extraction buffer and extraction procedure varies, depending on the plant material and the proteins to be analyzed.
3. The final SDS concentration of the protein sample should not be more than 0.5% (w/v). The SDS concentration can be minimized by dialysis.
4. Keep the flask containing the first-dimension gel solution at 37°C, with intermediate shaking if necessary, but never heat to temperatures higher than 50°C.
5. Do not equilibrate the first-dimension gel for more than 10 min. Longer equilibration leads to diffusion and eventually loss of proteins.
6. It is advisable to use transfer buffer of 50 m*M* boric acid and 50 m*M* Tris (*see* **Subheading 3.3.**) in order to avoid overtransfer.

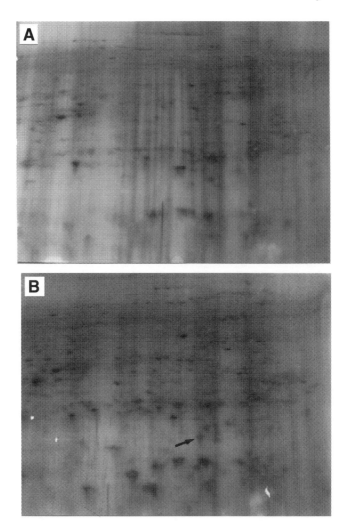

Fig. 5. Two-dimensional electrophoretic (IEF and SDS-PAGE) pattern of albumins from cultivar **(A)** White Martin and **(B)** IS 21702. The protein used for microsequence determination is indicated by an arrow.

Acknowledgments

This work was supported by the European Economic Community-Science and Technology Development (TS3 - CT91005) programme "Voies génétiques et moléculaires d'une maitrise de la qualité du grain chez le Sorgho." The authors would like to thank Marc Vauterin, Serge Vernaillen, and Tadesse Yohannes for their assistance in reviewing the manuscript.

References

1. Raymond, S. and Weintraub, L. S. (1959) Acrylamide gel as a supporting medium for zone electrophoresis. *Science* **130,** 711–721.
2. O'Farrel, P. Z. (1975) High resolution two-dimensional electrophoresis of proteins. *J. Biol. Chem.* **250,** 4007–4021.
3. Weiner, A. M., Platt, T., and Weber, K. (1972) Amino-terminal sequence analysis of proteins purified on a nanomole scale by gel electrophoresis. *J. Biol. Chem.* **247,** 3242–3251.
4. Hunkapiller, M. W., Lujan, E,. Ostrander, F., and Hood, L. E. (1983) Isolation of microgram quantities of proteins from polyacrylamide gels for amino acid sequence analysis, in *Enzyme Structure, part I: Methods in Enzymology,* vol. 91 (Hirs, C. H. W. and Timasheff, S. N., eds.), Academic, Orlando, pp. 227–236.
5. Vandekerckhove, J., Bauw, G., Puype, M., Van Damme, J., and Van Montagu, M. (1985) Protein-blotting on polybrene-coated glassfibre sheets: a basis for acid hydrolysis and gas-phase sequencing of picomole quantities of protein previously separated on sodium dodecyl sulphate/polyacrylamide gel. *Eur. J. Biochem.* **152,** 9–91.
6. Vandekerckhove, J., Bauw, G., Puype, M., Van Damme, J., and Van Montagu, M. (1986) Protein blotting from polyacrylamide gels on glass microfibre sheets: acid hydrolysis and gas-phase sequencing of glass-fibre immobilized proteins, in *Advanced Methods in Protein Microsequencing Analysis* (Wittman-Liebold, B., Salnikov, J., and Erdmann, V., eds.), Berlin, Springer-Verlag, pp. 179–193.
7. Towbin, H., Staehelin, T., and Gordon, J. (1979) Electrophoretic transfer of proteins from polyacrylamide gels to nitrocellulose sheets: procedure and some applications. *Proc. Natl. Acad. Sci. USA* **76,** 4350–4354.
8. Beisiegel, U. (1986) Protein blotting. *Electrophoresis* **7,** 1–18.
9. Gershoni, J. M. (1988) Protein blotting: a manual. *Methods Biochem. Anal.* **33,** 1–58.
10. Anderson, L. and Anderson, N. G. (1977) High resolution two-dimensional electrophoresis of human plasma proteins. *Proc. Natl. Acad. Sci. USA* **74,** 5421–5425.
11. Anderson, N. L., Nance, S. L., Pearson, T. W., and Anderson, N. G. (1982) Specific antiserum staining of two-dimensional electrophoretic patterns of human plasma proteins immobilized on nitrocellulose. *Electrophoresis* **3,** 135–142.
12. Bauw, G., De Loose, M., Inze, D., Van Montagu, M., and Vandekerckhove, J. (1987) Alternations in the phenotypes of plant cells studied by NH_2-terminal amino acid sequence analysis of proteins electroblotted from two-dimensional gel-separated total extracts. *Proc. Natl. Acad. Sci. USA* **84,** 4806–4810.
13. Bauw, G., Van den Bulcke, M., Van Damme, J., Puype, M., Van Montagu, M., and Vandekerckhove, J. (1988) Protein electroblotting on polybase-coated glassfibre and polyvinylidene difluoride membranes: an evaluation. *J. Prot. Chem.* **7,** 194–196.
14. Bauw, G., Van Damme, J., Puype, M., Vandekerckhove, J., Gesser, B., Ratz, G. P., Lauridsen, J. B., Celis J. E. (1989) Protein-electroblotting and microsequencing

strategies in generating protein databases from two-dimensional gels. *Proc. Natl. Acad. Sci. USA* **86,** 7701–7705.

15. Bauw, G., Van den Bulcke, M., Van Damme, J., Puype, M., Van Montagu, M., and Vandekerckhove, J. (1989) NH$_2$-terminal and internal microsequencing of proteins electroblotted on inert membranes, in *Methods in Protein Sequence Analysis* (Wittman-Liebold, B., ed.), Berlin, Springer-Verlag, pp. 220–233.

16. Bauw, G., Holm Rasmussen, H., Van den Bulcke, M., Van Damme, J., Puype, M., Gesser, B., Celis, J. E., and Vandekerckhove, J. (1990) Two-dimensional gel electrophoresis, protein electroblotting and microsequencing: a direct link between proteins and genes. *Electrophoresis* **11,** 528–536.

17. Garrels, J. I. (1979) Two-dimensional gel electrophoresis and computor analysis of proteins synthesized by clonal cell lines. *J. Biol. Chem.* **254,** 7961–7977.

18. Celis, J. E. and Bravo, R. (1984) *Two-Dimensional Gel Electrophoresis of Proteins. Methods and Applications.* Academic, New York.

19. Bradford, M. M. (1976) A rapid and sensitive method for the quantification of microgram quantities of protein utilizing the principle of protein-dye binding. *Anal. Biochem.* **72,** 248–254.

Index